Analytical Economics

ISSUES AND PROBLEMS

Analytical Economics

ISSUES AND PROBLEMS

NICHOLAS GEORGESCU-ROEGEN

HARVARD UNIVERSITY PRESS

CAMBRIDGE, MASSACHUSETTS · 1966

To OTILIA

Gratiam plenam

Foreword by Paul A. Samuelson

Professor Georgescu-Roegen has been a pioneer in mathematical economics. The times have almost caught up with him; but, unlike the hare, he moves ahead of his pursuers in a divergent series. For along with the gathering together of important papers already published, we have here in Part I a new, long, and profound essay that goes to the very foundations of the possibility of a purely quantitative economic science — or for that matter of a purely quantitative science applicable to the physical and biological world. Even without this invaluable bonus, the collecting together from scattered learned journals of certain important essays by Georgescu-Roegen would be welcomed by economic scholars and students. For in Georgescu-Roegen we have a scholar's scholar, an economist's economist.

On short airplane rides, when we tire of Agatha Christie, we can benefit much from dipping into Veblen or Rostow. The papers in this volume must be tackled in a very different manner. Each paragraph must be thoroughly chewed — and rechewed. The margin of the page is scarce large enough for our summaries and queries. One cannot, alas, read Agatha Christie twice; but a Georgescu-Roegen article is like the Widow's Cruse — dip into it again and again, and there remains yet more for the taking. Contemplating this assemblage of articles, I came to realize that the whole is greater than the sum of its parts: the 1936 classic on the pure theory of consumer's behavior is, to coin a phrase, *complementary* with the papers on choice of twenty years later.

The great economists of earlier generations — Alfred Marshall, A. C. Pigou, and Lord Keynes — thought that mathematical economics had a slim past and no future at all. Time makes fools even of great men. In the middle third of the twentieth century, mathematics has everywhere swept through economics like an epidemic of measles sweeping through a new continent. He who would occupy a chair at a great university, advise the Prince or the Chairman of the Board, must serve his apprenticeship in this difficult art. Deplore it or applaud it, the fact is there — like the Dead Sea or Pike's Peak. I shall not pronounce on the utility of this trend, but I must set the record straight on the aesthetic quality of modern mathematical economics.

Marshall, and Keynes and Pigou after him, insisted that mathematical economics lacked the intellectual grandeur of mathematical physics or of pure mathematics itself. And these were not second-rate minds, innocent of an art that was conspiring to deprive them of reputation and a living. Yet they should have known better. Consider the famous story told by Keynes — that Max Planck, the Nobel Laureate who started quantum physics, had quit economics early in life because it was too difficult! — and Keynes's quick antidote to the anecdote: "Professor Planck could easily master the whole corpus of mathematical economics in a few days." Or consider the dictum of Pigou: "But to the student of theoretical physics or the pure mathematician, watching from their Everest, the severest of so-called mathematical economists are merely flies crawling upwards to the towering summit of Primrose hill!" When I was a student, I believed such nonsense. For it is nonsense.

As these pages show, mathematical economics is difficult enough for anyone. Read the collected works of Max Planck — great man that he was — and you will realize that like anyone else he would have to spend months and years in learning the tools of our trade. Even the genius of Keynes did not throw off, on the train from Cambridge to London, any recorded solutions to the many unsolved conundrums of his day: had Keynes anticipated the correct theoretical solution to the problem of transferring German reparations to the Allies, his *Economic Consequences of the Peace* might have been a duller book — but it would not have been a less accurate one.

And when it comes to intellectual beauty, how can they appraise the beauty of advanced economics who don't know advanced economics? Anyone who has felt his skin bristle — A. E. Housman's test for perception of poetry — when contemplating Hamilton's Principle in dynamics or Gibbs's equilibrium of heterogeneous substances will recognize that Ricardian comparative cost *is* beautiful. The logic of rational choice, so deeply explored in Part II here, will interest minds when today's skyscrapers have crumbled back to sand.

Professor Georgescu-Roegen is more than a mathematical economist. He is first an economist, and the first to debunk the pretensions of symbolic boondoggles. The niceties of marginal productivity and of original utility do not escape his skeptical notice. In one sentence he quotes Aristotle and in the next Pareto or von Neumann. An abstract theorem concerning the possible nonexistence of a positive factor-price suddenly becomes the basis for an insight into the problem of disguised unemployment for overpopulated Rumania and agrarian economies.

If a man is an economist and an expert in mathematics, usually you can guess that he is an enthusiast for mathematical economics, often too pretentious in his claims for that subject. Georgescu-Roegen is an exception to this. Because he is so superlatively trained as a mathematician, he is quite immune to the seductive charms of the subject, being able to maintain an objective and matter-of-fact attitude toward its use. Coming from such a scholar, paradoxical views — like the following nuggets in the brilliant new essay — demand the attention of every serious scholar: ". . . we must accept that *in certain instances* at least, 'B is both A and non-A' is the case"; and his idea that the theoretical precision of Newton's and Einstein's physics, far from being the rule in the realm of physical science and the goal for less developed sciences, is actually the result of physicists' confining their attention to the easy subsets of experience. I defy any informed economist to remain complacent after meditating over this essay.

Wassily Leontief's input-output analysis — which statisticians are utilizing in the Soviet Union, the United States, and all over the world — has gained much from the papers in Part III of this collection. Thus, the basic theorem on nonsubstitution in a one-primary-factor system is merely one plum among many in the essay entitled "Some Properties of a Generalized Leontief Model" (1951).

This then is a book to own and to savor. My old master, Joseph Schumpeter, once hoped to collaborate with Nicholas Georgescu-Roegen on a definitive economic treatise. It has always been a source of regret to me that his prewar dream could not be fulfilled. These selected papers offer economists the best possible consolation.

Massachusetts Institute of Technology
January 1965

Author's Preface

If an author has the good fortune to have some of his papers reprinted together in a special volume, he does not need to instruct the reader on his gratitude to those who took the initiative and backed the project with their academic prestige. The reader knows perfectly well that this gratitude cannot be otherwise than immense, especially in cases — such as mine — where the author's contributions have at most a modest value. But since I have not done so earlier, I want to express now my heartfelt thanks to Paul A. Samuelson, Wassily W. Leontief, Edward S. Mason, and George W. Stocking, as well as to Rendigs Fels of Vanderbilt University, who in addition has helped me with the preparation of this volume beyond what one might expect from a friend. I wish also to thank Vanderbilt University, the institution with which I have been associated for the last fifteen years, for its enthusiastic support of the project.

The reader can also imagine without any prompting how gratified an author feels to see his intellectual endeavors receive such a high token of appreciation. However, since Harvard University Press asked me to write an introductory essay tracing the main thread of my thoughts and relating it to the lessons which I, at least, have learned from my probings and queryings, I feel it necessary to explain that my complying does not mean that I let my elation turn into false pride. I welcomed the suggestion only because it struck me that through this essay I might be able to perform some face-lifting which, in my opinion, the papers included in this volume needed badly. I felt that an essay on "Some Orientation Issues in Economics" (which constitutes Part I of the book) would cast the brightness necessary for espying the points of my previous contributions and also would correspond to the publisher's intention. Indeed, epistemological preoccupations have been the main source of inspiration for almost all my contributions, even though the relationship is not always conspicuous. A further reason for my choice is that the topic seemed most appropriate for rounding up some afterthoughts which for some time now I have presented at various invited seminar lectures. About these thoughts I wish to make a few prefatory remarks.

An essay on basic issues in any special science is, inevitably, philosophical in character. Differences in the scientific temperament, more than anything else, account for the widely divided opinions on the utility of such essays. Some specialists regard them as frivolous, if not as the hallmark of scientific immaturity. Others, on the contrary, maintain that the progress of science cannot dispense with a continual criticism of epistemological issues. The fact is that the greatest names in physics — I can think of no exception — have now and then indulged in some philosophizing about their science. On the other hand, it is beyond question that epistemological issues — or methodological issues as they are usually but improperly named — may be debated endlessly. Moreover, in economics their debating has been somewhat ineffective despite the brilliancy of many a participant. My justification for adding the introductory essay to the long list of previous contributions on the same general topic is twofold. In the first place, I have been able to illustrate some of my conclusions by concrete and highly topical cases. Secondly, my essay, I believe, explores the watershed of each issue more extensively, perhaps more intensively as well, than has been done in the past. Specifically, I have endeavored to follow each issue beyond the boundary not only of economics but also of the social sciences. As a result, I came to the conclusion that, contrary to what we generally seem to think, many issues which confront the economist are not specific to his particular domain. They spring up even in physical sciences, the only difference being that there they do not affect all special areas.

In all probability, the reason why the ubiquity of many epistemological issues has not been noticed earlier is to be found in a peculiar bent of contemporary philosophy of science. This philosophy completely ignores the existence of even some purely physical sciences — such as chemistry and material structure — let alone biological and social disciplines, as if no science besides physics contained any *scire* at all. To remedy in part the absence of any connecting bridge between theoretical physics and economics in modern philosophical literature, I had to venture into a large territory beyond the boundary of economics, a territory for which I possess no adequate knowledge. Nevertheless, I felt that the risk was worth taking. The adventure probably marks a beginning, and a beginning ought to be made by somebody. Moreover, I encourage myself with the thought that the accumulation of knowledge today — the pride but at the same time the burden of every special science — is so immense that the risk could not be appreciably reduced even by abler writers than myself. For the interpretation of facts outside the economic domain I had no other solution than to rely heavily on the opinion of other specialists. But, at least, I have always followed the rule of seeking counsel from the highest authorities in each field and of suppressing no reference to my sources.

A preface is also the place for acknowledging those debts that can be only acknowledged but never redeemed. Most of my contributions have been completed while I was free from ordinary duties thanks to research grants. Reasonably modest though these grants were, they have been extremely valuable to me, above all because they did not restrict my research activity to some pre-selected narrow area. It was thus possible for me to pursue freely whatever idea seemed to me most promising as ideas were budding accidentally in my mind. I wish to express my deep gratitude to those who have helped me in this liberal manner: the Rockefeller Foundation, the John Simon Guggenheim Memorial Foundation, and the Ford Foundation.

Finally, I wish to thank Harvard University Press for the many courtesies it has shown me in accepting and publishing this volume, and, especially, to acknowledge the excellent and painstaking editing of Mr. Max Hall, the editor for the social sciences.

Nicholas Georgescu-Roegen

Vanderbilt University
November 1964

Note on the Contents
of the Volume

As explained in the preface, Part I of this book is all new. The other three parts, II, III, and IV, contain twelve essays previously published. A note on the first page of each essay tells when and where it first appeared. The dates are also shown in the table of contents. Essay 9 is a full version of a paper which was previously published in an abbreviated form only. Postscripts have been added to Essays 7 and 9.

The reprinted papers, in comparison with the original versions, contain improvements of style. In a few instances, the argument has been rephrased so as to make it clearer, and, in one case, a slip in the original paper has been corrected. All these instances are noted in the proper places.

In the reprinted essays the numbered footnotes are those of the original versions. Only the very few footnotes marked with an asterisk or merely headed "Note" have been added in the present volume.

Table of Contents

PART I

---

*Introduction: Some Orientation
Issues in Economics*

ONE

SCIENCE: A BRIEF
EVOLUTIONARY ANALYSIS

1. *The Genesis of Science.* We can look at science from several viewpoints, for science is "a many splendored thing." However, science has not been in all places and at all times as we know it today. Nor has its modern form come to us by fiat as some specific commandments revealed in the shortness of a single blink to all men in every part of the globe. Science had a genesis and an evolution in the sense in which these terms are used in biology. The more we ponder how science has radically changed over the last three or four centuries, the more obvious it becomes that science is a living organism. This being so, we should not be surprised that every attempt to define it by one single trait has failed.

To proceed systematically, I shall search first for the reason why science came to be, that is, for its *causa efficiens* (in the Aristotelian sense). From what we can infer, this cause was the instinct of exploring the environment, an instinct man shares with all other animals. Here and there, some tribes came to realize, first, that knowledge gives controlling power over the environment (unfortunately, over men as well) and consequently makes life easier for him who possesses it; and second, that learning what others already know is far more economical than acquiring this knowledge by one's own experience. It was then that man began to value the aggregate knowledge of all individuals in the community and feel the need of storing and preserving this knowledge from generation to generation. Science, in its first form, came thus into being.

It is clear then that the *causa materialis* (again, in the Aristotelian sense) of science is stored communal knowledge, that is, the body of all descriptive propositions available to any member of a community and believed to be true according to the criteria of validity prevailing at the time of reference. To take this equation as a definition of science would be patently inept. On the other hand, we must agree that the equation is valid for all times and places, from the earliest cultures to those of today. Furthermore, the point disposes of the view that science is the opposite of description. On the contrary, science cannot exist without description.[1]

Furthermore, the equation set forth in the preceding paragraph applies not only to sciences of fact — like physics or political science, for instance — but also to sciences of essence, i.e., to mathematics and Logic.[2] Indeed, "p implies q and q implies r yields p implies r" is just as much a descriptive proposition as "an acid and a base yield a salt." Both propositions represent *gained* knowledge and, hence, their meaning is apt to change as this knowledge increases. By now we know that sciences of essence too have the privilege of discovering that not all swans are white. Bernhard Bolzano was perfectly right in cautioning us, more than a hundred years ago, that "many a fresh discovery remained to be made in logic." [3] Only the knowledge at which every individual *inevitably* arrives — such as "I am not you," for instance — does not change with time. Nor do such propositions form the *causa materialis* of a science.

2. *Evolution by Mutations.* As already intimated, the animal instinct of learning did not suffice for a community to develop science: the community had also to develop the utilitarian instinct to an appreciable degree so as to become conscious of the utility of storing all communal knowledge. There are examples of tribes which have survived to modern times and which have not developed science precisely because of their weak utilitarian instinct. This deficiency is responsible also for other cultural patterns that are common to these communities and which seem to us equally puzzling. We must observe also that the survival of science-less communities to our own time is due exclusively to their accidental isolation from others. For, otherwise, natural selection — as any Darwinist will instruct us — would have seen to their history's ultimately being brought to an end by the onslaught of other tribes which could put science in the service of war. History shows that even differences in the level of factual knowledge plays a paramount, if not decisive, role in the struggle between human societies. One can hardly doubt that had the European nations not acquired a vastly superior amount of factual knowledge in comparison with the rest of the world European colonialism would not have come about. In all probability China or India would have colonized the world, including Europe, if the Asian civilizations had first achieved this superiority.

Though the causes that could account for the birth of science seem to be the same everywhere, the evolution of science did not follow every-

---

1. Cf. P. W. Bridgman, *The Logic of Modern Physics* (New York, 1928), p. 175.

2. That fact as well as essence constitutes the object of description and, hence, of science represents the viewpoint of Edmund Husserl, *Ideas: General Introduction to Pure Phenomenology* (New York, 1931), pp. 61 ff.

3. Bernhard Bolzano, *Paradoxes of the Infinite* (New Haven, 1950), p. 42. See also P. W. Bridgman, *The Intelligent Individual and Society* (New York, 1938), p. 98.

where the same pattern. We may, with Veblen, impute the subsequent expansion and transformation of primitive science to the instinct of *idle curiosity*. But if we do, we must also admit that this instinct is not an innate one, as the instinct of learning is. This admission seems inevitable in view of the entirely different evolution of science in various parts of the world. The instinct of idle curiosity undoubtedly represents a later accidental mutation, which like any successful mutation was gradually diffused to larger and larger groups.

3. *Memory: The Earliest Store of Knowledge.* The problem of storing and preserving knowledge soon led to the profession of scholars and to the institution of teaching. As long as the list of descriptive propositions remained relatively short, memorizing it provided the easiest mode of storage. This mode was also perfect in the sense that it permitted almost instantaneous access to every bit of extant knowledge. For a long time, therefore, good memory was the only required ability of a scholar; it thus came to be regarded as one of the most valuable virtues of a people.[4]

Ultimately a point was reached when the memory of even the ablest individual could no longer serve as a filing cabinet for the growing amount of knowledge. Nonhuman cabinets had to be invented lest knowledge be lost. The impasse was resolved fortunately by the invention of writing and papyri. But as knowledge still continued to expand, a new and most troublesome problem arose: how to file countless propositions so as to find the one needed without having to search through the whole cabinet. Though we do not find the problem stated in these precise terms, we can nevertheless understand that the need must have continuously irritated the minds of the learned. They first fell upon the idea of taxonomic filing, as witnessed by the earliest codes of moral or legal conduct. However, good taxonomic filing in turn requires a readily applicable criterion, such as the chronological order for filing historical facts. At least one — probably the only one — of the known cultures, namely that of Ancient Greece, thus came to talk about classification and to debate its delicate issues hotly.[5]

Classification as a filing system has survived to this very day for the simple reason that we still have to file a great deal of our factual knowledge taxonomically. This is true not only for biology but also for the highest realm of physics: physicists are now preoccupied with classifying the ever growing number of different intra-atomic particles. It seems that

4. Plato, *Phaedrus*, 274–275, relates that a famous Egyptian king deplored the invention of writing because it would induce people to pay less attention to the training of the memory.

5. For instance, Plato, *Sophist*, 219, 253, *Statesman, passim*, argued that dichotomy is the rational principle of classification. Aristotle, *De Partibus Animalium*, I. 2–4, strongly disagreed with this, rightly pointing out that in most cases dichotomy is "either impossible or futile."

the commandment formulated by Georges Cuvier, *"nommer, classer, décrire,"* has a lasting value, even though the three commands cannot always be executed separately or in that order. Unfortunately, the basic issues of classification too have survived unresolved and still torment the scholarly world from the biological taxonomist to the logician. For indeed, most logical paradoxes grow out of classification.

4. *From Taxonomic to Logical Filing.* The search for a universal principle of classification caused the Greek philosophers to inquire into the nature of notions and their relationship. Out of these intellectual labors, Logic was born. This marked the end of a prolonged and diffused process. Logical proofs of geometrical propositions were used as far back as the beginning of the sixth century B.C. Yet even Plato, Aristotle's teacher, had no idea of syllogism. He did talk about scientific propositions following from some basic truths, but a clear picture of the logical edifice of knowledge did not appear before Aristotle.[6] And the important fact is that even Aristotle himself was inspired by some *Elements of Geometry* which existed in his time and have come down to us in highly polished form from the hands of Euclid.[7] Time and again, the coming into being of a thing — in this instance the first theoretical science — preceded its description.

It goes without saying that the theoretical edifice of geometry — in its etymological meaning — was not erected in one day. And since no one had a definite idea of what the final result was going to be, its bricklayers must have been guided by other purposes. The abstract thinkers, in the characteristic tradition of Greek thought, were searching for some First Principle. On the other hand, we may plausibly surmise that the *arpedonapts*, the land surveyors in ancient Egypt, must have sooner or later observed that once one can remember, for instance, that

A. *The sum of the angles in a triangle is two right angles,*

one need not memorize also the proposition

B. *The sum of the angles in a convex quadrangle is four right angles.*

Thus *arpedonapts* came to use, however unawares, the logical algorithm long before the first *Elements of Geometry* was written, simply because the device saved them memorizing efforts. Without this economical aspect, the logical algorithm would have in all probability remained a notion as esoteric as the First Cause, for example.

Today the relationship between the logical algorithm and theoretical science seems simple. By logical sorting, all propositions, $P_1, P_2, \ldots, P_n$, already established in any particular field of knowledge can be separated into two classes (α) and (β), such that

6. Plato, *Republic*, VII. 533; Aristotle, *Analytica Posteriora*, I. 1–6.
7. Cf. W. D. Ross, *Aristotle* (3rd edn., London, 1937), p. 44.

(1) *every β-proposition follows logically from some α-propositions*, and
(2) *no α-proposition follows from some other α-propositions.*[8]

This logical sorting represents the inner mechanism by which a scientific theory is constructed and maintained. Theoretical science, therefore, is a catalog which lists known propositions in a logical — as distinct from taxonomic or lexicographic — order. In other words, we have a first equation

"Theoretical science" = "Logically ordered description."

Actually, the logical economy does not always stop here. Often some speculative propositions are "thought up" and added to (α) with a view of shifting as many α-propositions to (β). Thus, (α) is replaced by (ω), the latter having the same properties and the same relation with the new (β) as (α) has had. The only difference is that (ω) contains some unobservable propositions, i.e., some first principles. But this does not affect the validity of the equation written above.

5. *Theoretical Science and Economy of Thought.* By filing knowledge logically we do not increase it; we only carry the economical advantage of the logical algorithm to its utmost limits. Clearly, the ω-propositions of any individual science contain, explicitly or implicitly, the entire extant knowledge in a particular domain. Strictly speaking, therefore, to store all that is already known in a domain we need only to memorize (ω), i.e., what we currently call the logical foundation of the respective science. To be sure, a scholar normally memorizes some β-propositions as well but only because he finds it convenient to have immediate access to those propositions most frequently needed in the daily exercise of his profession.

The Greek philosophers may appear to have been preoccupied with ethereally abstract issues and pragmatically idle problems. But in the deep waters of their intellectual struggle there was the need for a classification of knowledge in a form that could be grasped by one individual mind. The heroes of the battle might not have been aware of the economic implications of this need, nor always of the need itself, just as no one seems to have paid any attention to the immense economy brought about by the change from ideographic to alphabetical writings, either when the change happened or, still less, before the event. Generally speaking, needs generated by evolution guide us silently, as it were; seldom, if ever, are we aware of their influence upon our complex activity (or even of their existence). Only after a very long time do we realize why we labored and what we searched for. Only after the event can we say with Oswald

8. For (β) not to be a null set, the propositions $P_1, P_2, \ldots, P_n$ must not be entirely circular. For instance, our factual knowledge should not consist only of: Lightning is light; Light is electricity; Electricity is lightning. This necessity may account for the traditional aversion men of science display for circular arguments.

Spengler that "a task that historic necessity has set *will* be accomplished with the individual or against him." [9]

It is not surprising, therefore, that the economic aspect of theoretical science remained completely unnoticed until 1872 when Ernst Mach first argued that science "is experience, arranged in economical order." [10]

To speak of the economy of thought achieved through theoretical science we must first show that memorizing is a costlier intellectual effort than ratiocination. Certainly, this does not seem true for an overwhelming majority of humans: even university students in appreciable numbers prefer descriptive courses where knowledge being presented taxonomically has to be memorized rather than logically sorted. Besides, memory and ratiocination are abilities that training can improve; training, on the other hand, may place the accent on one or the other depending upon the cultural tradition. For years on end the memory of Chinese and Japanese students has been subject to a continuous training unparalleled in the West; this will continue as long as they have to learn by heart thousands of ideographic characters. Yet, in the end, even Chinese and Japanese scholars had to succumb to the pressure on memory. Nowadays no one, however narrow is his chosen field, can dream of memorizing the vast amount of factual knowledge, just as no one can hope to reach the moon in an ordinary balloon. By memorizing only a part of factual knowledge one can succeed as a craftsman, but certainly not as a scholar.

But in evolution nothing is general or definitive. Thinking beings from other solar systems may have their brains so constructed that for them memory is a relatively free factor; for such beings theoretical science might very well be uneconomical. On the other hand, we should expect even the structure of science on this planet to change with time. We can already catch some glimpses, however faint, of its next mutation after the electronic brains will have completely taken over memorizing and sorting.

9. Oswald Spengler, *The Decline of the West* (New York, 1928), II, 507. Parenthetically, but apropos, one may speculate that the present space programs might at some distant future prove to have corresponded to the need of taking care of an exploding population.

10. Ernst Mach, *Popular Scientific Lectures* (Chicago, 1895), p. 197 and *passim*. See also his *The Science of Mechanics* (La Salle, Ill., 1942), pp. 578–596. The same idea was elaborated with much greater insight by Karl Pearson, *The Grammar of Science* (Everyman's Library edn., London, 1937), pp. 21, 32, and *passim*.

Mach, however, made little if anything out of *logical* order. Rather, he emphasized the disburdening of memory through numerical tables and mathematical symbolism. However, ephemerides existed long before mechanics became a theoretical science; and the multiplication table has always been only a mnemonic. The economy of thought yielded by tables and symbols should be attributed to the invention of writing rather than to anything else.

6. *Significant Differences Between East and West.* Facts described in a logical order form a texture entirely different from that of taxonomic description. We are therefore justified in saying that with Euclid's *Elements* the *causa materialis* of geometry underwent a radical transformation; from a more or less amorphous aggregate of propositions it acquired an *anatomic* structure. Geometry itself emerged as a living organism with its own physiology and teleology, as I shall presently explain. And this true mutation represents not only the most valuable contribution of the Greek civilization to human thought but also a momentous landmark in the evolution of mankind comparable only to the discovery of speech or writing.

Looking back at the developments of Greek thought one is tempted to conclude that the emergence of theoretical science was a normal, almost necessary, upshot of Logic. One could not be more mistaken. Both Indian and Chinese civilizations arrived at a logic of their own — in some respects even more refined than Aristotle's[11] — but neither came to realize its utility for classifying factual knowledge. As a result, science in the East never went beyond the taxonomic phase. Greek culture, therefore, must have had some peculiar feature which the East lacked: otherwise we could not account for the difference in the development of science in the East and the West.

It is not difficult to convince ourselves that the distinctive birthmark of Greek philosophy is the belief in a First Cause of a nondivine nature. As early as the beginning of the sixth century B.C., Thales, the scholar of many and diverse talents, taught that "water is the material cause of all things." [12] To discover the First Cause one must sooner or later come to inquire for the proximate cause. And indeed, only one generation after Thales, we find Anaximander explaining in a quite modern vein that the earth "stays where it is [held by nothing] because of its equal distance from everything." [13]

Other civilizations may have arrived at the notions of cause and effect, but only that of Ancient Greece struck, and almost from the outset, on the idea of causality as a two-way algorithm: except the First Cause, everything has a cause as well as an effect. However, because of their paramount interest in the First Cause, the Greek thinkers focused their

11. For instance, Oriental logic required that the premise of the syllogism should include an example so as to eliminate vacuous truth: "Where there is smoke, there is fire, as in the kitchen." (See Chang Wing-Tsit, "The Spirit of Oriental Philosophy," in *Philosophy: East and West*, ed. Charles A. Moore, Princeton, 1944, p. 162.) However, the logic of the East developed mainly along highly dialectical lines. (Chang Wing-Tsit, "The Story of Chinese Philosophy," in the same volume, pp. 41 ff.)

12. J. Burnet, *Early Greek Philosophy* (4th edn., London, 1930), p. 47.

13. *Ibid.*, p. 64.

attention on cause rather than on effect.[14] As we know, Aristotle turned around the notion of cause until he discovered four forms of it.[15]

To search for the proximate cause, at least, came to be regarded as one of the noblest activities of the mind, second only to the search for the First Cause. Remembering facts — the Greeks held — is half-knowledge, i.e., mere *opinion*; true knowledge, i.e., *understanding*, includes also knowing *causa rerum*. This view had already gained such a strong ground by Plato's time that Aristotle had no difficulty in setting it up as an unquestionable dogma.[16] There is no exaggeration in saying that the distinctive feature of Greek thought was its obsessive preoccupation with "why?"

But this obsession still does not suffice by itself to explain the marriage of Logic and science in Greek thought. The marriage was possible because of one peculiar confusion: between "the why" and "the logical ground," that is, between *causa efficiens* and *causa formalis*. The symptom is obvious in Aristotle's bringing them together in his four types of causes,[17] and even more so in our using "explanation" in two distinct senses, each related to one of the *causae* just mentioned.[18] Had Logic by chance been applied first to constructing a theoretical science in a different field from geometry — where things neither move nor change, but merely are — the war now fought between logical positivists and realists would have very likely exploded soon after the first *Elements*.

As partakers of the Western mind we are apt to believe that causality represents, if not an *a priori* form in Kant's sense, at least one of the earliest notions inevitably grasped by the primeval man.[19] Yet the brute fact is that in contrast to Greek civilization the ancient cultures of Asia never developed the idea of causality.[20] It was thus impossible for them to link the logical syllogism with the causal algorithm and organize factual knowledge theoretically. However, we cannot blame only the absence of theoretical science for the very well-known fact that over the last two millennia or so factual knowledge in the East progressed little,

14. The first formulation on record of the principle of causality (by Leukippos, middle of the fifth century B.C.) speaks for itself: "Naught happens for nothing, but everything from a ground and of necessity." *Ibid.*, p. 340.

15. *Physics*, II. 3; *Metaphysics*, I. 3, 10; V. 2.

16. Cf. Plato, *Meno*, 81–86, *Theaetetus*, 201 ff; Aristotle, *Analytica Posteriora*, 78a 23, 88b 30, 94a 20, *Physics*, 194b 15.

17. The point is obvious in Aristotle's beginning section iii of *Physics*, II, with the remark that there are "many senses [in which] 'because' may answer 'why.' " See also *Metaphysics*, 1013b 3–4.

18. Cf. John Stuart Mill, *A System of Logic* (8th edn., New York, 1874), pp. 332 ff.

19. E.g., Mach, *Lectures*, p. 190.

20. Cf. Junjiro Takakusu, "Buddhism, as a Philosophy of 'Thusness,' " in *Philosophy: East and West*, already cited, pp. 75 ff.

if at all, despite the substantial advance it had over the West at the outset.[21] Other factors as well counted heavily in the balance.

While in Greece philosophers were searching for the First Cause, in India, for instance, they were bending their efforts to discover the Absolute Essence behind the veil of Māyā. While the Greeks believed that truth is reached by ratiocination, the Indians held that truth is revealed through contemplation. Now, contemplation has some unquestionable merits: without it we could not arrive even at pure description, nor strike upon the interpretative fictions of modern science. But a purely contemplative mind, even if it may see things that "have completely eluded us," [22] can hardly observe systematically the happenings of all sorts in the domain of natural phenomena; still less can such a mind think of devising and carrying out experiments. With the contemplative bent of the intellectual elite, the growth of factual knowledge remained in the East dependent solely upon the accidental discoveries made by the craftsman, whose mind is ordinarily less qualified than the scholar's to observe and evaluate.[23]

7. *Theoretical Science: A Continuous Source of Experimental Suggestions.* The last remarks raise a new question: had the Eastern scholars not shrunk from observing ordinary natural phenomena, could their factual knowledge have grown as much as that of the West? In other words, is theoretical science not only a more convenient storage of knowledge but also a more efficient instrument than crude empiricism in expanding knowledge? On this we often hear two contradictory views: that most revolutionary discoveries have been made accidentally, and that theory frees us from depending on accidental discoveries.

On closer examination, however, it is seen that the notion of entirely new factual knowledge completely divorced from accident is a contradiction in terms. Yet this does not mean that there is no way by which

21. Notice of the difference is not new: Cf. G. W. F. Hegel, *Lectures on the Philosophy of History* (London, 1888), pp. 141 ff.

22. Cf. William Ernest Hocking, "Value of the Comparative Study of Philosophy," in *Philosophy: East and West*, p. 3.

23. As I have already remarked, without theoretical science the storing of knowledge with easy access relies exclusively on good memory. This is not unrelated to the survival of the ideographic writing in the Far East. We should also note that this survival constitutes a tremendous intellectual handicap: ideographic writing narrows the number of actual scholars and, further, wastes much of their intellectual energy. Nowadays it prohibits the use of typewriters and linotype machines, a far more general loss for the communities involved.

In the case of China the survival may have some justification: the multiplicity of dialects within an otherwise unitary culture. But the survival in Japan of a hybrid writing that defies any systematic rule whatever constitutes a puzzle which appears all the more intriguing in view of the development miracle achieved by the Japanese economy.

the probability of "lucky" accidents might be increased. Clearly, the advice "do not wait for accidents to happen but cause them by continuous experimenting" is excellent, but in following it we are bound to be confronted sooner or later with the problem of what experiment to undertake next. Were we entirely to depend upon imagination to furnish us with new experimenting suggestions, we would not be able to comply with the advice, for imagination is moody and often bears no fruit for long stretches of time. Is there a way out?

Let us observe that although the work of imagination is indispensable in the process of logical discovery, it is not so at all times. Logic, in all its forms, has some automatic rules which can keep the process moving for quite long strides without any outside aid. As this has been put, more often than not the tip of the pen displays an intelligence greater than the writer's. Besides, the road of logical inquiry always branches into so many directions that it is highly improbable that all ratiocinating processes should stall simultaneously because all have reached the stage where imagination is needed to start them running again. Consequently, new propositions can be derived from the logical foundation of a science without interruption. One physiological function of theoretical science is precisely the continuous derivation of *new* propositions, i.e., of propositions not already included in (β). As a result, laboratories are never short of new ideas to be tested experimentally and no total eclipse of the sun, for instance, occurs without immense experimental stir. It is clear then that the second economic advantage of theoretical science consists in the fact that experimental resources are always fully employed.

8. *Theoretical Science and the Analytical Habit.* At this juncture it is important to observe that the greatest strides in knowledge are made when a logically derived proposition is refuted by experiment, not when it is confirmed. We must then ask the pertinent question whether the full employment of experimental resources in testing logically derived propositions enhances the chance of such a lucky accident, i.e., of a real discovery. It is rather strange that the dogma of the rationality of reality, intended exclusively for proving the superiority of theoretical science, cannot help us in answering the question in the affirmative. For if reality is rational, the nearer science gets to it the greater is the probability that a logically derived proposition shall pass the experimental test. On the other hand, if we argue that the facts suffice to answer the question affirmatively, then we must forcibly conclude that reality is antirational, not merely irrational. The problem is extremely delicate.

It was the Eleatics who first propounded that "the thing that can be thought and that for the sake of which the thought exists is the same." [24] The dogma reached its apogee in the rationalist philosophy of the eight-

24. From a fragment of *Parmenides;* Burnet, *Early Greek Philosophy*, p. 176.

eenth century, when it could with immense pride invoke the new theoretical science, Newtonian mechanics, in its own support. But we know equally well that with almost every great discovery of the last hundred years rationalism has received a decisive blow.

Indeed, if reality is rational there can be no logical contradiction between any two factual propositions; in particular, the logical foundation of a science must be not only nonredundant — as warranted by the logical algorithm by which it is constructed — but also noncontradictory. However, contradictions have come up periodically in physics. To mention the most eloquent ones: (1) though in mechanics motion is indifferent to direction, heat moves only from the hotter to the colder body; (2) the electron appears at times as a localized particle, at others, as a wave filling the whole space.[25] Even Einstein, who categorically refused to renege the rationalist dogma, had to admit that "for the time being, . . . we do not possess any general theoretical basis for physics, which can be regarded as its logical foundation." [26] And since "for the time being" seemed to perpetuate itself, Niels Bohr proposed a new epistemological tenet known as the Principle of Complementarity: "Only the totality of phenomena exhausts the possible information about the objects." [27] For example, two theories of the electron, a corpuscular and a wave theory — mutually contradictory but each noncontradictory within itself — must be accepted side by side: which one to use depends upon the particular phenomenon observed. Or as Bridgman put it more directly, "the only sort of theory possible is a partial theory of limited aspects of the whole." [28]

The interesting fact is that even those men of science who repudiate the rationalist dogma behave like many atheists: they reject the gospel but follow its teachings. Regardless of their metaphysical beliefs, all thus strive to arrange facts in logical order. The origins of this mental habit, for habit it is, go back to the time of the first theoretical science, that is, to the first *Elements of Geometry*. The way it came about is familiar to us from the attitude of a housewife after using a labor-saving gadget for the first time: men of science, too, after having experienced the economic advantages of theoretical science, refuse to do without it. By tasting knowledge in the theoretical form only once, the human mind becomes infected with an incurable virus, as it were, which produces an irresistible

25. For further details see, for instance, R. E. Peierls, *The Laws of Nature* (London, 1957), pp. 152, 182, 246, and *passim*.

26. Albert Einstein, *Out of My Later Years* (New York, 1950), p. 110; also p. 71. Max Planck, in *The Universe in the Light of Modern Physics* (New York, 1931), p. 94, is more categorical: relativity theory and quantum mechanics "are even antagonistic."

27. Niels Bohr, *Atomic Physics and Human Knowledge* (New York, 1958), pp. 40, 90. See also Werner Heisenberg, *Physics and Philosophy: The Revolution in Modern Science* (New York, 1958), pp. 162–163.

28. P. W. Bridgman, *The Nature of Physical Theory* (Princeton, 1936), p. 118.

craving for logical order. This is why, whenever a Spencerian tragedy — a theory killed by a fact — takes place, the minds of the scholarly world know no rest until a new logical foundation is laid out.

Though economy of thought is the reason why the human mind acquired the analytical habit, this habit in turn has a very important economic role. Thanks to this habit, experimenting ceases to be a routine procedure by which the factual truth-value of logically derived propositions is established. By stimulating the imagination of the experimenting scholar, the analytical habit is alone responsible for the fact that theoretical experimenting is far luckier than mere experimenting. In the experimental undertaking, as Pasteur once observed, chance favors only the prepared minds.

Moreover, the analytical mind creates what it craves for: logical order. During the centuries that elapsed between Euclid and Newton, it slowly created patches of logically ordered knowledge, gradually increased the area of each patch, and ultimately united some in a single unit: theoretical mechanics. As I have said, whenever a theory is destroyed the analytical mind immediately sets out to rebuild a new logical foundation on the ashes of the old. The most important work in this reconstruction is the building of entirely new concepts. These concepts then open up new experimenting grounds, thus extending the fields upon which we harvest new factual knowledge. Thanks to the analytical habit, every Spencerian tragedy is followed by a scientific bonanza. On the other hand, as Einstein cautioned us, we should not believe that this habit consists only of pure logical thinking: it requires above all "intuition, resting on *sympathetic* understanding of experience," [29] and — I may add — a consummate intellectual phantasy. Logic helps us only present thought already thought out, but it does not help us think out thoughts.[30]

From the preceding analysis it follows that the immense difference between East and West in the progress of factual knowledge constitutes no evidence in support of a rational reality. It does prove, however, that theoretical science is thus far the most successful device for learning reality *given the scarcity pattern of the basic faculties of the human mind.*

9. *Theoretical Science: A Living Organism.* The main thesis developed in this chapter is that theoretical science is a living organism precisely because it emerged from an amorphous structure — the taxonomic science — just as life emerged from inert matter. Further, as life did not appear everywhere there was matter, so theoretical science did not grow wherever taxonomic science existed: its genesis was a historical accident.

29. Albert Einstein, *Ideas and Opinions* (New York, 1954), p. 226; also p. 270. My italics.

30. A well-documented and highly penetrating discussion of this problem is offered by Jacques Hadamard, *An Essay on the Psychology of Invention in the Mathematical Field* (Princeton, 1945).

The analogy extends still further. Recalling that "science is what scientists do," we can regard theoretical science as a purposive mechanism that reproduces, grows, and preserves itself. It reproduces itself because any "forgotten" proposition can be rediscovered by ratiocination from the logical foundation. It grows because from the same foundation new propositions are continuously derived, many of which are found factually true. It also preserves its essence because when destructive contradiction invades its body a series of factors is automatically set in motion to get rid of the intruder.

To sum up: *Anatomically*, theoretical science is logically ordered knowledge. A mere catalog of facts, as we say nowadays, is no more science than the materials in a lumber yard are a house. *Physiologically*, it is a continuous secretion of experimental suggestions which are tested and organically integrated into the science's anatomy. In other words, theoretical science continuously creates new facts from old facts, but its growth is organic, not accretionary. Its anabolism is an extremely complex process which at times may even alter the anatomic structure. We call this process "explanation" even when we cry out "science does not explain anything." [31] *Teleologically*, theoretical science is an organism in search of new knowledge.

Some claim that the purpose of science is prediction. This is the practical man's viewpoint even when it is endorsed by such scholars as Benedetto Croce or Frank Knight.[32] Neo-Machians go even further. Just as Mach focused his attention on economy of thought without regard for the special role of logical order, they claim that practical success is all that counts; understanding is irrelevant. No doubt, if science had no utility for the practical man, who acts on the basis of predictions, scientists would now be playing their little game only in private clubs, like the chess enthusiasts. However, even though prediction is the touchstone of scientific knowledge — "in practice man must prove the truth," as Marx said —[33] the purpose of science in general is not prediction, but knowledge for its own sake. Beginning with Pythagoras' school, science ceased to serve exclusively the needs of business and has remained always ahead of these.[34] The practical man may find it hard to imagine that what animates science is a delight of the analytical habit and idle curiosity;

31. Alfred A. Lotka, *Elements of Physical Biology* (Baltimore, 1925), p. 389.
32. Frank H. Knight, *The Ethics of Competition* (New York, 1935), pp. 109–110.
33. F. Engels, *Ludwig Feuerbach and the Outcome of Classical German Philosophy* (London, 1947), p. 76.
34. Burnet, *Early Greek Philosophy*, p. 99; P. W. Bridgman, *Reflections of a Physicist* (2nd edn., New York, 1955), pp. 348–349. What might happen to this relation in the very immediate future is a matter of speculation. But the contention of F. Engels, in *On Historical Materialism* (New York, 1940), p. 14, that science did not exist before the bourgeois society because only this society could not live without it, is a far cry from the truth.

hence, he might never realize what is the source of his greatest fortune. The overwhelming majority of great scholars react like Jacques Hadamard, who confessed to losing all interest in a problem as soon as he solved it.[35]

Others say that science is experimenting. As far as theoretical science at least is concerned, this view confuses the whole organism with one of its physiological functions. Those who commit this error usually proclaim that "Bacon [is science's] St. John the Baptist." [36] Naturally, they also blame Aristotle's philosophy of knowledge with its emphasis on Logic for the marasmus of science until Francis Bacon's time. Facts have never been more ignored. To begin with, Aristotle never denied the importance of experience; one eloquent quotation will suffice: "If at any future time [new facts] are ascertained, then credence must be given rather to observation than to theories and to theories only if what they affirm agrees with the observed facts." [37] In relation to the time in which he lived he was one of the greatest experimenters and keenest observers. As Darwin judged, Linnaeus and Cuvier are "mere schoolboys to old Aristotle." [38] His teachings should not be blamed for what Scholasticism did with them. Finally, mechanics was already moving fast on Aristotelian theoretical tracks at the time Bacon's works appeared. Without the analytical habit which had been kept alive by Euclid's *Elements* and Aristotle's writings, Kepler, Galileo, and Newton, as well as all the great men of science that came later, would have had to join the Sino-Indians in contemplative and casual observation of nature.[39] To the extent to which we may turn history around in thought, we may reason that without the peculiar love the Greeks had for Understanding,[40] our knowledge would not by far have reached its present level; nor would modern civilization be what it is today. For better or for worse, we have not yet discovered one single problem of Understanding that the Greek philosophers did not formulate.

35. Hadamard, *Psychology of Invention*, p. 60.
36. J. S. Huxley, "Science, Natural and Social," *Scientific Monthly*, L (1940), 5.
37. *De Generatione Animalium*, 760^b 30–33. Also *Metaphysics*, 981^a.
38. Quoted by G. R. G. Mure, *Aristotle* (New York, 1932), p. 124.
39. Cf. Alfred North Whitehead, *Process and Reality: An Essay in Cosmology* (New York, 1929), p. 7.
40. Cf. Plato, *Republic*, V. 435–436. Also W. T. Stace, *A Critical History of Greek Philosophy* (London, 1934), pp. 17–18; Cyril Bailey, *The Greek Atomists and Epicurus* (Oxford, 1928), p. 5.

TWO

CONCEPTS, NUMBERS, AND QUALITY

1. *"No Science Without Theory."* Theoretical science having the marvelous qualities just described, we can easily understand the sanguine hopes raised by Newton's success in transforming mechanics into such a science. At last, some two thousand years after Euclid's *Elements*, Newton's *Principia Mathematica* proved that theoretical science can grow in other domains besides geometry, and equally well. But sanguine hopes are sanguine hopes: thoughts on the matter, especially of those fascinated most by the powers of Logic, became prey to the confusion between "some fields" and "all fields." In the end almost everybody interpreted the evidence as proof that knowledge in *all* fields can be cast into a theoretical mold. Especially after the astounding discovery of Neptune "at the tip of Leverrier's pen," spirits ran high in all disciplines, and one scientist after another announced his intention of becoming the Newton of his own science. François Magendie aspired to place even physiology "on the same sure footing" as mechanics.[1] "Thus the confusion of tongues" — as one economist lamented — "was propagated from science to science." [2]

On the whole, the scientific temper has not changed much. To be sure, the position that mechanics constitutes the only road leading to divine knowledge — as Laplace argued in his magnificent apology[3] — has been officially abandoned by almost every special science. Curiously, the move was not caused by the recognition of the failures following the adoption of this position outside physics, but induced by the fact that physics itself had to reject it.[4] In place of "all sciences must imitate mechanics," the battle cry of the scholarly army is now "no science without theory." But the change is rather skin deep, for by "theory" is commonly under-

1. J. M. D. Olmsted and E. H. Olmsted, *Claude Bernard and the Experimental Method in Medicine* (New York, 1952), p. 23.
2. S. Bauer, quoted in J. S. Gambs, *Beyond Supply and Demand* (New York, 1946), p. 29n. My translation.
3. P. S. Laplace, *A Philosophical Essay on Probabilities* (New York, 1902), p. 4.
4. See chapter "The Decline of the Mechanical View" in A. Einstein and L. Infeld, *The Evolution in Physics* (New York, 1938).

stood a logical file of knowledge as exemplified by geometry and mechanics.[5]

No other science illustrates better than economics the impact of the enthusiasm for mechanistic epistemology upon its evolution. Does the transforming of economics into "a physico-mathematical science" require a measure of utility which escapes us? *"Eh bien!"* — exclaimed Walras characteristically — "this difficulty is not insurmountable. Let us suppose that this measure exists, and we shall be able to give an exact and mathematical account" of the influence of utility on prices, etc.[6] Unfortunately this uncritical attitude has ever since constituted the distinct flavor of mathematical economics. In view of the fact that theoretical science is a living organism, it would not be exaggerating to say that this attitude is tantamount to planning a fish hatchery in a flower bed.

Jevons showed some concern over whether the new environment — the economic field — would contain the basic elements necessary for the theoretical organism to grow and survive. Indeed, before declaring his intention to rebuild economics as *"the mechanics of utility and self-interest,"* he took pains to point out that in the domain of economic phenomena there is plenty of quantitative "moisture" in "the private-account books, the great ledgers of merchants and bankers and public offices, the share list, price lists, bank returns, monetary intelligence, Custom-house and other Government returns." [7] But Jevons, like many others after him, failed to go on to explain how ordinary statistical data could be substituted for the variables of his mechanical equations. By merely expressing the hope that statistics might become "more complete and accurate . . . so that our formulae could be endowed with exact meaning," [8] Jevons set an often-followed pattern for avoiding the issue.

Certainly, after it was discovered that theoretical science can function properly in another domain besides geometry, scientists would have been derelict if they had failed to try out "a fish hatchery in a flower bed." For trying, though not sufficient, is as absolutely necessary for the advancement of knowledge as it is for biological evolution. This is why we cannot cease to admire men like Jevons and Walras, or numerous others who even in physics hurried to adopt a new viewpoint without first testing their ground.[9] But our admiration for such unique feats does not

5. The point has been repeatedly recognized by numerous scholars: e.g., Max Planck, *Scientific Autobiography and Other Papers* (New York, 1949), p. 152.

6. Léon Walras, *Éléments d'économie politique pure* (3rd edn., Lausanne, 1896), p. 97. My translation.

7. W. Stanley Jevons, *The Theory of Political Economy* (4th edn., London, 1924), p. 21 and p. 11.

8. *Ibid.*, p. 21.

9. Cf. P. W. Bridgman, *Reflections of a Physicist* (2nd edn., New York, 1955), p. 355.

justify persistence in a direction that trying has proved barren. Nor do
we serve the interest of science by glossing over the impossibility of
reducing the economic process to mechanical equations. In this respect,
a significant symptom is the fact that Carl Menger is placed by almost
every historian on a lower pedestal than either Walras or Jevons only
because he was more conservative in treating the same problem, the
subjective basis of value.[10] Moreover, in spite of the fact that no economy,
not even that of a Robinson Crusoe, has been so far described by a
Walrasian system in the same way in which the solar system has been
described by a Lagrange system of mechanical equations, there are voices
claiming that economics "has gone through its Newtonian revolution":
only the other social sciences are still awaiting their Galileo or Pasteur.[11]
Alfred North Whitehead's complaint that "the self-confidence of learned
people is the comic tragedy of [our] civilization" [12] may be unsavory but
does not seem entirely unfounded.

Opposition to Walras' and Jevons' claim that "economics, if it is to
be a science at all, must be a mathematical science," [13] has not failed
to manifest itself. But, in my opinion, during the ensuing controversies
swords have not been crossed over the crucial issue. For I believe that
what social sciences, nay, all sciences need is not so much a new Galileo
or a new Newton as a new Aristotle who would prescribe new rules for
handling those notions that Logic cannot deal with.

This is not an extravagant vision. For no matter how much we may
preen ourselves nowadays upon our latest scientific achievements, the
evolution of human thought has not come to a stop. To think that we
have even approached the end is either utter arrogance or mortifying
pessimism. We cannot therefore write off the possibility of striking one
day upon the proper mutant idea that would lead to an anatomy of
science capable of thriving equally well in natural as in social sciences.
On rare occasions we find this hope more clearly expressed with the
extremely pertinent remark that in such a unifying science physics will
be "swallowed up" by biology, not the reverse.[14] Or, as Whitehead put it
more sharply, "murder is the prerequisite for the absorption of biology

10. E.g., K. Wicksell, *Value, Capital and Rent* (London, 1954), p. 53; Joseph A.
Schumpeter, *History of Economic Analysis* (New York, 1954), p. 918. Among the few
exceptions: Frank H. Knight, "Marginal Utility Economics," *Encyclopaedia of the
Social Sciences* (New York, 1931), V, 363; George Stigler, *Production and Distribution
Theories* (New York, 1948), p. 134.
11. Karl R. Popper, *The Poverty of Historicism* (Boston, 1957), p. 60 and note.
12. Alfred North Whitehead, *Science and Philosophy* (New York, 1948), p. 103.
13. Jevons, *Theory*, p. 3.
14. Cf. J. S. Haldane, *The Sciences and Philosophy* (New York, 1929), p. 211. Also
Erwin Schrödinger, *What Is Life?* (Cambridge, Eng., 1944), pp. 68–69; R. E. Peierls,
The Laws of Nature (London, 1957), p. 277; L. von Bertalanffy, *Problems of Life*
(New York, 1952), p. 153.

into physics." [15] A historical precedent already exists: physicists and scientific philosophers had for a long time denied that "scientific" laws exist outside physics and chemistry, because only there do we find rigidly binding relations. Today they work hard to convince everybody that on the contrary the laws of nature are not rigid but stochastic and that the rigid law is only a limiting, hence highly special, case of the stochastic law. Somehow they usually fail to point out that the latter type of law is not a native of physical science but of the life sciences.

The history of human thought, therefore, teaches us that nothing can be extravagant in relation to what thought might discover or where. It is all the more necessary for us to recognize fully the source as well as the nature of our difficulty *at present*.

2. *Theoretical Science versus Science.* The first condition an environment must satisfy in order to sustain the life of a certain organism is to contain the chemical elements found in the anatomy of that organism. If it does not, we need not go any further. Let us, therefore, begin our inquiry by a "chemical" analysis of the anatomy of theoretical science.

As I have pointed out, the *causa materialis* of science, not only of theoretical science, consists of descriptive propositions. I have further explained that the distinctive feature of theoretical science is its logically ordered anatomy. Whoever is willing to look at the brute facts and accept some of their unpleasantness, will agree that in some phenomenal domains an overwhelming majority of descriptive propositions do not possess the "chemical" properties required by logical ordering.

I can hardly overemphasize the fact that Logic, understood in its current Aristotelian sense, is capable of dealing only with one distinct class of propositions, such as

A. *The hypotenuse is greater than the leg,*

but it is largely impotent when it comes to propositions such as

B. *Culturally determined wants are higher than biological wants,*

or

C. *Woodrow Wilson had a decisive influence upon the Versailles Peace Treaty.*

A logician would hardly deny this difference. But many, especially the logical positivists, would argue that propositions such as B or C are meaningless and, hence, the difference does not prove at all the limitation of Logic. This position is clearly explained by Max Black: *red* being a vague concept, the question "Is this color red?" has scarcely any mean-

15. Alfred North Whitehead, "Time, Space, and Material," in *Problems of Science and Philosophy*, Aristotelian Society, suppl. vol. 2, 1919, p. 45.

ing.[16] However, the use of the term "meaningless" for propositions that Logic cannot handle is a clever artifice for begging a vital question.

At bottom, the issue is whether knowledge is authentic only if it can be unified into a theory. In other words, is theoretical science the only form of scientific knowledge? The issue resolves into several questions: the first is what accounts for Logic's impotence to deal with "meaningless" propositions.

3. *Numbers and Arithmomorphic Concepts.* The boundaries of every science of fact are moving penumbras. Physics mingles with chemistry, chemistry with biology, economics with political science and sociology, and so on. There exists a physical chemistry, a biochemistry, and even a political economy in spite of our unwillingness to speak of it. Only the domain of Logic — conceived as *Principia Mathematica* — is limited by rigidly set and sharply drawn boundaries. The reason for this is that *discrete* distinction constitutes the very essence of Logic: perforce, discrete distinction must apply to Logic's own boundaries.

The elementary basis of *discrete* distinction is the distinction between two written symbols: between "m" and "n," "3" and "8," "excerpt" and "except," and so on. As these illustrations show, good symbolism requires perfect legibility of writing; otherwise we might not be able to distinguish without the shadow of a doubt between the members of the same pair. By the same token, spoken symbolism requires distinct pronunciation, without lisping or mumbling.

Symbols, as we know, have one and only one purpose: to represent concepts visually or audibly.[17] We also know that logic deals with symbols *qua* representatives of concepts. But we do not go, it seems, so far as to realize (or to admit if we realize) that the fundamental principle upon which Logic rests is that *the property of discrete distinction should cover not only symbols but concepts as well.*

As long as this principle is regarded as normative no one could possibly quarrel over it. On the contrary, no one could deny the immense advantages derived from following the norm whenever possible. But it is often presented as a general law of thought. A more glaring example of Whitehead's "fallacy of misplaced concreteness" than such a position would be hard to find. To support it some have gone so far as to maintain that we can think but in words. If this were true, then thoughts would become a "symbol" of the words, a most fantastic reversal of the relationship between means and ends. Although the absurdity has been repeatedly

16. Max Black, *The Nature of Mathematics* (New York, 1935), p. 100n.
17. This limitation follows the usual line, which ignores tactile symbolism: taps on the shoulder, hand shakes, etc. Braille and especially the case of Helen Keller prove that tactile symbolism can be as discretely distinct and as efficient as the other two. Its only shortcoming is the impossibility of immediate transmission at a distance.

exposed, it still survives under the skin of logical positivism.[18] Pareto did not first coin the word "ophelimity" and then think of the concept. Besides, thought is so fluid that even the weaker claim, namely, that we can coin a word for every thought, is absurd. "The Fallacy of the Perfect Dictionary" [19] is plain: even a perfect dictionary is molecular while thought is continuous in the most absolute sense. Plain also is the reason for and the meaning of the remark that "in symbols truth is darkened and veiled by the sensuous element." [20]

Since any particular real number constitutes the most elementary example of a discretely distinct concept, I propose to call any such concept *arithmomorphic*. Indeed, despite the use of the term "continuum" for the set of all real numbers, within the continuum every real number retains a *distinct individuality* in all respects identical to that of an integer within the sequence of natural numbers. The number π, for instance, is discretely distinct from any other number, be it 3.141592653589793 or 10^{100}. So is the concept of "circle" from "10^{100}-gon" or from "square," and "electron" from "proton." In Logic "is" and "is not," "belongs" and "does not belong," "some" and "all," too, are *discretely* distinct.

Every arithmomorphic concept stands by itself in the same specific manner in which every "Ego" stands by itself perfectly conscious of its absolute differentiation from all other "Egos." This is, no doubt, the reason why our minds crave arithmomorphic concepts, which are as translucent as the feeling of one's own existence. Arithmomorphic concepts, to put it more directly, *do not overlap*. It is this peculiar (and restrictive) property of the material with which Logic can work that accounts for its tremendous efficiency: without this property we could neither compute, nor syllogize, nor construct a theoretical science. But, as happens with all powers, that of Logic too is limited by its own ground.

4. *Dialectical Concepts.* The antinomy between One and Many with which Plato, in particular, struggled is well known. It arises from the fact that the quality of discrete distinction does not necessarily pass from the arithmomorphic concept to its concrete denotations. There are, however, cases where the transfer operates. Four pencils are an "even number" of pencils; a concrete triangle is not a "square." Nor is there any great difficulty in deciding that Louis XIV constitutes a denotation of "king." But we can never be absolutely sure whether a concrete quad-

---

18. For a discussion of the psychological evidence against the equation "thought = word," see Jacques Hadamard, *An Essay on the Psychology of Invention in the Mathematical Field* (Princeton, 1945), pp. 66 ff. For what it might be worth, as one who is multilingual I can vouch that I seldom think in any language, except just before expressing my thoughts orally or in writing.

19. Alfred North Whitehead, *Modes of Thought* (New York, 1938), p. 235. See also P. W. Bridgman, *The Intelligent Individual and Society* (New York, 1938), pp. 69–70.

20. G. W. F. Hegel, *Hegel's Science of Logic*, 2 vols. (London, 1951), I, 231.

rangle is a "square." [21] In the world of ideas "square" is One, but in the world of the senses it is Many.

On the other hand, if we are apt to debate endlessly whether a particular country is a "democracy" it is above all because the concept itself appears as Many, that is, it is not discretely distinct. If this is true, all the more the concrete cannot be One. A vast number of concepts belong to this very category; among them are the most vital concepts for human judgments, like "good," "justice," "likelihood," "want," etc. They have no arithmomorphic boundaries; instead, *they are surrounded by a penumbra within which they overlap with their opposites.*

At a particular historical moment a nation may be both a "democracy" and a "nondemocracy," just as there is an age when a man is both "young" and "old." Biologists have lately realized that even "life" has no arithmomorphic boundary: there are some crystal-viruses that constitute a penumbra between living and dead matter.[22] Any particular want, as I have argued along well-trodden but abandoned trails, imperceptibly slides into other wants.[23]

It goes without saying that to the category of concepts just illustrated we cannot apply the fundamental law of Logic, the Principle of Contradiction: "B cannot be both A and non-A." On the contrary, we must accept that *in certain instances* at least, "B is both A and non-A" is the case. Since the latter principle is one cornerstone of Hegel's Dialectics, I propose to refer to the concepts that may violate the Principle of Contradiction as *dialectical.*[24]

In order to make it clear what we understand by dialectical concept, two points need special emphasis.

First, the impossibility mentioned earlier of deciding whether a concrete quadrangle is "square" has its roots in the imperfection of our senses and of their extensions, the measuring instruments. A *perfect* instrument would remove it. On the other hand, the difficulty of deciding whether a particular country is a democracy has nothing to do — as I shall explain in detail presently — with the imperfection of our sensory organs. It

---

21. Strangely, logicians do not argue that because of this fact, "square" is a *vague* concept and "Is this quadrangle a square?" has no meaning. Cf. Black as cited in my note 16.

22. On the arithmomorphic definition of life, see Alfred A. Lotka, *Elements of Physical Biology* (Baltimore, 1925), chap. i and p. 218n.

23. My essay entitled "Choice, Expectations and Measurability" (1954), reprinted in this book.

24. The connection between dialectical concepts thus defined and Hegelian logic is not confined to this principle. However, even though the line followed by the present argument is inspired by Hegel's logic, it does not follow Hegel in all respects. We have been warned, and on good reasons, that one may ignore Hegel at tremendous risks. To follow Hegel only in part might very well be the greatest risk of all; yet I have no choice but to take this risk.

arises from another "imperfection," namely, that of our thought which cannot always reduce an apprehended notion to an arithmomorphic concept. Of course, one may suggest that in this case too the difficulty would not exist for a *perfect* mind. However, the analogy does not seem to hold. For while the notion of a perfect measuring instrument is sufficiently clear (and moreover indispensable even for explaining the indeterminacy in physical measurements), the notion of a perfect mind is at most a verbal concoction. There is no direct bridge between an imperfect and the perfect measuring instrument. By the same token, the imperfect mind cannot know how a perfect mind would actually operate. It would itself become perfect the moment it knew how.

The second point is that a dialectical concept — in my sense — does not overlap with its opposite *throughout the entire range of denotations.* To wit, in most cases we can decide whether a thing, or a particular concept, represents a living organism or lifeless matter. If this were not so, then certainly dialectical concepts would be not only useless but also harmful. Though they are not *discretely distinct*, dialectical concepts are nevertheless *distinct*. The difference is this. A penumbra separates a dialectical concept from its opposite. In the case of an arithmomorphic concept the separation consists of a void: *tertium non datur* — there is no third case. The extremely important point is that the separating penumbra itself is a dialectical concept. Indeed, if the penumbra of A had arithmomorphic boundaries, then we could readily construct an arithmomorphic structure consisting of three discretely distinct notions: "proper A," "proper non-A," and "indifferent A." The procedure is most familiar to the student of consumer's choice where we take it for granted that between "preference" and "nonpreference" there *must* be "indifference." [25]

Undoubtedly, a penumbra surrounded by another penumbra confronts us with an infinite regress. But there is no point in condemning dialectical concepts because of this aspect: in the end the dialectical infinite regress resolves itself just as the infinite regress of Achilles running after the tortoise comes to an actual end. As Schumpeter rightly protested, there is "no sense in our case in asking: 'Where does that type [of entrepreneur] begin then?' and then to exclaim: 'This is no type at all!' " [26] Far from being a deadly sin, the infinite regress of the dialectical penumbra constitutes the salient merit of the dialectical concepts: as we shall see, it reflects the most essential aspect of Change.

5. *Platonic Traditions in Modern Thought.* To solve the perplexing prob-

25. Cf. my essay "Choice, Expectations and Measurability" (1954), reprinted in this book.

26. Joseph A. Schumpeter, *The Theory of Economic Development* (Cambridge, Mass., 1949), p. 82n.

lem of One and Many, Plato taught that ideas live in a world of their own, "the upper-world," where each retains "a permanent individuality" and, moreover, remains "the same and unchanging." [27] Things of the "lower-world" partake of these ideas, that is, resemble them.[28] The pivot of Plato's epistemology is that we are born with a latent knowledge of all ideas — as Kant was to argue later about some notions — because our immortal soul has visited their world some time in the past. Every one of us, therefore, can learn ideas by reminiscence.[29]

Plato's extreme idealism can hardly stir open applause nowadays. Yet his mystical explanation of how ideas are revealed to us in their purest form underlies many modern thoughts on "clear thinking." The Platonic tenet that only a privileged few are acquainted with ideas but cannot describe them publicly, is manifest, for example, in Darwin's position that "species" is that form which is so classified by "the opinion of naturalists having sound judgment and wide experience." [30] Even more Platonic in essence is the frequently heard view that "constitutional law" has one and only one definition: it is the law pronounced as such by the U.S. Supreme Court if and when in a case brought before it the Court is explicitly asked for a ruling on this point.

There can be no doubt about the *fact* that a consummate naturalist or a Supreme Court justice is far more qualified than the average individual for dealing with the problem of species or constitutional law. But that is not what the upholders of this sort of definition usually mean: they claim that the definitions are operational and, hence, dispose of the greatest enemy of clear thought — vagueness. It is obvious, however, that the claim is specious: the result of the defining operation is not One but Many.[31] Not only is the operation extremely cumbersome, even wholly impractical at times, but the definition offers no enlightenment to the student. Before anyone becomes an authority on evolution, and even thereafter, he needs to know what "fitness" means without waiting until natural selection will have eliminated the unfit. Science cannot be satisfied with the idea that the only way to find out whether a mushroom is poisonous is to eat it.

Sociology and political science, in particular, abound in examples of

27. *Phaedo*, 78, *Philebus*, 15. Plato's doctrine of ideas being "fixed patterns" permeates all his Dialogues. For just a few additional references, *Parmenides*, 129 ff, *Cratylus*, 439–440.

28. *Phaedo*, 100 ff. It is significant that although Plato (*Phaedo*, 104) illustrates the discrete distinction of ideas by referring to integral numbers, he never discusses the problem why some things partake fully and others only partly of ideas.

29. *Meno*, 81–82, *Phaedo*, 73 ff, *Phaedrus*, 249–250.

30. Charles Darwin, *The Origin of Species* (6th edn., London, 1898), p. 34.

31. As Charles Darwin himself observes in a different place, *The Descent of Man* (2nd edn., New York, n.d.), p. 190: Thirteen eminent naturalists differed so widely as to divide the human species into as few as two and as many as sixty-three races!

another form of disguised Platonic rationale. For instance, arguments often proceed, however unawares, from the position that the pure idea of "democracy" is represented by one particular country — usually the writer's: all other countries only partake of this idea in varying degrees.

Plato's *Dialogues* leave no doubt that he was perfectly aware of the fact that we know concepts either by definition or by intuition. He realized that since definition constitutes a public description, anyone may learn to know a concept by definition. He also realized that we can get acquainted with some concepts only by direct apprehension supplemented by Socratic analysis.[32] Plato's difficulty comes from his belief *that regardless of their formation all concepts are arithmomorphic*, that "everything resembles a number," as his good friend Xenocrates was to teach later. One Dialogue after another proves that although Plato was bothered by the difficulties of definition in the case of many concepts, he never doubted that in the end all concepts can be defined. Very likely, Plato — like many after him — indiscriminately extrapolated the past: since all defined concepts have at one time been concepts by intuition, all present concepts by intuition must necessarily become concepts by definition.

The issue may be illustrated by one of our previous examples. Should we strive for an arithmomorphic concept of "democracy," we would soon discover that no democratic country fits the concept: not Switzerland, because Swiss women have no voting right; not the United States, because it has no popular referendum; not the United Kingdom, because the Parliament cannot meet without the solemn approval of the King, and so on down the line. The penumbra that separates "democracy" from "autocracy" is indeed very wide. As a result, "even the dictatorship of Hitler in National-Socialist Germany had democratic features, and in the democracy of the United States we find certain dictatorial elements."[33] But this does not mean that Hitlerite Germany and the United States must be thrown together in the same conceptual pot, any more than the existence of a penumbra of viruses renders the distinction between "man" and "stone" senseless.

Furthermore, the efforts to define democracy are thwarted by a more general and more convincing kind of difficulty than that just mentioned. Since "democracy" undoubtedly implies the right to vote but not for all ages, its definition must necessarily specify the *proper* limit of the voting age. Let us assume that we agree upon L being this limit. The natural question of why $L-\epsilon$ is not as good a limit fully reveals the impossibility

---

32. *Republic*, VI. 511. In all probability, it was this sort of analysis that Plato meant by "dialectics," but he never clarified this term.

33. Max Rheinstein in the "Introduction" to Max Weber, *On Law in Economy and Society* (Cambridge, Mass., 1954), p. xxxvi.

of taking care of all the imponderables of "democracy" by an arithmomorphic concept.

Of "democracy" as well as of "good," "want," etc., we can say what St. Augustine in essence said of Time: if you know nothing about it I cannot tell you what it is, but if you know even vaguely what it means let us talk about it.[34]

6. *Dialectical Concepts and Science*. No philosophical school, I think, would nowadays deny the existence of dialectical concepts as they have been defined above. But opinions as to their relationship to science and to knowledge in general vary between two extremes.

At one end we find every form of positivism proclaiming that whatever the purpose and uses of dialectical concepts, these concepts are antagonistic to science: knowledge proper exists only to the extent to which it is expressed in arithmomorphic concepts. The position recalls that of the Catholic Church: holy thought can be expressed only in Latin.

At the other end there are the Hegelians of all strains maintaining that knowledge is attained only with the aid of dialectical notions in the strict Hegelian sense, i.e., notions to which the principle "A is non-A" applies *always*.

There is, though, some definite asymmetry between the two opposing schools: no Hegelian — Hegel included — has ever denied either the unique ease with which thought handles arithmomorphic concepts or their tremendous usefulness.[35] For these concepts possess a built-in device against most kinds of errors of thought that dialectical concepts do not have. Because of this difference we are apt to associate dialectical concepts with loose thinking, even if we do not profess logical positivism. The by now famous expression "the muddled waters of Hegelian dialectics" speaks for itself. Moreover, the use of the antidialectical weapon has come to be the easiest way for disposing of someone else's argument.[36] Yet the highly significant fact is that no one has been able to present an argument against dialectical concepts without incessant recourse to them.

34. Saint Augustine, *Confessions*, XI. 17.
35. That Hegel's philosophy has been made responsible for almost every ideological abuse and variously denounced as "pure nonsense [that] had previously been known only in madhouses" or as "a monument to German stupidity," need not concern us. (Will Durant, in *The Story of Philosophy*, New York, 1953, p. 221, gives E. Caird, *Hegel*, London, 1883, as the source of these opinions; all my efforts to locate the quotation have been in vain.) But I must point out that the often-heard accusation that Hegel denied the great usefulness of mathematics or theoretical science is absolutely baseless: see *The Logic of Hegel*, tr. W. Wallace (2nd edn., London, 1904), p. 187.
36. Precisely because I wish to show that the sin is not confined to the rank and file, I shall mention that Knight within a single article denounces the concept of instinct as arbitrary and unscientific but uses the concept of want freely. Frank H. Knight, *The Ethics of Competition* (New York, 1935), p. 29 and *passim*.

We are badly mistaken if we believe that the presence of such terms as "only if" or "nothing but" in a sentence clears it of all "dialectical nonsense." As an eloquent example, we may take the sentence "A proposition has a meaning only if it is verifiable," and the sentence "When we speak of verifiability we mean *logical* possibility of verification, and nothing but this," [37] which together form the creed of the Vienna positivism. If one is not a positivist, perhaps he would admit that there is some sense in these tenets, despite the criticism he may have to offer. But if one is a full-fledged positivist, he must also claim that "the dividing line between logical possibility and impossibility is *absolutely sharp and distinct*; there is no gradual transition between meaning and nonsense." [38] Hence, for the two previous propositions to have a meaning, we need to describe "the logical possibility of [their] verification" in an absolutely sharp and distinct manner. To my knowledge, no one has yet offered such a description. Positivism does not seem to realize at all that the concept of verifiability, or the position that "the meaning of a proposition is the method of its verification" [39] is covered by a dialectical penumbra in spite of the apparent rigor of the sentences used in the argument. Of course, one can easily give examples of pure nonsense — "my friend died the day after tomorrow" is used by Moritz Schlick — or of pure arithmomorphic sense. However — as I have argued earlier — this does not dispose of a dialectical penumbra of graded differences of clearness between the two extreme cases. I hope the reader will not take offense at the unavoidable conclusion that most of the time all of us talk some nonsense, that is, express our thoughts in dialectical terms with no clear-cut meaning.

Some of the books written by the very writers who — like Bertrand Russell or Bridgman, for example — have looked upon combatting vagueness in science as a point of highest intellectual honor, constitute the most convincing proof that correct reasoning with dialectical concepts is not impossible. [40] Such a reasoning is a far more delicate operation than syllogizing with arithmomorphic concepts. Long ago, Blaise Pascal pointed out the difference between these two types of reasoning as well as their correlation with two distinct qualities of our intellect: *l'esprit géométrique* and *l'esprit de finesse*. [41] To blame dialectical concepts for any muddled thinking is, therefore, tantamount to blaming the artist's colors

37. Moritz Schlick, "Meaning and Verification," *Philosophical Review*, XLV (1936), 344, 349.

38. *Ibid.*, 352. My italics.

39. *Ibid.*, 341.

40. E.g., Bertrand Russell, *Principles of Social Reconstruction* (London, 1916), and P. W. Bridgman, *The Intelligent Individual and Society* (New York, 1938).

41. Pensées, 1–2, in Blaise Pascal, *Oeuvres complètes*, ed. J. Chevalier (Paris, 1954), pp. 1091 ff.

for what the artless — and even the talented at times — might do with
them.

Now, both *l'esprit géométrique* and *l'esprit de finesse* are acquired (or
developed) through proper training and exposure to as large a sample
of ideas as possible. And we cannot possibly deny that social scientists
generally possess enough *esprit de finesse* to interpret correctly the proposition "democracy allows for an equitable satisfaction of individual
wants" and to reason correctly with similar propositions where almost
every term is a dialectical concept. (And if some social scientists do not
possess enough *esprit de finesse* for the job, God help them!) The feat is
not by any means extraordinary. As Bridgman once observed, "little
Johnnie and I myself know perfectly well what I want when I tell him
to be good, although neither of us could describe exactly what we meant
under cross-examination." [42] But, no sooner has Bridgman recognized
this than he concludes that "we will not have a true social science until
eventually mankind has educated itself to be more *rational*." [43] These
two remarks taken together can mean only one thing: like many other
scientists, Bridgman equates "science" with "theoretical science." But
he, more than any other, has let us see his position clearly: even though
not all concepts are arithmomorphic and even though we can operate
successfully with dialectical concepts, these concepts have no place in
science. The question is whether this advice would not increase another
sort of muddled thinking which already plagues social sciences.

7. *Science and Change.* As I explained in the preceding chapter, Greek
philosophy began by asking what causes things to change. But the recognition of Change soon raised the most formidable question of epistemology. How is knowledge possible if things continuously change, if "you
cannot step twice into the *same* rivers," as the obscure Herakleitos
maintained?[44] Ever since, we have been struggling with the issue of what
is *same* in a world in flux. What is "same" in, say, a sodium vapor which,
as its temperature increases, turns from violet to a yellow glow, or in a
tumbler of water that continuously evaporates?[45] If we rest satisfied with
the argument of the continuity in time of the *things* observed, then we
must necessarily accept as perfectly scientific also the procedure by which
Lamaism decides who is the *same* Dalai Lama through death and birth.

On the other hand, if there were no Change at all, that is, if things have

42. Bridgman, *Intelligent Individual and Society*, p. 72; also pp. 56 ff.
43. Bridgman, *Reflections of a Physicist*, p. 451. My italics.
44. Fragment 41 in J. Burnet, *Early Greek Philosophy* (4th edn., London, 1930),
p. 136. My italics.
45. Ernst Mach, *Popular Scientific Lectures* (Chicago, 1895), p. 202; P. W.
Bridgman, *The Logic of Modern Physics* (New York, 1928), p. 35. Bridgman adds that
even $2 + 2 = 4$ collapses if applied to "spheres of a gas which expand and inter-
penetrate."

been and will always be as they are, all science would be reduced to a sort of geometry: *ubi materia, ibi geometria* — where there is matter, there is geometry — as Kepler thought.

The knot was cut but not untied by the distinction, introduced quite early, between change of nature and change of place.[46] And since, as Aristotle was to express it straightforwardly, "place is neither a part nor a quality of things," [47] it appeared expedient to resolve that all Change is locomotion, change of nature being only appearance. To avoid any reference whatever to quality, the ancient atomistic doctrine originated by Leukippos held that Change consists only of the locomotion of atomic particles of a *uniform* and *everlasting* matter. The first systematic criticism of monistic atomism came from Aristotle who opposed to it the doctrine of matter and form. This led him to analyze Change into change (1) of place, (2) of quantity (related to change by generation or annihilation), and (3) of quality.[48] Though we have ever since abided by this analysis in principle, the attitude of science toward Change has had a most uneven history.

To begin with, atomism suffered a total eclipse for some two thousand years until Dalton revived it at the beginning of the last century. It then gradually came to rule over almost every chapter of physics. However, the recent discoveries of one intra-atomic particle after another, all qualitatively different, have deprived monistic atomism of all its epistemological merit. Quality, being now recognized as a primary attribute of elementary matter, is no longer reducible to locomotion. For the time being, one point of Aristotle's doctrine is thus vindicated.

For quite a while change by generation and annihilation lingered in Scholastic speculations. But after the various principles of conservation discovered by physics during the last hundred years, we became convinced that this type of change was buried for good. Only cosmologists continued to talk about the creation of the universe. However, the idea that matter is continuously created and annihilated in every corner of the universe seems to have acquired increasing support recently. If it turns out to be a helpful hypothesis, then it will not only revolutionize cosmology but also solve the greatest mystery of physics, that of gravitation.[49] The universe also will become more intelligible because its laws will then become truly invariant with respect to Time. It is thus quite possible that we shall return to Aristotle's views and reconsider the modern axiom that "the energy concept without conservation is meaningless." [50]

46. See, for instance, Plato, *Parmenides*, 138.
47. Physics, 209[b] 26–27, 221[a] 1.
48. Physics, 190[a] 33–37, 260[a] 27–29.
49. Cf. Reginald O. Kapp, *Towards a Unified Cosmology* (New York, 1960), pp. 57, 104, *passim*, and the works of H. Bondi, T. Gold, F. Hoyle, and W. H. McCrea there quoted.
50. Bridgman, *Logic of Modern Physics*, p. 127.

Qualitative change has never ceased to be a central theme of the life sciences. But, time and again, the admiration produced by the operational successes of physics in almost every direction — in spite of its decision to ignore Change — misled us into thinking that science cannot study Change. Social scientists, in particular, continue to pay frequent lip service to this principle.[51] In spite of all these professions and the repeated arguments in their support, we may as well recognize that the highest ambition of any science is to discover the laws of whatever Change is manifest in its phenomenal domain. The task is extremely difficult, but challenge is the very soul of scientific activity.

8. *Change and Dialectical Concepts.* The undeniably difficult problem of describing qualitative change stems from one root: qualitative change eludes arithmomorphic schematization. The leitmotiv of Hegel's philosophy, "wherever there is movement, wherever there is life, wherever anything is carried into effect in the actual world, there Dialectic is at work," [52] is apt to be unpalatable to a mind seasoned by mechanistic philosophy. Yet the fact remains that Change is the fountainhead of all dialectical concepts. "Democracy," "feudalism," "monopolistic competition," for instance, are dialectical concepts because political and economic organizations are continuously evolving. The same applies to "living organism": biological life consists of a continuous and insinuating transformation of inert into living matter. What makes "want" a dialectical concept is that the means of want satisfaction can change with time and place: the human species would have long since vanished had our wants been rigid like a number. Finally, "species" is dialectical because every species "includes the unknown element of an act of creation." [53]

The reason that compelled Plato to exclude all *qualitative change* from his world of arithmomorphic ideas is obvious. The issue of whether motion too is excluded from this world is not discussed by Plato. But we can be almost certain that he had no intention — for there was no need for it — of conceiving that world as *motionless.* He thus implicitly recognized that an arithmomorphic structure is incompatible with qualitative change but not with locomotion, even though he admitted that Change consists of either.[54] As a result, Plato was as puzzled as the generation before him by Zeno's paradoxes, and could not crack them.

Through his paradoxes Zeno aimed to expose the flaws of the Pythagorean doctrine of Many as opposed to Parmenides' doctrine of One. The Arrow Paradox, in particular, intends to prove that even locomotion is incompatible with a molecular (i.e., arithmomorphic) structure of Space

---

51. E.g., Knight, *The Ethics of Competition*, p. 21.
52. *The Logic of Hegel*, p. 148. One page earlier he says that "the Dialectical principle constitutes the life and soul of scientific progress."
53. Darwin, *Origin of Species*, p. 30.
54. Plato, *Parmenides*, 139.

and Time. For, to reinterpret Zeno, if at any given instant the arrow is in some *discretely distinct* place, how can it move to another such place? Some argue that the paradox is resolved by defining locomotion as a relation between a time-variable and a space-coordinate.[55] The fact that this "mathematical" solution is good enough for physics is beyond question. However, in one respect the paradox is simpler, while in another more intricate, than this solution suggests.

It is simpler, because all that Zeno does is to ignore the qualitative difference between "to be in a place" and "to move through a place," i.e., between "rest" and "motion." To recall, locomotion is only change of place; locomotion by itself can change none of the qualities of the object, including the quality of being in motion.

But the paradox is more intricate than its mathematical solution leads us to believe, in that it discloses the perplexities surrounding the idea that Space and Time are not *continuous wholes* but mere *multiplicity of indivisible points*. As has been repeatedly pointed out by many mathematical authorities, these issues are still with us in spite of the splendid achievements of Dedekind, Weierstrass, and Cantor in connection with the arithmetical continuum.[56] No doubt, what these famous mathematicians mainly sought was a mathematical formalization of the intuitive continuum. Dedekind, in particular, constantly referred his argument to the intuitive aspects of the line continuum.[57] But Bertrand Russell's claim, still heard now and then, that "no other continuity [other than that of the arithmetical continuum] is involved in space and time," [58] lacks any basis. The truth is that the proposition that there exists a one-to-one correspondence between the real numbers and the points on a line is either an axiom or a mathematical definition of line.[59]

Developments in mathematics — later than Russell's statement quoted above — prove that Aristotle's tenet, point is the limit of line not *part* of it,[60] is far from groundless.

In the first place, the modern theory of measure is a belated admission that at least the tenet is not concerned with a pseudo problem. Still more telling is Ernst Zermelo's famous theorem that the arithmetical continuum can be well ordered, which means that every real number has an immediate successor. Even though this immediate neighbor cannot be

55. E.g., Bertrand Russell, *The Principles of Mathematics* (Cambridge, Eng., 1903), chap. liv.
56. E.g., Hermann Weyl, *Das Kontinuum* (Leipzig, 1918), p. 16; Henri Poincaré, *The Foundations of Science* (Lancaster, Pa., 1946), p. 52.
57. For an enlightening and well-balanced discussion of the above points, see Black (note 16, above), pp. 85–97.
58. Russell, *Principles of Mathematics*, p. 260.
59. See G. D. Birkhoff, "A Set of Postulates for Plane Geometry, Based on Scale and Protractor," *Annals of Mathematics*, XXXIII (1932), 329.
60. Aristotle, *Physics*, 231ª 25–29.

named, the proof of its existence reveals a point made earlier, namely, that a number has a perfectly isolated individuality. Whatever properties the arithmetical continuum might have, its structure is still that of beads on a string, but *without the string*. For the time being this is, probably, all the light mathematics can throw upon the issue of the arithmetic vs. intuitive continuum.[61]

Shifting to a more direct approach, we may observe that the intuitive continuum, whether of Space, Time, or Nature itself, constitutes a *seamless* whole. "Things that are in one world are not divided nor cut off from one another with a hatchet." [62] The world continuum has no joints where, as Plato thought, a good carver could separate one species from another.[63] Numbers more than anything else are artificial slits cut by us into this whole. Of course, given any whole we can make as many slits into it as we please. But the converse claim, implicit in arithmetical positivism, that the whole can be reconstructed from the slits alone rests on the thinnest air.

One cannot pass lightly over the fact that none other than a coauthor of *Principia Mathematica*, Alfred North Whitehead, has centered his entire philosophical system upon the essential difference between the continuum of the world and that of mathematics. Time, as is obvious from Whitehead's writing, supplies the best basis for illustrating the point. But the essence of Whitehead's philosophical position is not altogether new.

It was again Aristotle who argued that time is not made of point-instants succeeding each other like the points on a line.[64] The message has been variously echoed throughout the subsequent centuries. In modern times, it has been revived not only by philosophers, such as Henri Bergson or Whitehead, but also by prominent physicists: the "now" of our experience is not the point of separation in mathematics.[65] Bergson

61. Modern logicians have acquired a rather peculiar habit: each time a paradox springs up they legislate new rules outlawing one of the steps by which the paradox is reached. Clearly, the procedure means nothing less than the hara-kiri of reason. (Cf. H. Weyl, *Philosophy of Mathematics and Natural Science*, Princeton, 1949, p. 50. See also the sharp criticism by Henri Poincaré, *Foundations*, pp. 472 ff, esp. p. 485.) In any case, it does not *resolve* the paradox; it merely *shelves* it. As to Zermelo's theorem, the proposal is to outlaw choosing a member of a set without actually naming it. To use the highly instructive analogy of Bertrand Russell, *Introduction to Mathematical Philosophy* (New York, 1930), p. 126, though it would be legal to choose the left boot from a pair of boots, choosing one sock out of a pair of identical socks would be an illegal operation. I completely fail to see why the latter choice would be improper in a domain like mathematics where far more bizarre operations are performed all the time. Is not marrying nobody to nobody — as in the mapping of the null set onto itself — a most bizarre idea?

62. Anaxagoras, Fragment 8, in Burnet, *Early Greek Philosophy*, p. 259.

63. Plato, *Phaedrus*, 265.

64. Aristotle, *Physics*, 231[b] 6–10, 234[a] 23.

65. Bridgman, *Intelligent Individual and Society*, p. 107.

and Whitehead, however, go much further and explain why the difference matters in science in spite of the fact that physics has fared splendidly without any overt concern for the intuitive continuum. As Whitehead admits,[66] they both insist that the ultimate fact of nature is Change. Whether we prefer to use instead the word "happening," "event," or "process," Change requires time to be effected or to be apprehended. Nature at an instant or state of change at an instant are most forbidding abstractions. To begin with, there is no answer to the question "what becomes of velocity, at an instant?" Even "iron at an instant" is unintelligible without the temporal character of an event. "The notion of an instant of time, *conceived as a primary simple fact*, is nonsense." [67] The ultimate facts of nature vanish completely as we reach the abstract concept of point of time. An instant has an arithmomorphic structure and, hence, is indifferent to "whether or no there be any other instant." [68]

The ultimate fact of nature, Bergson's becoming or Whitehead's event, includes a *duration* with a temporal extension.[69] But "the immediate duration is not clearly marked out for our apprehension." It is rather "a wavering breadth" between the recalled past and the anticipated future. Thus, the time in which we apprehend nature is not "a simple linear series of durationless instants with certain mathematical properties of serial [arithmetic] continuity," [70] but a *sui generis* seriation of durations. Durations have neither minimum nor maximum extension. Moreover, they do not follow each other externally, but each one passes into others because events themselves "interfuse." No duration is discretely distinct from its predecessor or its successor, any more than an event can be completely isolated from others: "an isolated event is not an event." [71] Durations overlap durations and events overlap events in a peculiar complexity, which Whitehead attempted with relative success to analyze through the concept of extensive abstraction and abstractive

66. Alfred North Whitehead, *The Concept of Nature* (Cambridge, Eng., 1930), p. 54.

67. Whitehead, *Modes of Thought*, pp. 199, 207 (my italics); Whitehead, *An Enquiry concerning the Principles of Natural Knowledge* (2nd edn., Cambridge, Eng., 1925), p. 23. The same ideas occur as a leitmotif in all Whitehead's philosophical works, although they are more clearly stated in the early ones. See his *Enquiry*, pp. 1–8, and his "Time, Space, and Material" (note 15, above). See also Erwin Schrödinger, *Science, Theory and Man* (New York: Dover Publications, 1957), p. 62.

For Bergson's approach, see Henri Bergson, *Time and Free Will* (3rd edn., London, 1913), and his *Creative Evolution* (New York, 1913).

68. Whitehead, *Modes of Thought*, p. 199; Whitehead, "Time, Space, Material," p. 45.

69. Bergson, *Time and Free Will*, pp. 98 ff; Bergson, *Creative Evolution*, pp. 1–7; Whitehead, "Time, Space, Material," pp. 45–46; Whitehead, *Enquiry*, chap. ix.

70. Whitehead, *Concept of Nature*, p. 69 and *passim;* Whitehead, "Time, Space, Material," p. 44; Bergson, *Creative Evolution*, pp. 21–22. Also P. W. Bridgman, *The Nature of Physical Theory* (Princeton, 1936), p. 31.

71. Whitehead, *Concept of Nature*, p. 142.

classes.[72] However, everything he says in "vague" words leaves no doubt that both "duration" and "event" as conceived by Whitehead are concepts surrounded by dialectical penumbras, in our sense.[73]

With regard to the opposition between Change and arithmomorphic structure, Whitehead's position is essentially the same as Hegel's. Perhaps in no other place does Hegel state his thought on the matter more clearly than in the following passage: "Number is just that entirely inactive, inert, and indifferent characteristic in which every movement and relational process is extinguished." [74] The statement has generally been criticized as Hegelian obscurantism and anti-scientism. Yet, as I have already intimated, Hegel did not intend to prove anything more than Whitehead, who maintained that no science can "claim to be founded on observation" if it insists that the ultimate facts of nature "are to be found at durationless instant of time." [75] Whitehead only had the benefit of a far greater knowledge in mathematics and sciences of fact than was available in Hegel's time.

Whitehead's censure of durationless events may not seem as important for physics as for the life sciences. Physics may very well argue that $s = vt$, for instance, is shorthand for $\Delta s = v\Delta t$. But this explanation is not always available; one of the many examples is provided by the phenomenon of surface stress.[76] Consequently, even physics cannot plead completely not guilty.

9. *A Logistic Solution.* Even though the onus of proof rests with him who affirms the operationality of an idea, no one among those who claim that Change can be completely described by means of arithmomorphic concepts seems to have shown how this can be done in all instances. (Merely to point at physics would be obviously insufficient even if physics were a model of perfection in this respect.) To my knowledge, there is only one exception, which thus is all the more instructive. In an *oeuvre de jeunesse*, Bertrand Russell asserted that any qualitative change can be represented as a relation between a time variable and the *truth-value* of a set of propositions "concerning the *same* entity." [77] The assertion raises several questions.

72. Whitehead, *Enquiry*, Part III. In my opinion, his analysis represents rather a *simile*, for in the end his operations of extensions, intersection, etc., imply discrete distinction, as is obvious from the diagrammatical analysis on his pp. 103, 105.

73. See the following works by Whitehead: "Time, Space, Material," p. 51; *Enquiry*, p. 4 and *passim; Concept of Nature*, pp. 55, 59, 72–73, 75; *Process and Reality: An Essay on Cosmology* (New York, 1929), p. 491; *Science and the Modern World* (New York, 1939), pp. 151, 183 ff.

74. G. W. F. Hegel, *The Phenomenology of Mind* (2nd edn., New York, 1931), p. 317.

75. Whitehead, *Enquiry*, p. 2, and *Concept of Nature*, p. 57.

76. Whitehead, *Enquiry*, pp. 3–4.

77. Russell, *Principles of Mathematics*, p. 469. My italics.

Perhaps, we ought to ask first what *"same"* is in such a complex changing structure; however, it appears expedient to beg this question for a while. Therefore, let E denote "the same entity." To take the simplest possible case of a continuous change, what Russell further means is that (1) for every value of the time variable t, there is one proposition "E is $A(t)$" that is true, and (2) this very proposition is false for any other value of the time variable. Obviously, the set of all propositions "E is $A(t)$" and, hence, the set $[A(t)]$ have the power of continuum. There are now two alternatives.

First, $[A(t)]$ represents a range of a *quantified* quality. In this case $A(t)$ is a number, and Russell's solution is no better but no worse than the mathematical representation of locomotion. Its operationality, however, is confined to the domain of measurable qualities.

The second alternative, upon which the matter hinges, is the case where $A(t_1)$ and $A(t_2)$ for $t_1 \neq t_2$ represent two distinct pure qualities, like "feudalism" and "capitalism," for example. In this case, Russell's solution is purely formal, nay, vacuous. On paper, we can easily write that E is $A(t)$ at the time t, but if $A(t)$ is a pure quality it must be defined independently of the fact that it is an attribute of E at t. Obviously, to say that on January 1, 1963, the United States economic system is "the United States economic system on January 1, 1963" is the quintessence of empty talk. What we need is a proposition in which $A(t)$ is replaced by, say, "free enterprise under government vigilance." If $A(t)$ is a pure quality, i.e., if it cannot be represented by a number, then the representation of continuous change by Russell's scheme runs against a most elementary stumbling block: any vocabulary is a finite set of symbols. We may grant, at most, that the structure of vocabulary is that of a countable infinity, but certainly it does not have the power of continuum. Russell's proposal thus breaks down before we can ask any question of the sort that a logistic philosopher would dismiss as "metaphysical."

10. *What Is Sameness?* There are indeed other issues which cannot be pinpointed by the simple illustration I have used in the preceding discussion. The most relevant case of qualitative change is where for any value of t there is more than one true proposition concerning E. To take the simplest case, Russell's scheme tells us only this: given $t \neq t'$, there exists a pair of propositions, "E is A" and "E is B," true at t and false at t', and another pair, "E is C" and "E is D," true at t' and false at t. Nothing is said about whether the pairs are *ordered* or not. But without the condition that they are ordered, the scheme is inadequate for describing even a quantitative change. For what would become of any physical law if the observer were unable to ascertain which member in each pair, (A, B) and (C, D), represents, say, gas pressure and which represents temperature? To order each pair by using the Axiom of Choice would

not do, even if we regard the axiom as perfectly legitimate. Therefore, if the scheme is to be operational at all, it must include from the outset the condition that one member of every pair, such as (A, B), belongs to a set $[P_1(t)]$, and the other to another set $[P_2(t)]$. Clearly, this additional information corresponds to the fact that the observer must know beforehand whether or not two attributes observed at two different times belong to the *same* qualitative range. An operational Russell's scheme, therefore, requires the concept of *sameness* not only in relation to E but to each attribute as well. To establish *sameness* of attribute we need to know what "*same* quality" is. Therefore, Russell's exercise in formal logic does not do away with what intuition posits; on the contrary, on closer examination it is seen that it cannot function without what it purports to destroy.[78]

Perhaps nothing illustrates more aptly the staggering issues raised by "sameness" than one of Bridgman's remarks. With the discovery of relativity in physics, it is perfectly possible that two observers traveling in different directions through space may record a signal from a third source as two different facts. For instance, one observer may see "a flash of yellow light" while the other may only feel "a glow of heat on his finger." How can they be sure then that they have reported the same event, since they cannot turn to simultaneity in the absence of absolute time?[79] Bridgman's point is that even relativity physics presupposes sameness in some absolute sense although it fails to show how it could be established. The upshot is that we have to recognize once and for all that sameness is an internal affair of a single mind, whether an individual one or one embracing several individual minds. We have gone too far, it appears, in believing that natural phenomena can be reduced to signal registrations alone and hence that mind has no direct role in the process of observation. Mind, on the contrary, is as indispensable an *instrument* of observation as any physical contrivance. The point is of paramount importance for social sciences, and I shall return to it later on.

On the philosophical problem of "sameness," one can only say that it is as thorny as it is old. How thorny it is can be shown in a brief sum-

78. The fallacy of believing that the weapon of pure logic suffices by itself to kill any creature of intuition is not uncommon. An instance of this fallacy is discussed in the author's paper, "The End of the Probability Syllogism?" *Philosophical Studies*, February 1954, pp. 31–32. An additional example is the refutation of historical laws on "*strictly* logical reasons." (Karl R. Popper, *The Poverty of Historicism*, Boston, 1957, pp. ix–xi.) The very first premise of Popper's argument, "the course of human history is strongly influenced by the growth of human knowledge," is plainly a historical law. That is, the conclusion that [L] is empty is derived from the proposition that [L] is not empty! We should observe, however, that in a new footnote (*The Logic of Scientific Discovery*, New York, 1959, p. 279n2) Popper takes a more softened line, somewhat agnostic.

79. Bridgman, *Nature of Physical Theory*, p. 77, and especially his *Reflections of a Physicist*, p. 318 ff.

mary of Whitehead's ideas on the subject. According to Whitehead, we apprehend nature in terms of *uniform* objects and *unique* events, the former being ingredients of the latter. "Objects are elements in nature which do not pass." Because they are "out of time," they "can be 'again,' " so that we can say "Hullo, there goes Cleopatra's Needle again." Events, on the contrary, once passed "are passed, and can never be again." At most, we may recognize that one event is *analogous* to another.[80] One cannot help feeling that this dualist view is far from settling the issue, and that "analogous events" stand in the same relation to one another as two objects recognized as being the same. Moreover, one is still baffled by the question of whether any object, such as Cleopatra's Needle, is really out of time so that thousands of years from now we could still say "there it goes again." And if we think in millions of years, we should doubt whether the universe itself is "out of time." Besides, in describing nature we are interested as much in *uniform* objects as in *analogous* events. That is, keeping within Whitehead's system, we know that science is equally concerned with whether we can say "there goes *another* 'King of England' again" and whether we can say "there goes *another* 'coronation' again." Actually, science may even dispense with objects, but not with events. The electron, for instance, "cannot be identified, it lacks 'sameness.' " [81] We cannot therefore say "there goes the same electron again," but only that "there goes another electron-event again."

But then, why should we distinguish between object, i.e., Being, and event, i.e., Becoming? In the end, we verify what we have known of old, that dualism is full of snags. The only way out is to recognize that the distinction between object and event is not discrete but dialectical, and probably this is Whitehead's message too.[82] Any further discussion of this point, however, would involve us too deeply in pure Hegelian Dialectics.

11. *How Numerous Are Qualities?* In my first discussion of Russell's formalization of change (§ 10), I have shown that an impasse is reached because words are not as numerous as pure qualities. But perhaps the impasse might be cleared by using numbers instead of words for labeling qualities. An example of such a *continuous* catalogue is readily available: each color in the *visual* spectrum can be identified by the wave length of the equivalent unmixed color. As is almost needless to add, such a cataloguing does not necessarily imply the measurability of the range of the qualities involved. However, the question whether the cataloguing

---

80. Whitehead, *Concept of Nature*, pp. 35, 77–78, 143 ff, 169 ff; Whitehead, *Enquiry*, pp. 61 ff, 167–168.

81. Schrödinger, *Science, Theory and Man*, p. 194; Bridgman, *Intelligent Individual and Society*, pp. 32–33.

82. Cf. Whitehead, *Concept of Nature*, pp. 166–167.

is possible forms a prerequisite to that of measurability, although for some reason or other the point has not been recognized, at least in economics, until recently. Clearly, there is no reason why the cardinal power of all the qualities we can think of even in a simple set-up should not exceed that of the arithmetical continuum. On the contrary, as I have argued in relation to individual expectations and preferences,[83] there are good reasons for the view that real numbers are not always sufficient for cataloguing a set of qualities. In other words, the manifold of our thoughts differs from the arithmetical continuum not only by its indivisible continuity — as I have explained in some of the preceding sections — but also by its dimensionality.[84] As we say in mathematics, the continuum of the real number system forms only a simple infinity.

The suggestion, natural at this juncture, of using more than one real number, i.e., a vector, for labeling qualities would still not reduce quality to number. For, as set theory teaches us, no matter how many coordinates we add, no set of vectors can transcend simple infinity. There is an intimate connection between this mathematical proposition and the well-known difficulties of biological classification.

It was Linnaeus who first struck upon the idea of using a two-word name for each species, the first word representing the genus, the second the species within the genus. By now all naturalists agree that any taxonomical term, in spite of its two-dimensionality, does not cover one immutable, arithmomorphic form but a dialectical penumbra of forms. The fact that they still use Linnaeus' *binary* system clearly indicates that the manifold of biological species is in essence more complex than simple, linear infinity. The problem of biological classification therefore is not equivalent to that illustrated by the continuous cataloguing of colors, and hence the predicament of naturalists would not come to an end even if a numerical vector would be used for labeling species.

One naturalist after another has intuitively apprehended that — as Yves Delage put it — "whatever we may do we will never be able to account for all affinities of living beings by classifying them in classes, orders, families, etc." [85] Many have argued that this is because in the domain of living organisms only *form* (shape) counts and shape is a fluid concept that resists any attempt at classification.[86] Some have simply asserted that form cannot be identified by number.[87] Even Edmund

83. See "Choice, Expectations and Measurability" (1954), reprinted in this book.

84. I am not at all sure that these two aspects do not boil down to a single one.

85. Quoted in G. G. Simpson, "The Principles of Classification and a Classification of Mammals," *Bulletin of the American Museum of Natural History*, LXXXV (1945), 19, except that I have translated Delage's words into English.

86. E.g., Theodosius Dobzhansky, *Evolution, Genetics, and Man* (New York, 1955), chap. x, and especially the eloquent picture on p. 183.

87. E.g., P. B. Medawar, *The Uniqueness of the Individual* (New York, 1958), pp. 117 ff.

Husserl, though educated as a mathematician, thought the point to be obvious: "The most perfect geometry" — he asserts — cannot help the student to express in precise concepts "that which in so plain, so understanding, and so entirely suitable a way he expresses in the words: notched, indented, lens-shaped, umbelliform, and the like — simple concepts which are *essentially and not accidentally inexact*, and are *therefore* also unmathematical." [88] Yet a simple proposition of the theory of cardinal numbers vindicates the gist of all these intuitive claims. It is the proposition that the next higher cardinal number mathematics has been able to construct after that of the arithmetical continuum is represented by the set of all functions of a real variable, i.e., by a set of forms. Clearly, then, *forms cannot be numbered*.

12. *The Continuity of Qualities.* The peculiar nature of most qualitative structures leads to a somewhat similar sort of difficulty in connection with their ordering. I can best illustrate this difficulty by an example from my own work. Thirty years ago, as I was trying to unravel the various thoughts underlying early and contemporary writings on utility and to map them out as transparent "postulates," I became convinced of the logical necessity of settling first of all one issue, that with which Postulate A of one of my early papers is concerned. [89] This postulate states that given a preferential set $[C_\alpha]$ — where α is a real number and $[C_\alpha]$ is preferentially ordered so that C_α is preferred to C_β if $\alpha > \beta$ — and C not belonging to $[C_\alpha]$, there exists an i such that C and C_i are indifferent combinations. At the time, the postulate bothered me; intuitively I felt that the accuracy of human choice cannot be compared with that of a physical instrument, but I was unable to construct a *formal* example to convince myself as well as the few colleagues with whom I discussed the matter, i.e., that Postulate A can be negated. The most I could do was to introduce a stochastic factor in choice — which, I believe, was a quite new idea. But this still did not settle my doubts, nor those of my colleagues, about my Postulate A.

In retrospect, the objections of my colleagues and my inability — due to a deficiency in my mathematical knowledge — to meet these objections are highly instructive and also apropos. My critics generally felt that Postulate A is entirely superfluous: some argued that it is impossible to pass from nonpreference to preference without effectively reaching a stage of indifference; [90] others held that since $[C_\alpha]$ is continuous there is no room in it for other items, not even for one. An example which I offered as a

88. Edmund Husserl, *Ideas: General Introduction to Pure Phenomenology* (New York, 1931), p. 208. Italics are Husserl's.

89. See my essays, "The Pure Theory of Consumer's Behavior" (1936) and "Choice, Expectations and Measurability" (1954), reprinted in this volume.

90. From recent discussions I learned that even mathematicians are apt to raise this objection.

basis for discussion was too clumsy for everyone concerned: a hypothetical wine lover who always prefers more to less wine but has nevertheless a very slight preference for red wine, so that between two equal quantities of wine he prefers the red. I denoted by x_r and x_w the quantities of red and white wine respectively, but as I came to write $x_r > x_w$, I invited the objection that "x is x." Today the connection between the example and the old notion of a hierarchy of wants may seem obvious, but I was unable to clarify my own thoughts on the matter until much later, after I came across an objection raised by a reviewer to one of Harold Jeffrey's propositions. Learning then for the first time of lexicographic ordering, I was able to solve my problem.[91] However, my initial difficulties with the example of the wine lover bear upon a point I wish to make now.

Either set, $[x_r]$ or $[x_w]$, taken by itself, is continuous in the mathematical sense. Consequently, no brutal offense is committed by regarding, say, $[x_r]$ as the arithmomorphic representation of the preference continuum in case only red wine is available. However, if both red and white wine are introduced into the picture the arithmomorphic representation of the wine lover's preference suddenly becomes discontinuous: in the corresponding lexicographic ordering (with respect to the *subscript*) there is no element between x_w and x_r, or alternatively, x_r is the immediate successor of x_w. On the other hand, there is no reason why preference itself should become discontinuous because of qualitative variations in the object of preference. To argue that preference is discontinuous because its arithmomorphic simile is so, is tantamount to denying the three-dimensionality of material objects on the ground that their photographs have only two dimensions. The point is that an arithmomorphic simile of a qualitative continuum displays spurious seams that are due to a peculiar property of the medium chosen for representing that continuum. The more complex the qualitative range thus formalized, the greater the number of such artificial seams. For the variety of quality is continuous in a sense that cannot be faithfully mirrored by a mathematical multiplicity.

13. *A Critique of Arithmomorphism.* Like all inventions, that of the arithmomorphic concept too had its good and its bad features. On the one hand, it has speeded the advance of knowledge in the domain of inert matter; it has also helped us detect numerous errors in our thinking, even in our mathematical thinking. Thanks to Logic and mathematics in the ultimate instance, man has been able to free himself of most animistic superstitions in interpreting the wonders of nature. On the other hand, because an arithmomorphic concept has absolutely no relation to life,

91. "Choice, Expectations and Measurability" (1954), below. Perhaps this bit of personal history suffices to show how indispensable to the student of economics is a substantial familiarity with every branch of mathematics.

to *anima*, we have been led to regard it as the only sound expression of knowledge. As a result, for the last two hundred years we have bent all our efforts to enthrone a superstition as dangerous as the animism of old: that of the Almighty Arithmomorphic Concept. Nowadays, one would risk being quietly excommunicated from the modern *Akademia* if he denounced this modern superstition too strongly. The temper of our century has thus come to conform to one of Plato's adages: "He who never looks for numbers in anything, will not himself be looked for in the number of the famous men." [92] That this attitude has also some unfortunate consequences becomes obvious to anyone willing to drop the arithmomorphic superstition for a while: today there is little, if any, inducement to study Change unless it concerns a measurable attribute. Very plausibly, evolution would still be a largely mysterious affair had Darwin been born a hundred years later. The same applies to Marx and, at least, to his analysis of society. With his creative mind, the twentieth-century Marx would have probably turned out to be the greatest econometrician of all times.

Denunciations of the arithmomorphic superstition, rare though they are, have come not only from old-fashioned or modern Hegelians, but recently also from some of the highest priests of science, occasionally even from exegetes of logical positivism. Among the Nobel laureates, at least P. W. Bridgman, Erwin Schrödinger, and Werner Heisenberg have cautioned us that it is the arithmomorphic concept (indirectly, Logic and mathematics), not our knowledge of natural phenomena, that is deficient.[93] Ludwig Wittgenstein, a most glaring example in this respect, recognizes "the bewitchment of our understanding by the means of our [rigidly interpreted] language." [94] The arithmomorphic rigidity of logical terms and symbols ends by giving us mental cramps. We can almost hear Hegel speaking of "the dead bones of Logic" and of "the battle of Reason . . . to break up the rigidity to which the Understanding has reduced everything." [95] But even Hegel had his predecessors: long before him Pascal had pointed out that "reasoning is not made of *barbara* and *baralipton*." [96] The temper of an age, however, is a peculiarly solid phenomenon which advertises only what it likes and marches on undisturbed by the self-criticism voiced by a minority. In a way, this is only natural:

92. Plato, *Philebus*, 17.

93. Bridgman, *Logic of Modern Physics*, p. 62, and *Nature of Physical Theory*, p. 113; Erwin Schrödinger, *What Is Life?* (Cambridge, Eng., 1944), p. 1; Werner Heisenberg, *Physics and Philosophy: The Revolution in Modern Science* (New York, 1958), pp. 85 ff.

94. L. Wittgenstein, *Philosophical Investigations* (New York, 1953), I. 109. My translation.

95. *The Logic of Hegel*, p. 67.

96. Blaise Pascal, "De l'esprit géométrique et de l'art de persuader," in *Oeuvres complètes*, ed. J. Chevalier (Paris, 1954), p. 602.

as long as there is plenty of gold dust in rivers why should one waste time in felling timber for gold-mine galleries?

There can be no doubt that all arguments against the sufficiency of arithmomorphic concepts have their roots in that "mass of unanalyzed prejudice which Kantians call 'intuition,' " [97] and hence would not exist without it. Yet, even those who, like Russell, scorn intuition for the sake of justifying a philosophical flight of fancy, could not possibly apprehend or think — or even argue against the Kantian prejudice — without this unanalyzed function of the intellect. The tragedy of any strain of positivism is that in order to argue out its case it must lean heavily on something which according to its own teaching is only a shadow. For an excellent illustration in point we may turn to a popular treatise which aims to prove that if "no possible sense-experience" can determine the truth or falsehood of a nontautological proposition then "it is metaphysical, . . . neither true nor false, but literally senseless." [98] After reading this statement on the first page of the preface one cannot help wondering in what way the rest of the book can support it if the statement is true — as its author claims. Certainly, the subsequent argument has no relation whatever to sense-experience — except, of course, the visual perception of black letters, nay, spots on a white background.

The frequent diatribes against this or that particular dialectical concept are guilty of the same sin. Cornelius Muller, for example, preaches the abolition of the concept of community. The reasoning is that since "the several examples of one class of communities are not identical and [since] two adjacent classes of communities are not distinct from one another . . . the word has no meaning." [99] But the argument is obviously self-destroying, for the meaning of the premise is negated by its own conclusion. We have not learned, it seems, everything from the legendary Cretan liar of the ancient sophist school.

The propounders of the views such as those just mentioned — or this author for that matter — would not go to the trouble of discussing the issues of dialectical concepts if we thought that these issues have no bearing upon scientific orientation. It is therefore not surprising that Muller, who argues that there are no "real entities" — whatever this might mean — unless we can distinguish them in the same way we distinguish one carbon isotope from another, begins his attack on "community" by asking "Is there a mechanistic theory that . . . conforms to the true nature of communities?" [100] The moral is plain: social sciences

97. Russell, *Principles of Mathematics*, p. 260.
98. A. J. Ayer, *Language, Truth and Logic* (2nd edn., New York, 1946), p. 31.
99. Cornelius H. Muller, "Science and Philosophy of the Community Concept," *American Scientist*, XLVI (1958), 307–308.
100. *Ibid.*, 298.

and biology should cling to the universality of mechanics, that is, to a retrograde position long since abandoned even by physics.

Unfortunately for everyone concerned, life phenomena are not as simple as that, for not all their aspects are as pellucid as an arithmomorphic concept. Without dialectical concepts life sciences could not fulfill their task. As I have argued earlier, there is no way of defining "democracy" or "competition," for instance, so as to comply with Muller's criterion of real entity. The most we can do for a greater precision is to distinguish species within each genus, as in biology: "American democracy," "British democracy," "monopolistic competition," "workable competition," etc. Let us observe that even the familiar and apparently simple notion of the struggle for existence has many shades of meaning "which pass into each other" [101] and, hence, is dialectical. Finally, let us observe that the only proof of evolution is the dialectical relation of species in their phylogenetic classification. Should we one day succeed in constructing an arithmomorphic concept of species (or of something equivalent), that very day biology will have to return to the pre-Lamarckian views: species were created immutable and by fiat. A self-identical species, a self-identical community, anything self-identical, cannot account for biological or social evolution: "self-identity has no life." [102] More explicitly, no process of change can be completely decomposed into arithmomorphic parts, themselves devoid of change.[103] And it is because society and its organization are in constant flux that genuine justice cannot mean rigid interpretation of the words in the written laws. Only bitter and unnecessary conflict, as Bridgman correctly observed, can result from ignoring the dialectical nature of "duty" and using the term as if it has the "sharpness and uniqueness of a mathematical concept." [104]

From whatever angle we look at the issue illustrated by the examples cited in this section, we reach the same inescapable conclusion: thinking, even mathematical thinking, would come to a standstill if confined only to self-identical notions. "As soon as you leave the beaten track of vague clarity, and trust to exactness, you will meet difficulties." [105] Infinitely continuous qualities, dialectical penumbras over relations and concepts, a halo of varying brightness and contour, this is thought: a gaseous medium, as Wittgenstein pictured it.[106] By fighting this obvious fact, positivism sets itself on the path to self-defeat. The idea of pretending to be color-blind in order to argue that those who insist upon their seeing

101. Darwin, *Origin of Species*, 6th edn., p. 46.
102. *Hegel's Science of Logic*, II, 68.
103. Whitehead, *Modes of Thought*, pp. 131–132. See also *Hegel's Science of Logic*, II, 251–252.
104. Bridgman, *Intelligent Individual and Society*, p. 116.
105. Whitehead, *Science and Philosophy*, p. 136.
106. Wittgenstein, *Philosophical Investigations*, I. 109.

something that cannot be reduced to tone are either "blind" to tone or have metaphysical hallucinations, will never work.

Robert Mayer's outcry that "a single number has more true and permanent value than a vast library of hypotheses" was perfectly in place. He spoke as a physicist addressing physicists, and hence there was no need for him to add, "provided that that number helps us describe reality adequately." Omissions such as this one have allowed similar statements by the greatest authorities in science to be interpreted as applying to *any* number. The fascination of our intellect by number is not easily conquered. It is responsible also for the fact that Galileo's advice to astronomers and physicists has been transformed into a definition of essence: "science is measurement." The consequences of these gratifying generalizations have not always been fortunate.

Planck, for example, observed that by exaggerating the value of measure we might completely lose touch with the real object. Of the many examples that could illustrate the point, one is particularly piquant. From as far back as we can go in history a man's degree of ageing has been measured by his age. Because of this biologists have simply thought little, if at all, of ageing. So, recently they suddenly discovered "an unsolved problem in biology": age may be an average measure of ageing, but ageing is something entirely different from growing old in years.[107]

Undoubtedly, for the sciences concerned with phenomena almost devoid of form and quality, measure usually means expanded knowledge. In physics, which has quite appropriately been defined as the quantitative knowledge of nature, there is no great harm if measurement is regarded as an end in itself. But in other fields the same attitude may lead to empty theorizing, at the very least. The advice "look for number" is wise only if it is not interpreted as meaning "you must find a number in everything." *We do not have* to represent beliefs by numbers just because our mind feels similarly embarrassed if it has to predict the outcome of a coin-tossing or the political conditions in France ten years from now. The two events are not instances of the same phenomenon. A measure for all uncertainty situations, even though a number, has absolutely no scientific value, for it can be obtained only by an intentionally mutilated representation of reality. We hear people almost every day speaking of "calculated risk," but no one yet can tell us how he calculated it so that we could check his calculations. "Calculated risk" if taken literally is a mere parade of mathematical terms.[108]

It was under the influence of the idea "there is a number in everything"

107. See Medawar, *The Uniqueness of the Individual*, chap. ii.

108. For the argument I have offered against the measurability of even documented belief, see below, my article "Choice, Expectations and Measurability" (1954), and, especially my article "The Nature of Expectation and Uncertainty" (1958).

that we have jumped to the conclusion "where there is 'more' and 'less' there is also quantity," and thus enslaved our thoughts to what I have called "the ordinalist's error." In fact, the sin unfortunately affects a far larger circle than the writers on utility. There seems to be some need for pointing out — as I have done in some of the papers included in this volume and also in the present chapter — that "there is a limit to what we can do with numbers, as there is a limit to what we can do without them." [109]

109. Below, p. 275.

THREE

SOME OBJECT LESSONS
FROM PHYSICS

1. *Physics and Philosophy of Science.* A social scientist seeking counsel and inspiration for his own activity from the modern philosophy of science is apt to be greatly disappointed, perhaps also confused. For some reason or other, most of this philosophy has come to be essentially a praise of theoretical science and nothing more. And since of all sciences professed today only some chapters of physics fit the concept of theoretical science, it is natural that almost every modern treatise of critical philosophy should avoid any reference to fields other than theoretical physics. To the extent to which these other fields are mentioned (rarely), it is solely for the purpose of proving how unscientific they are.

Modern philosophy of science fights no battle at all. For no one, I think, would deny that the spectacular advance in some branches of physics is due entirely to the possibility of organizing the description of the corresponding phenomenal domain into a theory. But one would rightly expect more from critical philosophy, namely, a nonprejudiced and constructive analysis of scientific methodology in all fields of knowledge. And the brutal fact is that modern works on philosophy of science do not even cover fully the whole domain of physics.

The result of this uncritical attitude is that those who have worked inside the edifice of physics do not always agree with those who admire it only from the outside. The insiders admit, to their regret, that the crown of physics has lost some of the sparkling jewels it had at the time of Laplace. I have already mentioned one such missing jewel: the impossibility, which becomes more convincing with every new discovery, of a noncontradictory logical foundation for all properties of matter. For the biologist or social scientist this constitutes a very valuable object lesson, but there are other lessons at least equally significant. In what follows I shall attempt to point them out.

I begin by recalling an unquestionable fact: the progress of physics has been dictated by the rhythm with which attributes of physical phenomena have been brought under the rule of measure. More interesting still for our purpose is the correlation between the development of various compartments of physics and the *nature* of the attributes conquered by measure.

As we may find it natural *ex post*, the beginning was made on those variables whose measure, having been practiced since time immemorial, raised no problem. Geometry, understood as a science of the *timeless* properties of bodily objects, has only one basic attribute: length, the prototype of a quality-free attribute. Mechanics was the next chapter of physics to become a complete theoretical system. Again, measures for the variables involved had been in practical use for millennia. It is very important for us to observe that what mechanics understands by "space" and "time" is not *location* and *chronological time*, but *indifferent distance* and *indifferent time interval*. Or, as the same idea is often expressed, mechanical phenomena are independent of Place and Time. The salient fact is that even the spectacular progress achieved through theoretical mechanics is confined to a phenomenal domain where the most transparent types of measure suffice. The space, the time, and the mass of mechanics all have, in modern terminology, a *cardinal* measure.

The situation changed fundamentally with the advent of thermodynamics, the next branch of physics after mechanics to acquire a theoretical edifice. For the first time *noncardinal* variables — temperature and chronological time, to mention only the most familiar ones — were included in a theoretical texture. This novelty was not a neutral, insignificant event. I need only mention the various scales proposed for measuring temperature, i.e., the level of heat, and, especially, the fact that not all problems raised by such a measure have been yet solved to the satisfaction of all.[1]

The extension of theoretical structure to other fields met with still greater difficulties. This is especially clear in the case of electricity, where all basic variables are *instrumentally* measured and none is connected directly with a sense organ — as are most variables in other branches of physics. It is perfectly natural that the invention of the special instruments for measuring electrical variables should have taken longer. Electricity, more than other branches, advanced each time only to the extent to which each measuring instrument could clear additional ground. The opposite is true for mechanics; its progress was not held up by the problem of measure.

We usually stop the survey of physics at this point and thus miss a very important object lesson from such fields as structural mechanics or

1. For example, P. W. Bridgman, in *The Logic of Modern Physics* (New York, 1928), p. 130, observes that "no physical significance can be directly given to the flow of heat, and there are no operations for measuring it."

metallurgy. The complete story reveals that these fields — which are as integral a part of the science of matter as is atomic theory — are still struggling with patchy knowledge not unified into a single theoretical body. The only possible explanation for this backwardness in development is the fact that most variables in material structure — hardness, deformation, flexure, etc. — are in essence *quantified qualities*. Quantification in this case — as I shall argue presently — cannot do away completely with the peculiar nature of quality: it always leaves a qualitative residual which is hidden somehow inside the metric structure. Physics, therefore, is not as free from metaphysics as current critical philosophy proclaims, that is, if the issues raised by the opposition between number and quality are considered — as they generally are — metaphysical.

2. *Measure, Quantity, and Quality*. As one would expect, man used first the most direct and transparent type of measure, i.e., he first measured *quantity*. But we should resist the temptation to regard this step as a simple affair. Quantity presupposes the abstraction of any qualitative variation: consequently, only after this abstraction is reached does the measure of quantity become a simple matter, in most instances. Undoubtedly it did not take man very long to realize that often no qualitative difference can be seen between two instances of "wheat," or "water," or "cloth." But an immense time elapsed until weight, for instance, emerged as a general measurable attribute of any palpable substance. It is this type of measure that is generally referred to as *cardinal*.

In view of the rather common tendency in recent times to deny the necessity for distinguishing cardinal from other types of measure, one point needs emphasis: cardinal measurability is the result of a series of specific *physical* operations without which the paper-and-pencil operations with the measure-numbers would have no relevance.[2] Cardinal measurability, therefore, is not a measure just like any other, but it reflects a particular physical property of a category of things. Any variable of this category always exists as a *quantum* in the strict sense of the word (which should not be confused with that in "quantum mechanics"). Quantum, in turn, possesses simple but specific properties.

Whether we count the number of medicine pills by transferring them one by one from the palm into a jar, or whether we measure the amount of water in a reservoir by emptying it pail by pail, or whether we use a Roman balance to weigh a heap of flour, cardinal measure always implies

---

2. For an axiomatic analysis of how cardinal measure is derived from these physical operations, see the author's "Measure, Quality, and Optimum Scale," in *Essays on Econometrics and Planning Presented to Professor P. C. Mahalanobis on the Occasion of His 70th Birthday*, ed. C. R. Rao (Oxford, 1964), pp. 232–246.

indifferent subsumption and *subtraction* in a definite physical sense. It is precisely because of the fact that cardinality is an elemental physical property that we have had no great difficulty in devising a pointer-reading instrument for every instance of cardinal measure.

As I have intimated, quantity cannot be regarded as a notion prior to quality, either in the logical or evolutionary order. Undoubtedly, before the thought of measuring quantities of, say, wheat, man must have first come to recognize that one pile of wheat is greater than another *without weighing them*. For a long time "colder" and "hotter" had no measure. Distinctions, such as between "more friendly" and "less friendly" and, especially, between "earlier" and "later," which reflect qualitative differences, must have long preceded the practice of quantity measure. The things to which terms such as these apply came to be arranged in a definite mental order. Only later was a ranking number assigned to each of them, as must have happened first with events in time and, probably, with kinship. This "ranking" step represents the basis of the modern concept of *ordinal* measure. But the precedence of the ranking concept over that of quantity had a lasting influence upon the development of our ideas in this domain. Bertrand Russell rightly observed that philosophers are wholly mistaken in thinking that quantity is essential to mathematics; wherever it might occur quantity is not *"at present* amenable to mathematical treatment." [3] But even nowadays, order, not quantity, occupies the central place in pure mathematics.

Old as the basic principles of measure are and frequently as they have been discussed in recent years, we have been rather slow in perceiving the essential difference between cardinal and purely ordinal measure. Specifically, from the fact that cardinal measure presupposes ordinality we have often concluded that distinguishing between cardinal and purely ordinal measure is irrelevant hairsplitting. This position completely ignores the shadow that quality casts over purely ordinal measure. The things encompassed by a purely ordinal measure must necessarily vary qualitatively, for otherwise there would be absolutely nothing to prevent us from subsuming and subtracting them physically and, hence, from constructing a cardinal measure for them. To take a most elementary example: we can subsume by a physical operation a glass of "water" and a cup of "water" in another instance of the same substance, "water"; and, we can reverse the operation, if we so choose. But there is no sense in which we can subsume two historical dates into another historical date meaningfully, not even in a paper-and-pencil operation. "Historical date" thus is not cardinally measurable.

3. Bertrand Russell, *The Principles of Mathematics* (Cambridge, Eng., 1903), p. 419. Italics mine, to emphasize that the mathematical theory of measure was yet rather an esoteric topic at the time of Russell's statement.

On the other hand, we must recognize that cardinal and purely ordinal measurability represent two extreme poles and that between these there is room for some types of measure in which quality and quantity are interwoven in, perhaps, limitless variations. Some variables, ordinally but not cardinally measurable, are such that what appears to us as their "difference" has an indirect cardinal measure. Chronological time and temperature are instances of this. There is only one rule for constructing a measuring scale for such variables that would reflect their special property. Because of its frequency among physical variables, I proposed to distinguish this property by the term *weak cardinality*.[4] For self-evident reasons, a weak cardinal measure, like a cardinal one, is readily transformed into an instrumental one.

At this juncture a thorny question inevitably comes up: are there ordinally measurable attributes that could not possibly be measured by a pointer-reading instrument? Any definitive answer to this question implies at least a definite epistemological, if not also a metaphysical, position. The prevailing view is that all attributes are capable of instrumental measure: with time we will be able to devise a pointer-reading instrument for every attribute. F. P. Ramsey's faith in the eventual invention of some sort of psychogalvanometer for measuring utility, for example, clearly reflects this position.[5] In Ramsey's favor, one may observe that nowadays a meter of an electronic computer could show the I.Q. of an individual within a fraction of a second after he has pushed a system of buttons related to the truth-falsehood of a series of questions. And if one is satisfied with the idea that the I.Q. measures intelligence, then intelligence *is* measured by a pointer-reading instrument. On the other hand, there is the fact that hardness has so far defied the consummate ingenuity of physicists, and its scale is still exclusively qualitative. But probably the most salient illustration in this respect is supplied by entropy: basic though this variable is in theoretical physics, there is no *entropometer* and physicists cannot even suggest how it might be designed. Thus, while the evidence before us shows that physics has been able to devise measuring instruments for an increasing number of measurable attributes, it does not support the view that potentially all measures are reducible to pointer readings.

3. *The Qualitative Residual.* Variables in all equations of physics, whether in mechanics or in material structure, represent numbers. The only way in which quality appears explicitly in these equations is through a differentiation of symbols, as in $E = mc^2$ where E, m, and c stand for discretely distinct categories or constants. Ordinarily a physicist is not

---

4. Cf. the author's "Measure, Quality, and Optimum Scale," p. 241.
5. F. P. Ramsey, *The Foundations of Mathematics and Other Logical Essays* (New York, 1950), p. 161.

at all preoccupied by the fact that some variables are quantity measures while others measure quantified qualities. However, the quantification of a qualitative attribute — as I argued in the preceding section — does not change the nature of the attribute itself. Nor can quantification therefore destroy the quality ingredient of a phenomenon involving such an attribute. It stands immediately to reason that, since quantification does not cause quality to vanish, it leaves a qualitative residual which perforce must be carried over into the numerical formula by which the phenomenon is described. Otherwise this formula would not constitute an adequate description. The problem is to find out under what form the qualitative residual is hidden in a purely numerical pattern.

An examination of the basic laws of classical mechanics will show us the direction in which the answer lies. As already pointed out, this oldest branch of physics covers only cardinal variables. Newton's Second Law states, first, that the effect of a force upon a given body, the acceleration of that body's motion, is *proportional* to the quantum of force, and second, that the effect of a given force upon any particular body is *proportional* to the latter's mass. Furthermore, the essence of Newton's Law of Gravitation can be formulated in a similar manner: the attraction exerted by one body upon a unit of mass is *proportional* to the mass of the body and uniformly diffused in all directions.

One could cite other basic laws in physics that also affirm the proportional variation of the variables involved: the various transformation laws of energy, or such famous laws as Planck's ($E = h\nu$) and Einstein's ($E = mc^2$). The point I wish to make is that this simple pattern is not a mere accident: on the contrary, in all these cases the proportional variation of the variables is the inevitable consequence of the fact that every one of these variables is free from any qualitative variation. In other words, they all are cardinal variables. The reason is simple: if two such variables are connected by a law, *the connection being immediate in the sense that the law is not a relation obtained by telescoping a chain of other laws*, then what is true for one pair of values must be true for all succeeding pairs. Otherwise, there would be some difference between the first and, say, the hundredth pair, which could mean only a qualitative difference.

There is therefore an intimate connection between cardinality and the homogeneous linearity of a formula by which a direct law is expressed. On the basis of this principle, nonhomogeneous linearity would generally indicate that some of the variables have only a weak cardinality. Indeed, a nonhomogeneous linear relation is equivalent to a linear homogeneous relation between the finite differences of all variables.

A counter-proof of the principle just submitted is even more enlightening. For this we have to turn to the least advertised branch of physics, that of material structure. This field abounds in quantified qualities: tensile strength, elastic limit, flexure, etc. We need only open at random any treatise on material structure to convince ourselves that no law involving such variables is expressed by a linear formula. (In fact, in some cases there is no formula at all but only an empirically determined graph.) The reason is, again, simple. Successive pounds of load may be regarded as equal causes, but their individual effect upon the shape of a beam is not the same. Deformation being a measurable quality, the nth degree of deformation is not qualitatively identical to any of the preceding degrees. Nor does "n degrees of deformation" represent the subsumption of n times "one degree of deformation." We thus reach the correlative principle to the one stated in the preceding paragraph: nonlinearity is the aspect under which the qualitative residual appears in a numerical formula of a quality-related phenomenon.

One may think of refuting this conclusion by *implicit measuring*, i.e., by choosing an ordinal scale for the quantified quality so as to transform the nonlinear into a linear relation. Joan Robinson once tried this idea for labor efficiency.[6] The reason why her attempt failed is general: we have to establish an implicit measure for *every* situation to which the relation applies. That would be no measure at all. Moreover, most quality-related phenomena have a sort of climax, followed by a rapid breakdown; such a nonmonotonic variation cannot possibly be represented by a linear function.

The situation is not as limpid in the case of homogeneous linearity. Some laws covering quantified quality are nevertheless expressed as proportional variations. An example is Robert Hooke's law: elastic stress is proportional to the load strain. But the contradiction is purely superficial, for in all such cases the linear formula is valid only for a limited range and even for this range it represents only a convenient approximation, a rule of thumb. Such cases suggest that some of the other laws now expressed by linear formulae may be only a rule of thumb. One day we may discover that linearity breaks down outside the range covered by past experiments. The modern history of physics offers several examples of such discoveries. Perhaps the most instructive is the famous formula proposed by H. A. Lorentz for velocity addition. In the classical formula, which proceeds from the principle that equal causes produce equal effects on velocity, we have $V = v + v + \cdots + v = nv$, which is a homogeneous linear function of n, that is, of scale. But for the same situation, as is easily proved, the Lorentz law yields $V = c[(c + v)^n - (c - v)^n]/[(c + v)^n + (c - v)^n]$. In this case, the effect of each additional v decreases with the scale n. We can then understand

6. Joan Robinson, *Economics of Imperfect Competition* (London, 1938), p. 109 and *passim*.

why physicists lose no opportunity to decry the extrapolation of any known law outside the range of actual experiments.[7] Even though the protests should not be taken literally, their ground is unquestionable. It would seem, therefore, that if we take cardinality to be a physical property we should also admit that this property too might be limited to a certain range of the quantum. This would vindicate Hegel's dictum, that quantitative changes in the end bring about qualitative changes,[8] over the entire domain of physics — and perhaps to an extent not intended even by Hegel. Indeed, if the dictum applies to quantity itself then it loses all meaning.[9]

To banish completely the notion of quantity — hence, of cardinality — from physics would be fatal. For then all laws of physics would be reduced to nonmetric, topological propositions and its success in practice would almost come to an end. This is perhaps why no physicist — to my knowledge — has denounced the extrapolation of cardinality. On the contrary, there is at least one memorable situation where the choice between two alternative descriptions was made so as to save the cardinality of the most basic coordinate in physics. I am referring to the replacement of the Lorentz contraction of "length" by Einstein's relativity theory.

We may discover that some of the variables presently considered cardinal are not really so — as happened with velocity — but it does not seem possible for any quantitative science to get rid of quality altogether. The point is important, and I shall presently illustrate it with examples from economics.

4. *The Problem of Size.* Without forgetting the caveats I have inserted into the preceding analysis, we can generally expect that if the variables *immediately connected* by a phenomenon are cardinally measurable, then they can all be increased in the same proportion and still represent the same phenomenon. The formula describing the phenomenon then must be homogeneous and linear, or more generally, a homogeneous function of the first degree. On the other hand, if some variable is a quantified quality, nothing seems to cast doubt over our expectation that the formula will be nonlinear.

Since the first situation characterizes a phenomenon (or a process) *indifferent to size*, it is clear that the problem of size arises only for processes involving quantified qualities, and conversely. Needless to add, the

7. Bridgman, *Logic of Modern Physics*, p. 203; P. W. Bridgman, *The Intelligent Individual and Society* (New York, 1938), p. 13; Werner Heisenberg, *Physics and Philosophy: The Revolution in Modern Science* (New York, 1958), pp. 85–86.

8. *The Logic of Hegel*, tr. W. Wallace (2nd edn., London, 1904), pp. 203 and *passim*.

9. "In quantity we have an alterable, which in spite of alterations still remains the same." *Ibid.*, p. 200.

same would apply with even greater force to processes involving non-quantifiable qualities. The conclusion is that the problem of size is strictly confined to quality-related processes.

The point I wish to emphasize is that in support of this conclusion I have not invoked one single piece of evidence outside the domain of inert matter. The fact that it is physics which teaches us that size is indissolubly connected with quality, therefore, deserves special notice on the part of students of life phenomena. Indeed, the prevailing opinion regarding size, which constitutes one of the most important chapters in biology and social sciences, has been that the problem arises only in these sciences because they alone have to study organisms.

I know only too well the endless controversy surrounding the problem of optimum size in economics, and also how ingenious are some of the arguments against the existence of such a size. The nature of these arguments proves that more often than not economists fail to see the relation between that problem and physical laws. The same can be said about most counter-arguments, for they resort to biological analogies, from the mosquito to the elephant. In fact, the optimum size of the elephant, just like that of a manufacturing plant, is determined by physical laws having to do with quantified qualities. If some biologists — such as Herbert Spencer, especially[10] — have been able to perceive the connection between biological size and the laws of material structure, whereas economists merely referred the problem back to biology, it is because biologists alone have been interested in what happens inside the individual organism. The common flow-complex of the economist leads to the position that what happens inside a production unit concerns exclusively the engineer, that economics is concerned only with the flows observed at the plant gate, i.e., with *inter-unit flows*. And this flow-complex is responsible for many myopic views of the economic process.

One can hardly overemphasize the point that the problem of size is indissolubly connected with the notion of sameness, specifically, with the notion of "same phenomenon" or "same process." In economics we prefer the term "unit of production" to "same process," in order to emphasize the abstract criterion by which sameness is established. Whatever term we use, in this case too sameness remains basically a primary notion which is not susceptible to complete formalization. "The same process" is a class of analogous events, and it raises even greater difficulties than "the same object." But we must not let our analysis — in this case any more than in others — run aground on this sort of difficulty. There

10. Herbert Spencer, *The Principles of Biology*, 2 vols. (New York, 1886), I, 121 ff. For a summary of the present views on this matter see L. von Bertalanffy, *Problems of Life* (New York, 1952), pp. 136–137 and P. B. Medawar, *The Uniqueness of the Individual* (New York, 1958), pp. 110 ff.

are many points that can be clarified to great advantage once we admit
that "sameness," though unanalyzable, is in most cases an operational
concept.

Let then P_1 and P_2 be any two distinct instances of a process. The
problem of size arises only in those cases where it is possible to subsume
P_1 and P_2 *in vivo* into another instance P_3 of the same process. If this is
possible we shall say that P_1 and P_2 are added internally and write

$$(1) \qquad\qquad P_1 \oplus P_2 = P_3$$

We may also say that P_3 is *divided* into P_1 and P_2 or that the corresponding process (P) is *divisible*. For an illustration, if the masses m_1
and m_2 are transformed into the energies E_1 and E_2 respectively, by two
distinct instances P_1 and P_2, these individual instances can be added
internally because there exists an instance of the same process which
transforms $m_1 + m_2$ into $E_1 + E_2$. We can also divide P_3 into P_1 and P_2
or even into two half P_3's — provided that P does not possess a natural,
indivisible unit. Needless to say, we cannot divide (in the same sense of
the word) processes such as "elephant" or even "Harvard University."

It is obvious that it is the internal addition *in vivo* that accounts for
the linearity of the corresponding paper-and-pencil operation. For even
if the subsumption of P_1 and P_2 is possible but represents an instance of
a *different* process, our paper-and-pencil operations will reveal a nonlinear
term.[11]

Another point that deserves emphasis is that processes can also be
added externally. In this case, P' and P'' need not even be instances of
the same process. In the external addition,

$$(2) \qquad\qquad P' + P'' = P''',$$

P' and P'' preserve their individuality (separation) *in vivo* and are lumped
together only in thought or on paper. External and internal addition,
therefore, are two entirely separate notions.

When an accountant consolidates several balance sheets into one balance sheet, or when we compute the net national product of an economy,
we merely add all production processes externally. These paper-and-pencil
operations do not imply in the least any real amalgamation of the processes involved. In bookkeeping all processes are additive. This is why we
should clearly distinguish the process of a unit of production (plant or
firm) from that of *industry.* The point is that an industry may expand
by the accretion of *unconnected* production processes, but the growth of

11. This term reflects what we may call the interaction generated by the merging
of two distinct individual phenomena. For an instructive illustration the reader may
refer to Erwin Schrödinger's interpretation of the nonlinear term in the equation of
wave mechanics. E. Schrödinger, "Are There Quantum Jumps?" *British Journal for
the Philosophy of Science,* III (1952), 234.

a unit of production is the result of an internal morphological change.

It follows that if the units which are externally added in the book-keeping process of industry are *identical*, then proportionality will govern the variations of the variables involved — inputs and outputs. The constancy of returns to scale therefore is a tautological property of a granular industry.[12] To the extent that an actual industry represents an accretion of practically identical firms, no valid objection can be raised against the assumption of constant coefficients of production in Wassily Leontief's system.

One point in connection with the preceding argument is apt to cause misunderstanding. Since I have argued that phenomena involving only cardinally measurable variables necessarily are indifferent to scale, one may maintain that I thereby offered the best argument against the existence of the optimum scale of the plant, at least. Indeed, a critic may ask: by and large, are not plant inputs and outputs cardinally measurable?

Such an interpretation would ignore the very core of my argument, which is that only if the cardinally measurable variables are immediately connected — as cause and effect in the strictest sense of the terms are — can we expect the law to be expressed by a homogeneous linear formula. To return to one of the examples used earlier, we can expect acceleration to be proportional to force because force affects acceleration directly: to our knowledge there is no intermediary link between the two. I have not even hinted that cardinality by itself suffices to justify homogeneous and linear law formulae. I visualize cardinality as a physical property allowing certain definite operations connected with measuring, and, hence, as a property established prior to the description of a phenomenon involving cardinal variables. Precisely for this reason, I would not concur with Schrödinger's view that energy may be in some cases "a 'quantity-concept' (Quantitätsgrösse)," and in others "a 'quality-concept' or 'intensity-concept' (Intensitätsgrösse)." [13] As may be obvious by now, in my opinion the distinction should be made between internally additive and nonadditive processes instead of saying that the cardinality of a variable changes with the process into which it enters.

As to the case of a unit of production, it should be plain to any economist willing to abandon the flow-complex that inputs and outputs are not *directly connected* and, hence, there is no *a priori* reason for expecting the production function to be homogeneous of the first degree. The familiar plant-production function is the expression of an external addi-

12. I am using the term "granular industry" instead of "atomistic industry," not only because nowadays the latter may cause confusion, but also because the property of constant returns to scale does not necessarily require the number of firms to be extremely large.

13. Schrödinger, "Are There Quantum Jumps?" p. 115.

tion of a series of physical processes, the addition being justified by the unitary nature of management whose authority begins and ends at the gate. It is because most of these intermediary processes are quality-related that no plant process can be indifferent to scale. We know that the productive value of many inputs that are unquestionably cardinally measurable does not reside in their material quantum. Although materials are bought by weight or volume, what we really purchase is often resistance to strain, to heat, etc., that is, quality, not quantity. This is true whether such materials are perfectly divisible or not. Consequently, the so-called tautological thesis — that perfect divisibility of factors entails constant returns to scale — is completely unavailing. If, nevertheless, it may have some appeal it is only because in the course of the argument "divisibility of factors" is confused with "divisibility of processes." Whenever this is the case the argument no longer applies to the unit of production; with unnecessary complications it only proves a tautological feature of a molecular industry.

5. *Cardinality and the Qualitative Residual.* Perhaps the greatest revolution in modern mathematics was caused by Evariste Galois' notion of group. Thanks to Galois' contribution, mathematics came to realize that a series of properties, which until that time were considered as completely distinct, fit the *same abstract* pattern. The economy of thought achieved by discovering and studying other abstract patterns in which a host of situations could be reflected is so obvious that mathematicians have turned their attention more and more in this direction, i.e., towards formalism. An instructive example of the power of formalism is the abstract pattern that fits the point-line relations in Euclidean geometry and at the same time the organization of four persons into two-member clubs.[14] Time and again, the success of formalism in mathematics led to the epistemological position that the basis of knowledge consists of formal patterns alone: the fact that in the case just mentioned the pattern applies to points and lines in one instance, and to persons and clubs in the other, is an entirely secondary matter. By a similar token — that any measuring scale can be changed into another by a strictly monotonic transformation, and hence the strictly monotonic function is the formal pattern of measure — cardinality has come to be denied any epistemological significance. According to this view, there is no reason whatsoever why a carpenter should not count one, two, . . . , 2^n, as he lays down his yardstick once, twice, . . . , n-times.

There are also economists who have propounded the relativity of measure. Apparently, they failed to see that this view saps the entire foundation upon which the economic science rests. Indeed, this founda-

14. Cf. R. L. Wilder, *Introduction to the Foundations of Mathematics* (New York, 1956), pp. 10–13.

tion consists of a handful of principles, all stating that some particular phenomenon is subject to increasing or decreasing variations. There is no exception, whether the principle pertains to consumption or production phenomena: decreasing marginal utility, decreasing marginal rate of substitution, increasing internal economies, and so on.

It is a relatively simple matter to see that these principles lose all meaning if cardinality is bogus. Clearly, if there is no epistemological basis for measuring corn one way or the other, then marginal utility may be freely increasing or decreasing over any given interval. Surprising though it may seem, the relativity of measure would cause a greater disaster in the study of production than in that of consumption. Isoquants, cost curves, scale curves could then be made to have almost any shape we choose.[15] The question whether theoretical physics needs a cardinal basis is beyond the scope of this essay, but there can hardly be any doubt that economic activity, because of its pedestrian nature, cannot exist without such a basis.

We buy and sell land by acres, because land is often homogeneous over large areas; and because this homogeneity is not general, we have differential rent. How unimaginably complicated economic life would be if we adopted an ordinal measure of land chosen so as to eliminate differential rent, let alone applying the same idea to all economic variables involving qualitative variations!

Since cardinality is associated with the complete absence of qualitative variation, it represents a sort of natural origin for quality. To remove it from the box of our descriptive tools is tantamount to destroying also any point of reference for quality. Everything would become either "this" or "that." Such an idea would be especially pernicious if followed by economics. Any of the basic principles, upon which a good deal of economic analysis rests, is at bottom the expression of some qualitative residual resulting from the projection of quality-related phenomena upon a cardinal grid. The principle of decreasing elasticity of factor substitution, to take one example, is nothing but such a residual. A critical examination of its justification would easily disclose that substitutable factors belong to a special category mentioned earlier: they participate in the production process through their qualitative properties. The other category of factors, which are only carried through the process as mere substances of some sort, cannot, strictly speaking, cause any qualitative residual and, hence, give rise to substitution. For an illustration one can cite the inputs of copper and zinc in the production of a particular brass. We thus come to the conclusion that every relation between two inputs, or an input and the output, may or may not show a qualitative residual

15. For details, see my article cited above, "Measure, Quality, and Optimum Scale," pp. 234, 246.

depending on the kind of role the corresponding factors play in the production process. This difference is responsible for the great variety of patterns which a production function may display and which is covered by the general notion of limitationality.[16]

Many economists maintain that economics is a deductive science. The preceding analysis of the nature of the basic principles pertaining to the quantitative variations of cardinally measurable economic goods justifies their position, but only in part. To be sure, to affirm the existence of a qualitative residual is an *a priori* synthetic judgment rather than an empirical proposition. But only by factual evidence can we ascertain whether the qualitative residual is represented by increasing or decreasing variations. The point seems obvious enough. Nevertheless, I wish to illustrate it by a particularly instructive example.

Still groping towards the idea that the basic feature of the preference map in *a field of cardinally measurable commodities* reflects a qualitative residual, in a 1954 article I replaced the principle of decreasing marginal rate of substitution by a new proposition which brings quality to the forefront. To put it elementarily, my point of departure was that if *ten pounds* of potatoes and *ten pounds* of corn flour happen to be equivalent incomes to a consumer, then an income of *ten pounds* of any mixture of potatoes *and* corn flour could not be equivalent to either of the initial alternatives. This negative statement simply acknowledges the existence of a qualitative residual in the preference map and, hence, needs no empirical support: the "axiom" that choice is quality-related suffices. But under the influence of the tradition-rooted notion that indifference curves are *obviously* convex, I went one step further and asserted that the ten-pound mixture is (generally) preferred to either of the other two. For obvious reasons, I called the postulate thus stated the *Principle of Complementarity*.[17] Carl Kaysen questioned the postulate on the ground that some ingredients may very well produce a nauseating concoction. At the time, I was hardly disturbed by the objection, for I was satisfied by the observation that my postulate compels the individual neither to actually mix the ingredients nor to consume them in a certain order. It was only later that I came to see the relevance of Kaysen's question, as I struck upon a simple counter-example of the postulate: a pet lover may be indifferent between having two dogs or two cats but he might find life unpleasant if he had one dog and one cat. The example shows that since some commodities may have an antagonistic effect the Principle

16. The above remarks may be regarded as some afterthoughts to my 1935 paper, "Fixed Coefficients of Production and the Marginal Productivity Theory," reprinted below, which in all probability represents the first attempt at a general analysis of limitationality in relation to the pricing mechanism.

17. See section II of "Choice, Expectations and Measurability" (1954), reprinted in this volume.

of Complementarity is not generally valid. As I have said, only factual evidence can determine in which direction the qualitative residual disturbs proportionality. And since without specifying this direction the basic principles of economics are practically worthless, the position that they are *a priori* synthetic truths is only half justified. Like all half-truths, this position has had some unfortunate effects upon our thoughts.

6. *Theory and Novelty.* Modern philosophy of science usually fails also to pay sufficient attention to the fact that the study of inert matter is divided between physics and chemistry. Probably it is thought that the separation of chemistry from physics is a matter of tradition or division of labor. But if these were the only reasons, chemistry would have long since become an ordinary branch of physics, like optics or mechanics, for instance. With the creed of unified science sweeping the intellectual world, why are the frontier posts still in place? The recent establishment of physical chemistry as an intermediary link between physics and chemistry clearly indicates that the complete merger is prevented by some deep-lying reason. This reason is that chemistry does not possess a theoretical code of orderliness. Hence, only harm could result from letting that Trojan horse inside the citadel of physics.

One may be greatly puzzled by the observation that there is no chemical theory. After all, chemistry, like physics, deals with quantities and quantified qualities. That two atoms of hydrogen and one atom of oxygen combine into a molecule of water is an example of a quantitative chemical proposition. True, chemistry does study some quantified qualities of substance: color, hardness, acidity, water repellence, etc. But in the end, even these qualitative properties are expressed by arithmomorphic propositions. *Prima facie*, therefore, nothing could prevent us from passing all chemical propositions through a logical sieve so as to separate them into an α-class and a β-class. Why then is there no chemical theory?

The key to the answer lies in the observation that no general formula exists for deducing the qualities of a substance from its chemical composition. We know quite a lot about every chemical element, but more often than not, this knowledge is of no avail in predicting all qualities of a new compound. From the point of extant knowledge, therefore, almost every new compound is a *novelty* in some respect or other. That is why the more chemical compounds chemistry has synthesized, the more baffling has become the irregularity of the relation between chemical structure and qualitative properties. If this historical trend teaches us anything at all, it is that nothing entitles us to expect that this increasing irregularity will be replaced by some simple principles in the future.

Let us suppose that we have taken the trouble to sift all *known* propositions of chemistry into an α-class and a β-class. But new chemical compounds are likely to be discovered almost every day. From the preceding paragraph it follows that with every such discovery, even a minor one, the α-class has to be increased by new propositions, at times more numerous than those to be added to the β-class. Known compounds being as numerous as they are, we do not actually have to determine today's α-class of chemistry in order to ascertain that it contains an immense number of propositions, perhaps greater than that of the β-class. It is now obvious why no one has attempted to construct a logical foundation for chemistry. As I argued in the first chapter of this essay, the *raison d'être* of a theoretical edifice is the economy of thought it yields. If novelty is an immanent feature of a phenomenal domain — as is the case in chemistry — a theoretical edifice, even if feasible at all, is uneconomical: to build one would be absurd.

It is not necessary to see in novelty more than a relative aspect of knowledge. In this sense, the concept is free from any metaphysical overtones. However, its epistemological import extends from chemical compounds to all forms of Matter: colloids, crystals, cells, and ultimately biological and social organisms. Novelty becomes even more striking as we proceed along this scale. Certainly, all the qualitative attributes which together form the entity called elephant, for example, are novel with respect to the properties of the chemical elements of which an elephant's body is composed. Combination *per se* — as an increasing number of natural scientists admit — contributes something that is not deducible from the properties of the individual components.[18] The obvious conclusion is that the properties of simple elements, whether atomic or intra-atomic, do not describe Matter exhaustively. The complete description of Matter includes not only the property of the molecule of, say, carbon, but also those of all organizations of which carbon is a constituent part. In this perspective, Matter has infinitely many properties, nay, infinitely many *potentiae*. If we wish, we may include thinking and even the feeling of being alive among these *potentiae*. But if we follow this line, we must also admit that novelty suffices to explain why thought or any other specific attribute of life organizations cannot be reduced to (or deduced from) the properties of elementary matter.[19]

7. *Novelty and Uncertainty.* There are several object lessons that students of life phenomena could derive from the emergence of novelty by combination. The most important one for the social scientist bears upon those doctrines of human society that may be termed "chemical" because

18. P. W. Bridgman, *The Nature of Physical Theory* (Princeton, 1936), p. 96; Bertalanffy, *Problems of Life*, chap. ii.

19. Many, probably most, students of "man" would meet this statement with strong objections. The contrary can be expected from most physicists; actually, many great names in physics have denounced the psycho-physical parallelism in most categorical terms. For an example, see C. N. Hinshelwood, *The Structure of Physical Chemistry* (Oxford, 1951), pp. 456, 471.

they overtly recognize chemistry as their source of inspiration and model. Since the problem raised by these doctrines is of crucial importance for the orientation of all social sciences, especially economics, it deserves to be discussed in detail. This will be done in a special section later on. But at this juncture, I propose to point out one object lesson which pertains to the difference between risk and uncertainty.

Since an exhaustive description of Matter implies the experimenting with, and study of, an essentially limitless set of material combinations (or organizations), it goes without saying that the fate of human knowledge is to be always incomplete. From the analysis in the preceding section, the meaning of "incomplete" should be perfectly clear. However, in the controversies over the difference between risk and uncertainty, *incomplete* knowledge has often been confused with what may be termed *imperfect* knowledge. The point is that — in the terminology here adopted — *incomplete* refers to knowledge as a whole, but *imperfect* refers to a particular piece of the extant knowledge. Some illustrations may help clarify the difference. Our knowledge is incomplete because, for instance, we have absolutely no idea what sort of biological species will evolve from *homo sapiens*, or even whether one will evolve at all. On the other hand, we know that the next birth (if normal) will be either a boy or a girl. Only we cannot know far in advance which it will be, because our knowledge concerning the sex of future individuals is imperfect.

Risk describes the situations where the exact outcome is not known but the outcome does not represent a novelty. Uncertainty applies to cases where the reason why we cannot predict the outcome is that the same event has never been observed in the past and, hence, it may involve a novelty.

Since in a paper included in the volume[20] I have insisted upon the necessity of this distinction probably more strongly than the authors who first broached the issue, further comments at this place may seem unnecessary. However, it may be instructive to illustrate by a topical problem the connection between novelty arising from new combinations and the nature of expectation.

Notation being the same as in the paper just cited, let $\mathcal{E}_1(E_1)$, $\mathcal{E}_2(E_2)$, ..., $\mathcal{E}_n(E_n)$ be the expectations of the n members of a committee *before* they meet on a given occasion. Let us also assume that the committee is not a pseudo committee. This rules out, among other things, the existence of a "best mind" (in all relevant respects) among members as well as their complete identity. In these circumstances, during the discussion preceding the vote, part of the evidence initially possessed by one member but not by another will combine with the initial evidence of the latter. In the end, everybody's evidence is increased and, hence, every-

20. "The Nature of Expectation and Uncertainty" (1958).

body will have a new expectation, $\mathcal{E}_k(E'_k)$. The new combination should normally produce some novelty: the decision adopted may be such that no one, whether a member or a poll-taker, could think of it prior to the meeting.[21]

8. *Hysteresis and History.* The unparalleled success of physics is generally attributed to the sole fact that physics studies only matter and matter is uniform. It would be more appropriate to say that physics studies only those properties of Matter that are uniform, that is, independent of combination. But perhaps even this clarification would not suffice. The current meaning of uniformity of matter is that its behavior at any moment depends exclusively upon *present conditions.* In other words, the behavior of matter is completely independent of *past history.* Undoubtedly, matter often behaves in this manner. For what would the world be like if drops of water or grains of salt behaved differently according to their individual histories? And if in addition matter remained indestructible — as it is believed to be — then a physical science would be quite impossible.

Yet, in some cases behavior, even physical, does depend upon past history as well. The most familiar case is the behavior of a magnet, or to use the technical term, the magnetic hysteresis. But hysteresis is not confined to magnetism: structural deformation and the behavior of many colloids too depend upon past history. According to a recent idea of David Bohm, shared also by Louis de Broglie, the Heisenberg indeterminacy would result from the fact that the past history of the elementary particle is not taken into account in predicting its behavior.[22] The case where all past causes work cumulatively in the present, therefore, is not confined to life phenomena.

There is, however, one important difference between physical hysteresis and the historical factor in biology or social sciences. A physicist can always find as many bits of nonmagnetized iron — i.e., *magnets without history* — as he needs for proving experimentally that magnets with an identical history behave identically. It is vitally important to observe that if we were unable to experiment with cases where *the level of history is zero*, we could not arrive at a complete law of magnetic hysteresis. But in the macro-biological and social world getting at the zero level of history seems utterly impossible. That is why in these two domains the

21. The fact that most, perhaps all, behavioristic models completely ignore this particular group effect is self-explanatory: a predicting model must keep novelty off the field. But it is highly surprising to find the point ignored by analyses of another sort. A salient example: N. Kaldor, in "The Equilibrium of the Firm," *Economic Journal*, XLIV (1934), 69n1, states that "the supply of coordinating ability could probably be enlarged by dismissing the Board and leaving the single most efficient individual in control."

22. Cf. Heisenberg, *Physics and Philosophy*, pp. 130–131.

historical factor invites endless controversy. We may recall in this connection the elementary remark of C. S. Peirce that universes are not as common as peanuts. Because there is only one Western civilization, the question of whether its historical development merely follows a trajectory determined entirely by the initial condition or whether it represents a hysteretic process can be settled neither by an effective experiment nor by the analysis of observed data. Unfortunately, the answer to this sort of question has an incalculable bearing upon our policy recommendations, especially upon those with a long-run target — such as the policies of economic development.

Physicists not only can determine the law that relates present behavior of a magnet to its history, but also can make history vanish by demagnetization. In other words, for any given history, $\mathcal{H}$, there is an $\mathcal{H}'$ such that $\mathcal{H} + \mathcal{H}' = 0$; moreover, $\mathcal{H}'$ is a very short history. Is this true also for the history of a society or of an individual, and if not exactly true, to what extent? Perplexing though this and other similar issues raised by human hysteresis seem to be, no search for a complete description of social phenomena can avoid them. Actually, the stronger our intention of applying knowledge to concrete practical problems — like those found in economic development, to take a topical example — the more urgent it is for us to come to grips with these issues.

The difficulties of all sorts which arise can be illustrated, though only in part, by the simple, perhaps the simplest, instance of the hysteresis of the individual consumer. The fact that the individual's continuous adjustment to changing price and income conditions changes his tastes seems so obvious that in the past economists mentioned it only in passing, if at all. Actually, there is absolutely no stand upon which this phenomenon could be questioned. In 1950 I attempted a sketchy formalization of the problem mainly for bringing to light the nasty type of questions that besiege the Pareto-Fisher approach to consumer's behavior as soon as we think of the hysteresis effect.[23] By means of a simple analytical example I showed that in order to determine the equilibrium of the consumer (for a fixed budget and constant prices) we need to know much more than his particular hysteresis law. Still worse, this law being expressed by a very complex *set function*, we can only write it on paper but not determine it in actual practice. Set functions cannot be extrapolated in any useful way. Consequently, however large the number of observations, the effect of the last experiment can be known only after we observe what we wish to predict. The dilemma is obvious. How nasty this dilemma may be is shown by the case where the order of observations too matters. In this case, even if it were possible to make the consumer

23. "The Theory of Choice and the Constancy of Economic Laws" (1950), reprinted below.

experiment with all possible situations we would not be able to know the general law of the hysteresis effect. All the more salient, therefore, are the contributions of James Duesenberry and Franco Modigliani on the hysteresis effect upon the saving ratio.

But the most unpleasant aspect of the problem is revealed in the ordinary fact that behavior suffers a qualitative shock, as it were, every time the individual is confronted with a *novel commodity*.[24] This is why we would be utterly mistaken to believe that technological innovations modify supply alone. The impact of a technological innovation upon the economic process consists of both an industrial rearrangement and a consumers' reorientation, often also of a structural change in society.

9. *Physics and Evolution.* The analysis of the two preceding sections leads to a few additional thoughts. The first is that history, of an individual or of a society, seems to be the result of two factors: a hysteresis process and the emergence of novelty. Whether novelty is an entirely independent element or only a consequence of the hysteresis process is perhaps the greatest of all moot questions, even though at the level of the individual it is partly tractable. Unquestionably, the invention of the carbon telephone transmitter was a novelty for all contemporaries of Edison. But what about Edison himself? Was his idea a novelty for him too or was it a result, partly or totally, of his own hysteresis process?

Be this as it may, we cannot avoid the admission that novel events, beginning with the novelty of chemical transformations, punctuate the history of the world. The several philosophical views which speak of "the creative advance of nature" [25] are not therefore as utterly metaphysical or, worse, as mystical, as many want us to believe. However, we need more than the existence of novelty to support the notion of a nature advancing from one creative act to another. Novelty, as I have tried to stress, need not represent more than a relative aspect of our knowledge; it may emerge for us without nature's advancing on a path marked by novel milestones. The masterpieces in a picture gallery are not being painted just as we walk from one to the next. On the other hand, geology, biology, and anthropology all display a wealth of evidence indicating that at least on this planet there has always been evolution: at one time the earth was a ball of fire which gradually cooled down; dinosaurs emerged, vanished and, undoubtedly, will never appear again; man moved from cave dwellings into penthouses. Impressive though all this evidence is, all efforts of biologists and social scientists to discover

24. *Ibid.*
25. The most outstanding representatives of this philosophy are Henri Bergson (*Creative Evolution*, New York, 1913, pp. 104–105 and *passim*) and Alfred North Whitehead (*An Enquiry concerning the Principles of Natural Knowledge*, 2nd edn., Cambridge, Eng., 1925, pp. 63, 98; *The Concept of Nature*, Cambridge, Eng., 1930, p. 178; *Process and Reality: An Essay in Cosmology*, New York, 1929, p. 31; etc.).

an evolutionary law for their phenomenal domains have remained utterly fruitless. But, perhaps, we should clarify this statement by explaining what an evolutionary law is.

An evolutionary law is a proposition that describes an ordinal attribute E of a given system (or entity) and also states that if $E_1 < E_2$ then the observation of E_2 is later in Time than E_1, and conversely.[26] That is, the attribute E is an *evolutionary index* of the system in point. Still more important is the fact that the ordinal measure of any such E can tell even an "objective" mind — i.e., one deprived of the anthropomorphic faculty of sensing Time — the direction in which Time flows. Or, to use the eloquent term introduced by Eddington, we can say that E constitutes a "time's arrow." [27] Clearly, E is not what we would normally call a cause, or the unique cause, of the evolutionary change. Therefore, contrary to the opinion of some biologists,[28] we do not need to discover a single cause for evolution in order to arrive at an evolutionary law. And in fact, almost every proposal of an evolutionary law for the biological or the social world has been concerned with a time's arrow, not with a single cause.

Of all the time's arrows suggested thus far for the biological world, "complexity of organization" and "degree of control over the environment" seem to enjoy the greatest popularity.[29] One does not have to be a biologist, however, to see that neither proposal is satisfactory: the suggested attributes are not ordinally measurable. We may also mention the interesting but, again, highly questionable idea of R. R. Marett that increasing charity in the broad sense of the word would constitute the time's arrow for human society.[30]

It is physics again that supplies the only clear example of an evolutionary law: the Second Law of Thermodynamics, called also the Entropy Law. But the law has been, and still is, surrounded by numerous controversies — which is not at all surprising. A brief analysis of entropy and a review of only its most important issues cannot avoid some technicalities. It is nevertheless worth doing, for it uncovers the unusual sort of epistemological difficulty that confronts an evolutionary law even

26. Needless to explain that $E_1 < E_2$ means that E_2 follows E_1 in the ordinal pattern of E.

27. A. S. Eddington, *The Nature of the Physical World* (New York, 1943), pp. 68–69.

28. Julian Huxley, *Evolution: The Modern Synthesis* (New York, 1942), p. 45.

29. For a comprehensive (but not wholly unbiased) discussion of these criteria see Huxley, *ibid.*, chap. x; also Theodosius Dobzhansky, in *Evolution, Genetics, and Man* (New York, 1955), pp. 370 ff, who argues that all sensible criteria of evolution must bear out the superiority of man. One should, however, appreciate the objection of J. B. S. Haldane, in *The Causes of Evolution* (New York, 1932), p. 153, that man wants thus "to pat himself on the back."

30. R. R. Marett, *Head, Heart, and Hands in Human Evolution* (New York, 1935), p. 40 and *passim*, and the same author's "Charity and the Struggle for Existence," *Journal of the Royal Anthropological Institute*, LXIX (1939), 137–149.

in the most favorable circumstances, those of the qualityless world of elementary matter. Although these difficulties were felt only in the later period and only serially, they are responsible for the agitated history of thermodynamics.

Thermodynamics sprang from a memoir of Sadi Carnot in 1824 on the efficiency of steam engines.[31] One result of this memoir was that physics was compelled to recognize as scientific an elementary fact known for ages: heat always moves by itself from hotter to colder bodies. And since the laws of mechanics cannot account for a unidirectional movement, a new branch of physics using nonmechanical explanations had to be created. Subsequent discoveries showed that all known forms of energy too move in a unique direction, from a higher to a lower level. By 1865, R. Clausius was able to give to the first two laws of thermodynamics their classic formulation:

The energy of the universe remains constant;
The entropy of the universe at all times moves toward a maximum.

The story is rather simple if we ignore the small print. According to Classical thermodynamics, energy consists of two qualities: (1) *free* or *available* and (2) *bound* or *latent*. Free energy is that energy that can be transformed into mechanical work. (Initially, free heat was defined roughly as that heat by which a hotter body exceeds the colder one, and which alone could move, say, a steam engine.) Like heat, free energy always dissipates by itself (and without any loss) into latent energy. The material universe, therefore, continuously undergoes a qualitative change, actually a qualitative degradation of energy. The final outcome is a state where all energy is latent, the Heat Death as it was called in the earliest thermodynamic theory.

For some technical reasons, which need not interest us, entropy was defined by the formula

(3)   Entropy = (Latent Energy)/(Absolute Temperature).[32]

The Entropy Law needs no further explanation. We should notice, however, that it is strictly an evolutionary law with a clearly defined time's arrow: entropy. Clausius seems to have thought of it in the same way, for he coined "entropy" from a Greek word equivalent in meaning to "evolution."

10. *Time: An Ambiguous Term.* A short word though it is, Time denotes a notion of extreme complexity. The problem of Time has tormented the

31. A full translation appears in *The Second Law of Thermodynamics*, ed. and tr. W. F. Magie (New York, 1899).

32. For which see J. Clerk Maxwell, *Theory of Heat* (10th edn., London, 1921), pp. 189 ff, and *Commentary on the Scientific Writings of J. Willard Gibbs*, ed. G. F. Donnan (New Haven, 1936), II, 19 ff.

minds of all great philosophers far more than its correlative, Space. To most of us Time does seem "so much more mysterious than space,"[33] and no one has yet proved that we are mistaken. To be sure, there are some who maintain that Einstein's relativity theory has proved that they are in fact one.[34] This sort of argument ignores, however, that Einstein's four-dimensional manifold is "a purely formal matter." A paper-and-pencil operation cannot possibly abolish the qualitative difference between the elements involved in it.[35] It is elementary that no observer can make proper records if he does not distinguish between time and space coordinates.

Moreover, as one eminent physicist after another has been led to admit, there is no other basis for Time than "the primitive stream of consciousness."[36] In fact, temporal laws in any science require a distinction between earlier and later, which only consciousness can make.[37] The Entropy Law is an excellent example in point. In formal terms, the law reads: let $E(T_1)$ and $E(T_2)$ be the entropies of the universe at two different moments in Time, T_1 and T_2, respectively; if $E(T_1) < E(T_2)$, then T_2 is later in Time than T_1 — and conversely. But, clearly, if we did not know already what "later" means, the statement would be as vacuous as "the gander is the male of the goose." The full meaning of the law is that the entropy of the universe increases *as Time flows through the observer's consciousness.* Time derives from the stream of consciousness, not from the change in entropy; nor, for that matter, from the movement of a clock. In other words, the relationship between Time and any "hour-glass" is exactly the reverse of what we generally tend to think.

If we know that Napoleon's death occurred later than Caesar's assassination it is only because the two events have been encompassed by the historical consciousness of humanity formed by the splicing of the consciousness of successive generations.[38] By going one step further and extrapolating in thought such a communal consciousness, we arrive at the notion of Eternity, without beginning and without end. This is the basis of Time.

But the word "time" is frequently used with many other meanings, some of which seem quite surprising. For example, the statement that

33. Eddington, *Nature of Physical World*, p. 51.

34. E.g., H. Margenau, *The Nature of Physical Reality* (New York, 1950), pp. 149 ff.

35. Cf. Bridgman, *Logic of Modern Physics*, p. 74; P. W. Bridgman, *Reflections of a Physicist* (2nd edn., New York, 1955), p. 254. In another place, *The Intelligent Individual and Society*, p. 28, Bridgman even denounces the thesis of the fusion as "bunkum."

36. H. Weyl, *Space, Time, Matter* (New York, 1950), p. 5.

37. On this point too we have not been able to go beyond Aristotle's teachings: *Physics*, 219ª 22 ff, 223ª 25 ff.

38. Cf. the remarks of Bridgman, *Reflections*, pp. 320–321, concerning the necessary ontinuity in observing a physical phenomenon.

"the time and the means for achieving [human] ends are limited," suggests that the term is used to represent not an endless flow but a scarce stock.[39] Economics abounds in similar loose uses of "time." A more stringent illustration is the use of "summation over time" to describe the operation by which the average age of a given population is computed. Surprising though it may seem, the terminological license originated in physics, where both a moment in Time and the interval between two such moments continued to be denoted, loosely, by the same term even after the distinction between the two meanings became imperative. The story of how this necessity was revealed is highly instructive.

The apparently innocuous admission that the statement "heat always moves by itself from hotter to colder bodies" is a physical law triggered one of the greatest crises in physics — which moreover is not completely resolved. The crisis grew out from the fact that mechanics cannot account for the unidirectional movement of heat, for according to mechanics all movements must be reversible. The earth, for instance, could have very well moved in the opposite direction on its orbit without contradicting any mechanical laws.[40] It is obvious that this peculiarity of mechanical phenomena corresponds to the fact that the equations of mechanics are invariant with respect to the sign of the variable t, standing for "time." The point led to the idea that in reality there are two Times: a reversible Time in which mechanical phenomena take place, and an irreversible Time related to thermodynamic phenomena. Obviously, the duality of Time is nonsense. Time moves only forward, and all phenomena take place in the same unique Time.[41]

Behind the idea of the duality of Time there is the confusion between the concepts I have denoted by T and t, a confusion induced by the practice of using the same term, "time," for both. In fact, T represents Time, conceived as the stream of consciousness or, if you wish, as a continuous succession of moments, but t represents the measure of an interval (T', T'') by a *mechanical clock*. Or to relate this description to our discussion of measurability (section 2 in this chapter), T is an ordinal variable, but t is a cardinal one. The fact that a weak cardinal scale can

39. Lionel Robbins, *An Essay on the Nature and Significance of Economic Science* (2nd edn., London, 1948), p. 12. The argument that "there are only twenty-four hours in the day" (*ibid.*, p. 15) increases the reader's difficulty in understanding Robbins position. Would the fact that there are one million microns in one meter make space (land) plentiful?

40. It is instructive to point out that, long before the crisis emerged in physics, G. W. F. Hegel, in *The Phenomenology of Mind* (2nd edn., New York, 1931), pp. 204–205, observed that the same scientific explanation would work for the inverted world.

41. Cf. Bridgman, *Logic of Modern Physics*, p. 79. Perhaps I ought to explain also that the impossibility of two observers to synchronize their clocks does not prove the multiplicity of Time. As anyone can verify it, this impossibility cannot be explained without referring events in both systems to a common time-basis.

be constructed for T on the basis of $t = $ Meas (T', T''), does not mean that it is not necessary to distinguish between t and T, even though we must reject the duality of Time.

It is the essential difference between the temporal laws which are functions of T and those which are functions of t that calls for a distinction between the two concepts. If we happen to watch a movie showing marshy jungles full of dinosaurs, we know that the event the movie intends to depict took place earlier than the founding of Rome, for instance. The reason invoked in this case is that the law governing such events — assuming that there is one — is, like the Entropy Law, a function of T. On the other hand, a movie of a purely mechanical phenomenon is of no help in placing the event in Time. For a pendulum moves and a stone falls in the same way irrespective of when the event occurs in Time. Mechanical laws are functions of t alone and, hence, are invariable with respect to Time. In other words, *mechanical phenomena are Timeless.*

Because only in thermodynamics, of all branches of physics, laws are functions of T, there was no strong compulsion for physics to eliminate the ambiguous use of "time." But it is hard to understand why other sciences, where the situation is not the same as in physics, have on the whole ignored the problem. All the greater is Schumpeter's merit for stressing, in his later writings, the difference between *historical* and *dynamic* time, by which he understood T and t respectively.[42] However, the root of the distinction does not lie in historical (evolutionary) sciences but — as we have seen — in the heart of physics, between mechanics and thermodynamics.

11. *Temporal Prediction and Clock-Time.* Ever since ancient astronomers succeeded in forecasting eclipses our admiration for the precision with which physics — the term being understood in the narrow sense, excluding thermodynamics — can predict future events, has steadily increased. Yet the reasons why only physics possesses this power are still obscure. The usual explanation that the future is determined exclusively by the initial (present) conditions, and that of all sciences physics alone has succeeded in ascertaining these conditions through measurements, raises more questions than it answers. In any case, it draws us into the muddled controversy over strict determinism.[43]

42. Joseph A. Schumpeter, *Essays*, ed. R. V. Clemence (Cambridge, Mass., 1951), p. 308, and, especially, Schumpeter's *History of Economic Analysis* (New York, 1954), p. 965n5.

43. The controversy turns upon whether the metaphysical doctrine of strict determinism meets with rebuttal in the domain of life phenomena. The issue is extremely complex. Without leaving firm ground, one may point out a piece of critical, though not entirely conclusive, evidence against strict determinism: at various stages embryos have been drastically mutilated but they nevertheless developed ultimately

The immediate reason why the temporal laws of physics are predictive is the fact that they are all functions of t, i.e., of the measure of Time-interval by a *mechanical clock*. What such a law tells us in essence is this: You set your mechanical clock to "zero" at the exact moment when you drop a coin from the top of the tower of Pisa; the tip of the clock hand will reach the mark t_0 at exactly the same moment as the coin reaches the ground. As this example shows, any temporal law of pure physics is nothing but the enunciation of a temporal parallelism between two mechanical phenomena, one of which is a mechanical clock. From this it follows that all mechanical phenomena, including that of a clock, are parallel in their ensemble. In principle therefore we could choose any such phenomenon to serve as the common basis for the enunciation of the parallelism. In part we have done so.

The point I wish to emphasize is that physical prediction is a symmetrical relation: we can very well say that the "falling body" predicts the "clock," or for that matter any other mechanical phenomenon. Why then do we prefer a clock mechanism over all other mechanical phenomena as the standard reference?

From the few physicists who cared to analyze the problem of "clock," we learn that the choice is determined by the condition that the concrete mechanism must be free, as much as possible, from the influence of factors that are not purely physical. This means that the clock must be almost Time-less, or in other words almost impervious to the march of entropy. As Eddington pointedly observed, the better the "clock" is, the less it shows the passage of Time.[44] That is why Einstein regarded the vibrating atom as the most adequate clock mechanism for physics.[45]

We can perfectly understand that if pure physics is to be a closed system, it needs a purely mechanical clock. But this internal necessity does not explain also why we associate the flow of Time with the movement of stars, of the sand in an hour-glass, or of a pendulum — all mechanical clocks. This association precedes by centuries the modern thoughts on clock. On the other hand, physics offers no proof that the clock hour just elapsed is equal to the one just beginning.[46] Time intervals

---

into normal individuals. The phenomenon, to which H. Driesch, in *The Science and Philosophy of the Organism* (London, 1908), I, 59 ff, 159–160, has given the name of *equifinality*, is better known under the form of restitution and regulation of living organisms. See also Bertalanffy, *Problems of Life*, pp. 59, 142–143, 188–189.

44. Eddington, *Nature of Physical World*, p. 99.

45. For various remarks on the problem of "clock," see Bridgman, *Nature of Physical Theory*, p. 73; Erwin Schrödinger, *What Is Life?* (Cambridge, Eng., 1944), pp. 84 ff; Weyl, *Space, Time, Matter*, pp. 7–8.

46. Karl Pearson, *The Grammar of Science* (Everyman Library edn., London, 1937), pp. 161–162; Henri Poincaré, *The Foundations of Science* (Lancaster, Pa., 1946), pp. 224–225.

cannot be superimposed so as to find out *directly* whether they are equal. Nevertheless, we have a strong feeling that they are, that Time flows at a constant rate hour by hour — as Newton taught. Perhaps the reason why we feel that the clock shows how fast Time flows is that suggested by Karl Pearson: in every clock hour there is packed "the same amount of consciousness." [47] The suggestion, however, could be accepted, if at all, only for two consecutive infinitesimal intervals. There is some evidence that the hours seem shorter as we grow older because — as has been suggested — the content of our consciousness increases at a decreasing rate. On the basis of the evidence available at present this is perhaps all we can say concerning the admiration scientists and laymen alike have for the prediction of future events by clock-time.

But, time and again, a legitimate admiration has turned into biased evaluation. Thus, at times, we can detect a deprecating intent in the statement that thermodynamics has no predictive power. The bare fact is that the Entropy Law tells us only that in, say, one clock-hour from now the entropy of the universe will be greater, but not by how much.[48] It is elementary that prediction by clock-time is impossible in a phenomenal domain — such as that of thermodynamics — where all temporal laws are functions of T exclusively. But I see no reason why we should deny such laws the power of predicting the future.

Indeed, let us suppose that we knew a Fourth Law of thermodynamics — which conceivably may be discovered any day. Let this law express the fact that some new variable of state, say, I, is a function of T. In this case, we could take either this new law or the Entropy Law as a "thermodynamic clock," and formulate the remaining law in exactly the same predictive form as we have cast earlier the law of falling bodies: When the thermodynamic clock will show I_0, the entropy of the system will simultaneously reach the level E_0. This example shows that, unless Pearson's suggested explanation of the constant rate of Time flow is substantiated, there can be no difference between prediction by clock-time and prediction by any time's arrow. (And even if it could be proved that Pearson's idea has a real basis, the superiority of prediction by clock-time would have only a purely anthropomorphic justification.) If we have nevertheless arrived at the contrary opinion, that thermo-

47. Pearson, *ibid.*, p. 159.
48. Cf. W. J. Moore, *Physical Chemistry* (2nd edn., Englewood Cliffs, N.J., 1955), p. 23; Margenau, *Nature of Physical Reality*, pp. 210–211; Philipp Frank, "Foundations of Physics," *International Encyclopedia of Unified Science* (Chicago, 1955), vol. I, part 2, p. 449. It may be contended that the First Law is nevertheless predictive by clock-time; however, the constancy of total energy represents a rather vacuous case of clock-time law. Perhaps, we ought to explain also that the Third Law, ordinarily called Nernst's Law, in essence states that the zero of absolute temperature can never be attained.

dynamics has no predictive value, it is no doubt because there the issue is obscured by another factor: in thermodynamics there is *only one* truly temporal law, the Entropy Law. But a single law, clearly, is useless for prediction: no law can be its own "clock." The difficulty is of the same nature as that inherent to any implicit definition.

Furthermore, there is absolutely no reason why in every domain of inquiry phenomena should be parallel to that of a mechanical clock. Only the dogma that all phenomena are at bottom mechanical could provide such a reason. But, as I have repeatedly emphasized, the mechanistic dogma has been abandoned even by physical sciences. We should therefore regard as a sign of maturity the reorientation of any science away from the belief that all temporal laws must be functions of clock-time. Wherever it has taken place, the reorientation paid unexpected dividends. For instance, many biological phenomena which appeared highly irregular as long as they were projected against a clock-time scale have been found to obey very simple rules when compared with some biological phenomenon serving as a "clock." [49]

Hoping that this essay will achieve one of its main objectives, namely, that of proving that the economic process as a whole is not a mechanical phenomenon,[50] I may observe at this juncture that the abandonment of Clément Juglar's formula for business cycles was a step in the right direction. Indeed, that formula implies the existence of a strict parallelism between business activity and a mechanical clock — the movement of sun spots. On the other hand the Harvard Economic Barometer, unhappy though its final fate was, reflects a more sound approach to the same problem. For in the ultimate analysis any similar type of barometer affirms a parallel relationship between some economic phenomena, one of which serves as a "clock" — an economic clock, that is. Most subsequent studies of business cycles have in fact adopted the same viewpoint. The palpable results may not be sufficiently impressive; hence doubts concerning the existence of an *invariant parallelism* between the various aspects of economic activity are not out of place. However, the alternative idea that the march of the entire economic process can be described by a system of differential equations with clock-time as the independent variable — an idea underlying many macro-dynamic models — is in all probability vitiated *ab ovo*.

12. *Mechanics and Probability.* In order to proceed systematically I have considered thus far only those lessons that a social scientist may learn from Classical thermodynamics. But the story has a very important epilogue.

---

49. Cf. P. Lecomte du Noüy, *Biological Time* (New York, 1937), pp. 156 ff.
50. And if on this point I am fighting a straw man, it is all the better for my other theses that depend upon the validity of the point.

It was quite difficult not only for physicists but also for other men of science to reconcile themselves to the blow inflicted on the supremacy of mechanics by the science of heat. Because the only way man can act upon matter directly is by pushing or pulling, we cannot easily conceive any agent in the physical universe that would have a different power. As Lord Kelvin, especially, emphasized, the human mind can comprehend a phenomenon clearly only if it can represent that phenomenon by a *mechanical model.* No wonder then that ever since thermodynamics appeared on the scene, physicists bent their efforts to reduce heat phenomena to locomotion. The result was a new thermodynamics, better known by the name of statistical mechanics.

First of all, we should understand that in this new discipline the thermodynamic laws have been preserved in exactly the same form in which Clausius had cast them. But the meaning of the basic concepts and the explanation of thermodynamic equilibrium have been radically changed. If technical refinements are ignored, the new rationale is relatively simple: heat consists of the *irregular* motion of particles, and thermodynamic equilibrium is the result of a *shuffling* process (of particles and their velocities) which goes on by itself. But I must emphasize one initial difficulty which still constitutes the stumbling block of statistical mechanics. The spontaneous shuffling has never been appropriately defined. Analogies such as the shuffling of playing cards or the beating of an egg have been used in an attempt at explaining the meaning of the term. In a more striking analogy, the process has been likened to the utter devastation of a library by an *unruly* mob.[51] Nothing is *destroyed* (The First Law of Thermodynamics), but everything is scattered to the four winds.

According to statistical mechanics, therefore, the degradation of the universe would be even more extensive than that envisaged by Classical thermodynamics: it covers not only energy but also material structures. As physicists put it in nontechnical terms,

In nature there is a constant tendency for order to turn into disorder.

Disorder, then, continuously increases: the universe thus tends toward Chaos, a far more forbidding picture than the Heat Death.

Within this theoretical framework, it is natural that entropy should have been redefined as a measure of the degree of disorder.[52] But as some philosophers and physicists alike have pointed out, disorder is a highly relative, if not wholly improper, concept: something is in disorder only in respect to some objective, nay, purpose.[53] All the less can we see how

51. Erwin Schrödinger, *Science, Theory, and Man* (New York, 1957), pp. 43–44.
52. For an authoritative discussion of this point, see P. W. Bridgman, *The Nature of Thermodynamics* (Cambridge, Mass., 1941), pp. 166 ff.
53. Bergson, *Creative Evolution*, pp. 220 ff and *passim;* Bridgman, *Nature of Thermodynamics*, p. 173.

disorder can be ordinally measurable. Statistical mechanics circumvents the difficulty by means of two basic principles:

A. *The disorder of a microstate is ordinally measured by the probability of the corresponding macrostate.*

B. *The probability of a macrostate is proportional to the number of the corresponding microstates.*[54]

A microstate is a state the description of which requires that each individual involved be named. "Mr. X in the parlor, Mr. and Mrs. Y in the living room," is an illustration of a microstate. The macrostate corresponds to a nameless description. Thus, the preceding illustration corresponds to the macrostate "One man in the parlor, one man and one woman in the living room." But it may equally well belong to the macrostate "One person in the parlor, two persons in the living room." This observation shows that the degree of disorder computed according to rule B — which is nothing but the old Laplacian rule — depends upon the manner in which microstates are grouped in macrostates. A second factor which affects the same measure is the criterion which determines whether or not a given microstate is to be counted. For example, in connection with the illustration used above it matters whether Emily Post rules that "Mrs. Y in the parlor, Mr. X and Mr. Y in the living room" is an unavailable microstate in a well-bred society.

The earliest but still the basic formula which connects the degree of disorder to the entropy of a microstate is Boltzmann's:

(4) $$\text{Entropy} = k \log n,$$

where k is a physical constant and n the number of *equivalent* microstates. Ludwig Boltzmann chose the criteria for the equivalence of microstates and for determining the set of permissible microstates with the intention that (4) should always give the same value as (3), i.e., as the ratio between latent energy and absolute temperature. But subsequently it was discovered that Boltzmann's statistics does not always achieve the coincidence between Classical and statistical entropy. Two new mechanical statistics, the Bose-Einstein and the Fermi-Dirac, had to be naturalized in order to fit new facts into (4).[55] This proves most eloquently that the double arbitrariness involved in rule B must in the end play havoc with any endeavor to establish microstates and macrostates by purely formal considerations.[56]

Though each problem discussed thus far uncovers some flaw in the foundation on which the measure of disorder rests, all are of an elementary

54. Cf. Margenau, *Nature of Physical Reality*, pp. 279 ff.
55. Schrödinger, *Science, Theory, and Man*, pp. 212 ff.
56. Cf. my criticism of Carnap's probability doctrine, in a paper reprinted below, "The Nature of Expectation and Uncertainty" (1958), section IV.

simplicity. They can hardly justify, therefore, the occasional but surprising admission that the concept of statistical entropy "is not easily understood even by physicists." [57] As far as mere facts are concerned, we know that ever since its conception statistical entropy has been the object of serious criticism; it still is. Although the risks of a layman's expressing opinions are all the greater in a situation such as this, I wish to submit that the root of the difficulty lies in the step by which statistical entropy is endowed with an additional meaning — other than a disorder index.

In order to isolate the issues let us beg the question of whether disorder is ordinally measurable. It is then obvious that nothing could be wrong with choosing that index of disorder which is computed according to the principles A and B, *provided that the index thus obtained fits the facts described by the Entropy Law* in the sense that it increases with T. But, as is well known, there are infinitely many rules which yield an equally adequate index of disorder. Therefore, it is not imperative that we should choose A and B. The choice, however, is a convenient one: it leads to a simple formula for disorder entropy. (Again, we must beg the question of whether the values of (3) and (4) necessarily coincide always.) The point I wish to emphasize now is elementary: from the fact that A and B serve as rules also for computing Laplacian probability *it does not follow that the index of disorder is a probability.* Clearly, nothing would be changed if in A and B the word "probability" were replaced by "relative measure" or by any other like term.[58] Statistical mechanics nevertheless takes the position that the disorder index computed according to A and B represents at the same time *the physical probability of the corresponding macrostate to occur.* It is this step, by which entropy acquires a double meaning, that constitutes the most critical link in the logical framework of new thermodynamics.

It is ultra-elementary that if a system is governed by rigid laws — such as those of mechanics — it passes from one microstate to another in a succession that is completely determined by those laws. Therefore there is absolutely no room in such a framework for a random agent, that is for physical probability. Yet, statistical mechanics does offer a justification for introducing probability in a system controlled by mechanical laws. It is the ergodic *theorem.* This theorem states that the relative frequency density with which a microstate occurs in the rigidly determined history of a mechanical system tends toward a definite limit as t

57. D. Ten Haar, "The Quantum Nature of Matter and Radiation," in *Turning Points in Physics*, R. J. Blin-Stoyle, *et al.* (Amsterdam, 1959), p. 37.
58. An earlier remark (section 5 of this chapter) illustrates the point just made. From the fact that the same propositions describe a geometrical as well as a social structure it does not follow that individuals and their associations *are* points and lines.

increases indefinitely.[59] The gist of the theorem is easily illustrated by a simple arithmetical example: the relative frequency of "5" in the decimal sequence of $\frac{1}{7}$ tends in the limit to $\frac{1}{6}$. Statistical mechanics equates this ergodic limit with physical probability. No objection could be raised against this procedure if "probability" were interpreted in the subjective sense — i.e., as an index of the mental incertitude resulting from our inability to measure the initial conditions exactly or, at times, to solve the motion equations completely. That, however, is not the case: the ergodic limit is taken to reflect the average intervention of a physical random factor. The arithmetical illustration offered above plainly shows that there is no sense in saying that there is a probability equal to $\frac{1}{6}$ for the next decimal digit to be "5"; this cipher will never appear after "1" and it always will after "8."

Let us now examine some of the most important consequences of the alleged equivalence of the index of disorder and physical probability. To begin with, a physical law, the Entropy Law, now becomes a tautology. If the higher the disorder the greater is the probability of its occurring, then obviously any closed system — such as nature — has a tendency to pass from any state to one of higher disorder. Naturally, the Entropy Law no longer states what will actually happen — as it does in Classical thermodynamics — but only what is likely to happen. The possibility that disorder will turn into order is not, therefore, denied. The event only has a very low probability.

But no matter how small the probability of an event is, over Time the event must occur infinitely many times. Otherwise the coefficient of probability would lose all physical significance. Consequently, over the limitless eternity, the universe necessarily reaches chaos and then rises again from ashes an infinite number of times.[60] The thought that the dead will one day rise again from their scattered and pulverized remains to live a life in reverse and find a new death in what was their previous birth, is likely to appear highly strange. Scientists, however, are used to each discovery being wholly surprising. If many physicists have questioned the explanatory value of statistical mechanics it is only because

59. G. D. Birkhoff, "Proof of the Ergodic Theorem," *Proceedings of the National Academy of Science*, XVII (1931), 656–660, and, especially, the same author's "What Is the Ergodic Theorem?" *American Mathematical Monthly*, XLIX (1942), 222–226. Obviously, there is an intimate connection between the ergodic theorem and the Poincaré-Zermelo theorem, which states that a mechanical system returns to any previous microstate if not exactly at least as close as one may wish. (Cf. Bridgman, *Reflections*, p. 262; Max Born, *Natural Philosophy of Cause and Chance*, Oxford, 1949, pp. 58–59.) Strange though it may seem, the defenders of statistical mechanics fight for the former but reject the relevance of the latter.

60. The writers who unreservedly endorse this view do not constitute a rare exception: e.g., P. Frank, "Foundations of Physics" (cited in note 48, above), pp. 452–453.

the idea of a universe periodically ageing and rejuvenating rests on a faulty foundation.

The most common and, perhaps, the most effective objection is that no factual evidence supports the idea that the rejuvenation of the universe has a non-null probability: no similar phenomenon has ever been observed in *nature* even at some smaller scale. Only factual evidence can endow a probability computed by a paper-and-pencil operation with physical significance.[61] The standard reply, that we have not yet witnessed a rejuvenating system because we have not observed nature long enough, may seem, though not decisive, at least acceptable.[62] In my opinion, the reply is nevertheless fallacious. Very likely, it is thought that the answer is justified by the proposition that if we wait long enough then a rare event must occur with quasi certainty. In fact the justification requires the converse proposition, namely, that a rare event cannot occur unless we wait a long time. But, as we know, this last proposition is false.

From whatever angle we look at statistical mechanics we discover what was plain from the outset: the impossibility of explaining unidirectional processes by laws that are indifferent to direction. In the new thermodynamics this impossibility is metamorphosed into a logical contradiction between the two basic hypotheses: (1) particles move according to rigid laws, and (2) states follow each other in some random fashion. No defender of statistical mechanics — to my knowledge — has ever denied this contradiction. Among the various mending proposals, one is highly instructive. The proposal is to adopt an additional "special hypothesis" by which to deny the contradiction between hypotheses (1) and (2).[63] However, if a special hypothesis is welcome, then it would be far less extravagant to assume — as some ancient atomists did — that particles can swerve freely from their normal course "at times quite uncertain and uncertain places." [64]

Many a physicist would probably dismiss the question of the logical consistency of statistical thermodynamics with the remark, reasonable to some extent, that after all its formulae work in practice. Yet, in a broader

61. The validity of this statement is in fact evidenced by the necessity of introducing the new statistics of Bose-Einstein and Fermi-Dirac, by which some microstates valid in Boltzmann's statistics are declared impossible. Formally, it can be illustrated by my earlier arithmetical example: it would be absurd to attribute a non-null ergodic limit to the microstate "9" in the decimal sequence of $\frac{1}{7}$ on reasoning *a priori* that all ciphers are equally probable. The point forms the basis of my criticism of the exclusively subjectivist doctrine of probability; see section V of my "The Nature of Expectation and Uncertainty" (1958), reprinted below.

62. Even an authority on probability such as Poincaré (cited in note 46, above), p. 304, seems satisfied with the answer. See also K. Mendelssohn, "Probability Enters Physics," in *Turning Points in Physics*, p. 53.

63. P. Frank, "Foundations of Physics," p. 452.

64. Lucretius, *De rerum natura*, II. 218-220.

intellectual perspective, the issue appears extremely important, for it pertains to whether the phenomenal domain where our knowledge is both the richest and the most incontrovertible supports or denies the existence of evolutionary laws. This is the reason why I felt it necessary to analyze the rationale of statistical mechanics in some detail even though to many students of social sciences such an analysis might at first appear to be a technical degression.

13. *Entropy and Purposive Activity.* The reorientation of physics from rigid to stochastic laws has been interpreted by many as proof of a "free will" in nature. It is argued that since we cannot say what particular atom of a radioactive substance will decay next, the decay of such a substance reflects some free will of the individual atom similar to that of an individual human to commit or not to commit suicide. Certainly, the issue of why both kinds of individuals seem to follow no rule is baffling. Yet, sociologists know that in a *given* society the frequency of suicides is as stable as that of the decaying atoms in a radioactive substance. It is in the stability of this frequency that the inexorability of stochastic laws lies. However, the fact that probability was introduced in thermodynamics by some logical legerdemain places the statistical Entropy Law on a different footing from all other stochastic laws of physics. In the opinion of many, statistical mechanics has thus deprived the Entropy Law of the inexorability all natural law must possess.

This view is related to a piquant fable of J. Clerk Maxwell's. He imagined a minuscule demon posted near a microscopic swinging door in a wall separating two gases, A and B, of equal temperature. The demon is instructed to open and close the door "so as to allow only the swifter molecules to pass from A to B, and only the slower ones to pass from B to A." Clearly, the demon can in this way make the gas in B hotter than in A. This means that it can unbind latent energy and, hence, defeat the Entropy Law of statistical thermodynamics.[65]

Ever since Maxwell wrote it (1871), the fable has been the object of a controversy which, I submit, is empty. Taken on its face value, the fable reveals a conflict between the tenet that physical laws are inexorable and the statistical explanation of thermodynamic phenomena. In this perspective, Maxwell's own point corresponds to eliminating the conflict by upholding the tenet and indicting the explanation. But one may equally well accept the statistical explanation and reject the tenet. This second alternative corresponds to the argument enthusiastically supported by all vitalists that a living being — as proved by Maxwell's demon — possesses the power of defeating the laws of matter. It is because of this last argument that the fable acquired a sweeping significance. However, like many other paradoxes, Maxwell's is still an intellectual riddle.

65. Maxwell, *Theory of Heat* (cited in note 32, above), pp. 338–339.

Hence, the fable cannot serve as a basis for any scientific argument.[66] Yet, Maxwell's demon was not to remain without glory. The fable had a decisive influence upon the orientation of the biological sciences. To begin with, it compelled us all to recognize the categorical difference between *shuffling* and *sorting*. In thermodynamics we do not ask ourselves whence comes the energy for the shuffling of the universe, even though we know only too well that it takes some work to beat an egg or to shuffle cards. The shuffling in the universe — like that of the gas molecules surrounding the demon — goes on by itself: it is automatic. But not so sorting: Maxwell invented a demon, not a mechanical device, for this job. "Sorting is the prerogative of mind or instinct," observed Eddington, and hardly anyone would disagree with him nowadays.[67]

Actually the more deeply biologists have penetrated into the biological transformations the more they have been struck by "the amazing specificity with which elementary biological units pick out of the building materials available just the 'right ones' and annex them just at the right places." [68] Irrespective of their philosophical penchant, all recognize that such orderly processes, which are "much more complex and much more perfect than any automatic device known to technology at the present time," occur only in life-bearing structures.[69] This peculiar activity of living organisms is typified most transparently by Maxwell's demon, which from its highly chaotic environment selects and directs the gas particles for some definite *purpose*. It is not surprising therefore that thermodynamics and biology have drawn continuously closer and that entropy now occupies a prominent place in the explanation of biological processes.[70]

I should hasten to add that this development does not vindicate the ultravitalist position that living structures can defeat the laws of *elementary matter*. These laws are inexorable. However, this very argument uncovers the real issue of the vitalist controversy. Given that even a simple cell is a highly ordered structure, how is it possible for such a structure to avoid being thrown into disorder instantly by the inexorable

66. Attempts at proving the absurdity of the fable itself have not been lacking. They usually concentrate on the point that a demon with a physical existence must consume free energy and thus increase the entropy of the whole system: it cannot accomplish its task "without any expenditure of work," as Maxwell (*ibid.*) assumed. But no *direct* proof exists that the demon cannot unbind more energy than it consume.s Merely to invoke the inexorability of the Entropy Law is to miss the point of the paradox. (For a very instructive discussion of Maxwell's fable, see Bridgman, *Nature of Thermodynamics*, pp. 155 ff, pp. 208 ff.)

67. Eddington, *Nature of Physical World*, p. 93.

68. Bertalanffy, *Problems of Life*, p. 29.

69. Ilya M. Frank, "Polymers, Automation and Phenomena of Life," *Izvestia*, Sept. 11, 1959. English translation in *Soviet Highlights*, I (1959), no. 3.

70. This intimate connection is admirably and with unique insight explained in a great little book already quoted: Schrödinger, *What Is Life?*

Entropy Law? The answer of modern science has a definite economic flavor: a living organism is a *steady going concern* which maintains its highly ordered structure by sucking low entropy from the environment so as to compensate for the entropic degradation to which it is continuously subject. Surprising though it may appear to common sense, life does not feed on mere matter and mere energy but — as Schrödinger aptly explained — on low entropy.[71]

Sorting, however, is not a natural process. That is, no law of elementary matter states that there is any sorting going on by itself in nature; on the contrary, we know that shuffling is the universal law of elementary matter. On the other hand, no law prohibits sorting at a higher level of organization. Hence, the apparent contradiction between physical laws and the distinctive faculty of life-bearing structures.[72]

Whether we study the internal biochemistry of a living organism or its outward behavior, we see that it continuously sorts. It is by this peculiar activity that living matter maintains its own level of entropy, although the *individual* organism ultimately succumbs to the Entropy Law. There is then nothing wrong in saying that life is characterized by the struggle against the entropic degradation of mere matter.[73] But it would be a gross mistake to interpret this statement in the sense that life can prevent the degradation of the entire system, including the environment. The entropy of the whole system must increase, life or no life.

Let me observe that the case, however, is not completely closed by the last explanation. A perhaps even more difficult question confronts us now: is the increase of entropy greater if life is present than if it is not?[74] *For if the presence of life matters, then life does have some effect upon physical laws.* Our ordinary knowledge of the change in the material environment brought about by the biosphere seems to lead to the belief that life speeds up the entropic degradation of the whole system. Actually, if examined closely, many occasional remarks by physicists on the life process tend to show that they too share, however unawares, this "vitalistic" belief.

71. *Ibid.*, chap. iv. The seed of this idea goes back to Ludwig Boltzmann who was first to point out that free energy is the object of the struggle for life.

72. Joseph Needham, "Contributions of Chemical Physiology to the Problem of Reversibility in Evolution," *Biological Reviews*, XII (1938), 248–249.

73. Bergson, *Creative Evolution*, pp. 245–246, is known for having presented this view more articulately and more insistently than any other philosopher. The multifarious accusations of mysticism directed against him are no longer in order, if they ever were.

74. Bergson, *ibid.*, maintains that life retards the increase, but offers no evidence in support of this view.

FOUR

EVOLUTION VERSUS MECHANICS

1. *Irreversible and Irrevocable Processes.* The idea that the life process can be reversed seems so utterly absurd to almost every human mind that it does not appear even as a myth in religion or folklore. The millenary evidence that life goes always in only one direction suffices as proof of the irreversibility of life for the ordinary mind but not for science. If science were to discard a proposition that follows logically from its theoretical foundation, merely because its factual realization has never been observed, most of modern technology would not exist. Impossibility, rightly, is not the password in science. Consequently, if one cornerstone of science is the dogma that all phenomena are governed by mechanical laws, science has to admit that life reversal is feasible. That the admission must cause great intellectual discomfort is evidenced by the fact that, apparently, no scholar of the Classical school made it overtly. Classical thermodynamics, by offering evidence — valid according to the code of scientific court procedure — that even in the physical domain there are irreversible processes, reconciled science's stand with generally shared common sense. However, after statistical mechanics began teaching, with even greater aplomb than Classical mechanics, that all phenomena are virtually reversible, universal reversibility became the object of a prominent controversy. From physics, the controversy spread into biology where the issue is far more crucial.

From the discussion in section 12 in the preceding chapter, we can expect the controversy to be highly involved. Unfortunately, it has been further entangled by the fact that "reversibility" has been used with different meanings by different authors and, hence, often with another meaning than in mechanics. There, a process is said to be *reversible* if and only if it can follow the same course phase by phase in the reverse order. It is obvious, however, that this is not the sense in which the term is used, for example, in Joseph Needham's argument that biological phenomena are reversible because protein micellae "are continually broken down and built up again." [1] Actually, the process of this illustration is irreversible according to the terminology of mechanics.

One source of this confusion is that only two terms, reversible and irreversible, are commonly used to deal with a situation that really is trichotomous. For the relevant aspects of a process call for the division of nonreversible phenomena — to use the stringent form of logical negation — into two categories.

The first category of "nonreversibility" consists of all processes which, though not reversible, can return to any previously attained phase. The flow of vehicles in a traffic circle comes immediately to mind, but the process of a tree growing and losing its leaves each year seems a more instructive illustration. Processes such as these are nonreversible but not irrevocable. We may refer to them simply as *irreversible*. No doubt, in the saying "history repeats itself," history is conceived as an irreversible process in this narrow sense.

The second category of "nonreversibility" consists of processes that cannot pass through a given state more than once. Of course, such a process is nonreversible, but *irrevocable* better describes its distinctive property. The entropic degradation of the universe as conceived by Classical thermodynamics is an irrevocable process: the free energy once transformed into latent energy can never be recuperated.

Another source of confusion about reversibility lies in the concept of process itself. Strange though it may seem, the process of the entire universe is a far more translucid concept than that of a single microorganism. The mere thought of a partial process necessarily implies some slits cut into the Whole. This, as we have already seen, raises inextricable problems. But at least we should not lose sight of where we intend the seams to be cut. It matters tremendously whether the process in Needham's illustration includes the life of a single protein micella or of an unlimited number. For in the first case, there are good reasons for regarding the process as irrevocable; however, the second process is unquestionably irreversible.[2]

2. *Evolution, Irrevocability, and Time's Arrow.* It is because science began to speak of evolution first in connection with biological phenomena that by evolution we generally understand *"the history of a system undergoing irreversible changes."* [3] (Actually the word should be "irrevocable.") The existence of evolutionary laws in nature depends then upon whether there are irrevocable phenomena: the existence of only irreversible

1. Joseph Needham, "Contributions of Chemical Physiology to the Problem of Reversibility in Evolution," *Biological Reviews*, XII (1938), 225.

2. The argument typified by Needham's article clearly refers to the latter process. Its fault is obvious: from the fact that this process is not irrevocable, it concludes that it is reversible.

3. Alfred A. Lotka, *Elements of Physical Biology* (Baltimore, 1925), p. 24.

phenomena — in the narrow sense — does not suffice. All the stronger, therefore, is the negation of evolutionary laws by the universal reversibility proclaimed by statistical mechanics. Many a scientist was thus induced to argue that evolution is appearance: a phenomenon may or may not appear evolutionary depending upon the angle from which we view it or upon the extent of our knowledge.

An epitome of this relativist position is Karl Pearson's argument that to an observer traveling away from the earth at a greater speed than light, events on our planet would appear in the reversed order to that in which they have actually occurred *here*.[4] The fact that since Pearson wrote we have learned that the speed of light cannot be exceeded does not destroy the gist of his argument. The gist is that evolution is appearance because any movie can be projected in two ways, "forward" and "backward." How can we then ascertain which is the right way?

Unquestionably, Pearson's argument implies the duality of Time. But a more serious fault is that it fails to explain what is the basis of "forward" and "backward." In other words, assuming that we are given the individual frames unconnected, *how can we splice them in the correct order?* If we cannot, then there is no Time at all in nature.

The truly unique merit of Classical thermodynamics is that of making perfectly clear the problem of Time in relation to nature. A basis of Time in nature requires this: (1) given two states of the universe, S_1 and S_2, there should be one general attribute which would indicate which state is later than the other, and (2) the temporal order thus established must be the same as that ascertained by a single or collective human consciousness assumed to be contemporary with both S_1 and S_2. It is elementary then that, since the stream of consciousness moves only "forward," the corresponding attribute must reflect an irrevocable process.

Actually, without a time's arrow even the concept of mechanical reversibility loses all meaning. The tables should, therefore, be turned. It behooves the side claiming that evolution is a relative aspect to show how, if there is no irrevocable process in nature, one can make any sense of ordinary temporal laws. To return to the movie analogy, a movie of a purely mechanical phenomenon — say, the bouncing of a *perfectly elastic* ball — can be run in either direction without anyone's noticing the difference. A biologist, however, will immediately become aware of the mistake if a movie of a colony of protein micellae is run in reverse. And everyone would notice the error if the movie of a plant germinating from seed, growing, and in the end dying, is run in reverse. However, that is not the whole difference. If the frames of each movie are separated and shuffled, only in the last case can we rearrange them in exactly the

4. Karl Pearson, *The Grammar of Science* (Everyman Library edn., London, 1937), pp. 343–344.

original order. This rearrangement is possible only because the life of a single organism is an irrevocable process. As to the other two processes mentioned, the first is reversible, the second irreversible.

Two important observations should now be made. First, if the movie of the micellae is irreversible it is because the process filmed consists of a series of overlapping irrevocable processes, the lives of the individual micellae. Second, if the first two movies have in the background an irrevocable process — say, that of a single plant — then their individual frames too can be rearranged immediately in the exact original order. The point is that only in relation to an irrevocable process do reversibility and irreversibility acquire a definite meaning.

3. *From Part to Whole.* An outsider may be surprised to see that the debate concerning the issue of Classical vs. statistical thermodynamics turns around the prediction each theory makes about the fate of the universe. The initiated knows that the reason is that no operational difference has yet been discovered between the two theories. Physicists work equally well with either, according to individual preferences: the literature covers both. But since an acid test of any prediction concerning the fate of the entire universe is well beyond our reach, opinions on which of the two theories is more verisimilar have been influenced mainly by the subjective intellectual appeal of each prediction. However, neither the picture of a universe irrevocably racing toward a Heat Death nor that of a universe devoid of temporal order seems particularly attractive. Undoubtedly, it is equally hard to admit that the Gods could create but a finite Existence or that, as Einstein once said, they only play dice continuously.[5]

Law extrapolation is the very soul of cosmology. However, the extrapolation of the Entropy Law — Classical or statistical — to the cosmic scale is particularly vulnerable because very likely the error thus committed is of a qualitative nature. Bridgman, who favors the Classical approach, has set forth some reasons to challenge the cosmic application of the Entropy Law. Moreover, he admitted — just as did Boltzmann, the founder of statistical mechanics — that in some sectors of the universe and for some periods of time entropy may very well decrease.[6] Perhaps still more interesting is one thought of Margenau's. He raised the question of whether even the Conservation Law applies to the entire universe: "If creation of matter-energy takes place . . . all our speculations [about the fate of the universe] are off." [7]

All these thoughts seem now prophetic. For they all concur with the

5. Reported by Niels Bohr, *Atomic Physics and Human Knowledge* (New York, 1958), p. 47.

6. P. W. Bridgman, *The Nature of Thermodynamics* (Cambridge, Mass., 1941), pp. 148 ff; Bridgman, *Reflections of a Physicist* (2nd edn., New York, 1955), pp. 263 ff.

7. H. Margenau, *The Nature of Physical Reality* (New York, 1950), p. 283.

recently ventilated hypothesis — mentioned earlier in section 7 of my Chapter Two, "Concepts, Numbers, and Quality" — that matter is continuously created and annihilated. From this hypothesis there emerges a universe that neither decays irrevocably nor is deprived of temporal order. It is a universe consisting of a congregation of *individual worlds*, each with an astronomically long but finite life, being born and dying at a constant average rate. The universe is then an everlasting steady state which, like any stationary population, does not evolve.[8] Not only its total energy but also its total entropy must remain constant, or nearly so.

In this picture, a time's arrow must come from some individual component if from anything. We are thus back to one of the oldest tenets. What is everlasting cannot evolve (change); evolution is a specific trait of that which is born and dies. In other words, evolution is the process that links birth to death: it is life in the broadest sense of the term. Witness the fact that even the whole universe must have a transient life between Creation and Heat Death if it is to be an evolving entity as pictured by the Classical Entropy Law.

The transparent principle that death is later in time than the birth of the *same* individual — be it a galaxy, a biological species, or a microscopic cell — does not suffice, however, to establish a complete chronology even if we beg such troublesome questions as whether birth and death can be operationally recognized as point-events. For a complete chronology we need a continuous time's arrow of at least one category of individuals the lives of which overlap without interruption.[9] If such a time's arrow can be found, then the cosmic panorama is as simple as our movie of protein micellae: the process of the entire universe is unidirectional, i.e., irreversible, because that of its individual members is irrevocable.

4. *Evolution: A Dialectical Tangle.* Paradoxical though it may seem, the evolution of the simplest micro-organism raises far more formidable issues than that of the whole universe. The Whole needs no boundaries to separate it from its Other, for there is no other Whole. And since there is no Other, we need not ask what sameness means for the Whole. A partial process, on the other hand, requires some conceptual cuts in the Whole. Cutting the Whole, as I have observed earlier, creates endless difficulties.

To begin with, across any boundary we may draw in space, time, or society, there is some traffic between the two domains thus separated. Hence, we get three partial processes instead of two, a contrariety to which little attention has been paid. The widespread practice is to ignore

8. Cf. Fred Hoyle, *Astronomy* (New York, 1962), pp. 300 ff.
9. This condition should be related to the manner in which the historical consciousness is formed, as explained above in section 10 of my preceding chapter

completely the processes one initially intended to separate and to reduce the whole picture to the traffic across the boundary. This flow-complex, as I have called it (section 4 in the preceding chapter), clearly throws away the baby with the bath water.[10]

Most social scientists are fully aware of a second difficulty. A separated (partial) process must have an individuality; otherwise we cannot ascertain which facts belong to one and the *same* process. The idea that an *individual* process is determined by the very setting of its boundaries, unfortunately, does not always work. If we still want to look at an oak's growing from an acorn as an individual process, we must admit that an ordinary feature of a partial process is the changing of its own boundary. Perhaps the point is more eloquently illustrated by the well-known difficulties of defining an individual plant or, especially, an individual firm so as to allow for variations in size. No doubt, the concept of an individual process is the most staggering of all.

The preceding remarks are borne out by the fact that the *isolated* system has ultimately become the unique reference for all propositions of theoretical physics. Of course, this manner of circumventing the difficulties of a partial process was made possible only because a physical universe can be reproduced in miniature and with some satisfactory approximation in the laboratory. Other disciplines are not as fortunate. Biologists too have experimented with isolated processes containing some organisms together with a part of environment. However, the great difference is that such an isolated process is far from being a miniature simulation of the actual process.

Experimenting with isolated systems in biology has reconfirmed — if reconfirmation was needed — that the evolution of the biosphere necessarily implies the evolution of the environment. To speak of biological or social evolution in a nonevolutionary environment is a contradiction in terms. *Ceteris paribus*, the indispensable ingredient of every physical law is poison to any science concerned with evolutionary phenomena. Evolutionary changes cannot be seen except in an isolated, at least quasi-isolated, system. Perhaps in some domains it might be unscientific to experiment with wholes because, as Popper argues, we cannot thus impute effects to individual causes.[11] That does not apply, however, to evolution which is inseparable from the Whole. Witness the fact that the only

10. In an earlier paper, "The Aggregate Linear Production Function and Its Application to von Neumann's Economic Model," in *Activity Analysis of Production and Allocation*, ed. T. C. Koopmans (New York, 1951), pp. 100–101, I touched briefly upon the inadequacy of representing a process only by either flows or stocks. A more detailed analysis of the issue is included in the author's forthcoming monograph *Process, Value, and Development*.

11. Karl R. Popper, *The Poverty of Historicism* (Boston, 1957), p. 89.

case in which we were able to formulate an evolutionary law is that of the whole universe.

All efforts to discover a time's arrow in the life (evolution) of a single organism or a species considered in isolation have remained vain. Beyond the intricate qualitative aspects of such lives, biology has found only a dualist principle: growth and decay, or anabolism and catabolism. To be sure, both anabolism and catabolism consist of physico-chemical processes, but the dualism comes from the fact that the two phases are not governed by the same category of laws.[12] And though we know that during growth anabolism exceeds catabolism and that the reverse happens during decay, there is no purely physico-chemical explanation of the reversal.[13]

As expected, entropy does enter into the picture but not as a time's arrow: it decreases during growth and increases during decay. Therefore, even if we were able to determine the entropy level of an organism we still could not say which of two states is earlier. We would also have to know whether entropy is increasing or decreasing in each situation. But this knowledge already presupposes a time's arrow.

The number of biochemical phenomena expressed by numerical formulae is continually increasing, but none of these formulae offers a basis for a biological time's arrow.[14] This, without much doubt, is why no description of an individual or collective organism is complete and meaningful unless these quantitative results are, first, related to the stream of consciousness and, then, cemented into a single picture by an immense dose of quality. For biology, and even more for a social science, to excommunicate dialectical concepts would therefore be tantamount to self-imposed paralysis.

5. *Evolution Is Not a Mystical Idea.* The preceding analysis was intended only to pinpoint the epistemological difficulties of the concept of evolution and their reflection upon the study of evolutionary processes. Nothing is further from my thought than to suggest thereby that evolution is a mystical concept. To make this point clear, let me return to the picture of the universe as a steady population of evolving individual worlds, a picture which, I believe, is intellectually far more satisfying than its alternatives. Certainly, this picture no longer compels us to believe in absolute novelty.

12. The burning of sugar in a biological structure is, no doubt, a physico-chemical process; yet only in such a structure can it take place without burning the whole structure at the same time. Moreover, some biochemical processes go in the "wrong" direction of the reaction. Cf. H. F. Blum, *Time's Arrow and Evolution* (Princeton, 1951), p. 33 and *passim;* L. von Bertalanffy, *Problems of Life* (New York, 1952), pp. 13–14.

13. Bertalanffy, p. 137. According to P. B. Medawar, *The Uniqueness of the Individual* (New York, 1958), chaps. i–ii, even death is a physico-chemical puzzle.

14. For a very instructive — by necessity somewhat technical — analysis of this problem see Blum, just cited.

For in a steady state nothing fundamentally new can happen: essentially the same story is told over and over again by each transient world. We need no longer assume that the laws of nature change over Time, some applying only before ylem turned into matter, others only thereafter. Complete knowledge no longer constitutes the exclusive privilege of a divine mind capable of discerning in the protogalactic ylem the distant emergence of man, nay, of superman. A demon having only an *ordinary* mind deprived of any clairvoyance, but lasting millions of eons and capable of moving from one galaxy to another, should be able to acquire a complete knowledge of every transient process, just as a biologist can arrive at a description of the typical life of a new strain of bacteria after observing a large number passing from birth to death. The principle "what holds on the average for one holds for all" would apply in both cases.

But, perhaps, the exceptional properties with which we have endowed our demon violate some (unknown) laws of nature, so that its existence is confined to our paper-and-pencil operations. Be this as it may, even the most optimistic expectations do not justify the hope that mankind might in the end fulfill the exceptional conditions with which we have endowed our demon. With a life span amounting to no more than a blink of a galaxy and restricted within a speck of space, mankind is in the same situation as a pupa destined never to witness a caterpillar crawling or a butterfly flying. The difference, however, is that the human mind wonders what is beyond mankind's chrysalis, what happened in the past and, especially, what will happen in the future. The greatness of the human mind is that it wonders: He "who can no longer pause to wonder and stand rapt in awe" — as Einstein beautifully put it — "is as good as dead." [15] The weakness of the human mind is the worshiping of the divine mind, with the inner hope that it may become almost as clairvoyant and, hence, extend its knowledge beyond what its own condition allows it to observe repeatedly.

It is understandable then why the notion of the unique event causes intellectual discomfort and is often assailed as wholly nonsensical. Understandable also is the peculiar attraction which, with "the tenacity of original sin" (as Bridgman put it), the scientific mind has felt over the years for all strains of mechanistic dogmas:[16] there is solace in the belief that in nature there is no other category of phenomena than those we know best of all. And, of course, if change consisted of locomotion alone, then evolution would be a mystical notion without place in scientific knowledge. However, as we have seen through some of the preceding

15. Quoted in *The Great Design*, ed. F. Mason (New York, 1936), p. 237.
16. P. W. Bridgman, *The Logic of Modern Physics* (New York, 1928), p. 47.

sections, it is far more mystical to believe that whatever happens around or within us is told by the old nursery rhyme:

> Oh, the brave old Duke of York
> He had ten thousand men;
> He marched them up to the top of the hill,
> And marched them down again.
> And when they were up, they were up,
> And when they were down, they were down,
> And when they were only half way up,
> They were neither up nor down.[17]

If evolution of large organizations and, especially, of our own species seems somewhat of a mystery, it is only for two reasons: first, not all natural phenomena follow the pattern of the rhyme, and second, the condition of mankind is such that we can observe nature only once, or more exactly, only in part. Some have seen an irreducible paradox of infinite regression in the problem of evolution. The study of the evolution of human society, it is argued, includes the study of that study itself.[18] That there is a contradiction in any self-evolution study is beyond question. But in the absence of absolute novelty the concept of evolution involves no paradox, as can be easily seen from the fact that any human can learn a lot about his own life by observing other humans during various phases of the same life pattern. The predicament of any evolutionary science derives from the fact that mankind has no access to observing other "mankinds" — of which there must be a multitude at all times in nature if the universe is a steady going concern subject to Time-less laws.

17. The *Oxford Dictionary of Nursery Rhymes* (Oxford, 1951), p. 442. My initial source is A. S. Eddington, *The Nature of the Physical World* (New York, 1943), p. 70.
18. E.g., Popper, *Poverty of Historicism*, p. 80 and *passim*.

FIVE

GENERAL CONCLUSIONS FOR
THE ECONOMIST

1. *From the Struggle for Entropy to Class Conflict.* One point of the agitated history of thermodynamics seems to have escaped notice altogether. It is the fact that thermodynamics was born thanks to a revolutionary change in the scientific outlook at the beginning of the last century. It was then that men of science ceased to be preoccupied almost exclusively with celestial affairs and turned their attention also to some pedestrian problems.

The most prominent product of this revolution is the memoir by Sadi Carnot on the efficiency of heat engines — of which I spoke earlier. In retrospect it is obvious that the nature of the problem in which Carnot was interested is economic: to determine the conditions under which one could obtain the highest output of mechanical work from a given input of free heat. Carnot, therefore, may very well be hailed as the first econometrician. But the fact that his memoir, the first spade work in thermodynamics, had an economic scaffold is not a mere accident. Every subsequent development in thermodynamics has added new proof of the bond between the economic process and thermodynamic principles. Extravagant though this thesis may seem *prima facie*, thermodynamics is largely a physics of economic value, as Carnot unwittingly set it going.[1]

A leading symptom is that purists maintain that thermodynamics is not a legitimate chapter of physics. Pure science, they say, must abide by the dogma that natural laws are independent of man's own nature, whereas thermodynamics smacks of anthropomorphism. And that it does so smack is beyond question. But the idea that man can think of nature in wholly non-anthropomorphic terms is a patent contradiction in terms. Actually, force, attraction, waves, particles, and, especially, *interpreted* equations, all are man-made notions. Nevertheless, in the case of thermo-

---

1. The avenues opened by this new vista are explored in the author's forthcoming monograph *Process, Value, and Development*, mentioned earlier. Only some of the highlights can be briefly discussed here.

dynamics the purist viewpoint is not entirely baseless: of all physical concepts only those of thermodynamics have their roots in economic value and, hence, could make absolutely no sense to a non-anthropomorphic intellect.

A non-anthropomorphic mind could not possibly understand the concept of order-entropy which, as I have argued earlier, cannot be divorced from the intuitive grasping of human purposes. For the same reason such a mind could not conceive why we distinguish between free and latent energy, should it see the difference at all. All it could perceive is that energy shifts around without increasing or decreasing. It may object that even we, the humans, cannot distinguish between free and latent energy at the level of a single particle where normally all concepts ought to be initially elucidated.

No doubt, the only reason why thermodynamics initially differentiated between the heat contained in the ocean waters and that inside a ship's furnace is that *we can use the latter but not the former*. But the kinship between economics and thermodynamics is more intimate than that. Apt though we are to lose sight of the fact, the primary objective of economic activity is the self-preservation of the human species. Self-preservation in turn requires the satisfaction of some basic needs — which are nevertheless subject to evolution. The almost fabulous comfort, let alone the extravagant luxury, attained by many past and present societies has caused us to forget, however, the most elementary fact of economic life, namely, that of all necessaries for life only the purely biological ones are absolutely indispensable for survival. The poor have had no reason to forget it.[2] But, as discussed earlier, biological life feeds on low entropy. We thus come across the first important indication of the connection between low entropy and economic value. For I see no reason why one root of economic value existing at the time when mankind was able to satisfy hardly any nonbiological needs should have dried out later on.

Casual observation suffices now to prove that *our whole economic life feeds on low entropy*, to wit, cloth, lumber, china, copper, etc., all of which are highly ordered structures. But this discovery should not surprise us. It is the natural consequence of the fact that thermodynamics developed from an economic problem and consequently could not avoid defining order so as to distinguish between, say, a piece of electrolytic copper — which is useful to us — and the same copper molecules when diffused so

---

2. The point is related to a consequence of the hierarchy of wants: what is always in focus for any individual is not the more vitally important; rather, it is the least urgent needs he can just attain. An illustration is the slogan, "what this country needs is a good five-cent cigar." Cf. section V of my article, "Choice, Expectations and Measurability" (1954), reprinted below.

as to be of no use to us.[3] We may then take it as a brute fact that low entropy is a *necessary* condition for a thing to be useful.[4]

But usefulness by itself is not accepted as a cause of economic value even by the discriminating economists who do not confuse economic value with price. Witness the keen arguments advanced in the old controversy over whether Ricardian land has any economic value. It is again thermodynamics which explains why the things that are useful have also an economic value — not to be confused with price. They are *scarce* in a sense that does not apply to land, because, first, the amount of low entropy within our environment (at least) decreases continuously and inevitably, and second, *a given amount of low entropy can be used by us only once.*

Clearly, both factors are at work in the economic process, but it is the last one that outweighs the other. For if it were possible, say, to burn the same piece of coal over and over again *ad infinitum*, or if any piece of metal lasted forever, then low entropy would belong to the same economic category as land. (That is, it could have only a scarcity value — a scarcity price, if you wish — and only after all environmental supply will have been brought under use.) Then, every economic accumulation would be everlasting. A country provided with as poor an environment as Japan, for instance, would not have to keep importing raw materials year after year, unless it wanted to grow in population or in income per capita. The people from the Asian steppes would not have been forced by the exhaustion of the fertilizing elements in the pasture soil to embark on the Great Migration. Historians and anthropologists, I am sure, could supply other, less known, examples of "entropy-migration."

Now, the explanation by Classical thermodynamics of why we cannot use the same amount of free energy twice and, hence, why the immense heat energy of the ocean waters has no economic value, is sufficiently transparent so as to be accepted by all of us. However, statistical thermodynamics — undoubtedly because of its ambiguous rationale — has failed to convince everyone that high order-entropy too is irremediably useless. Bridgman tells of some younger physicists who in his time tried to convince the others that one could fill "his pockets by bootlegging entropy," [5] that is, by reversing high into low entropy. The issue illustrates most vividly the thesis that thermodynamics is a blend of physics and economics.

3. By now the reader should know better than to suspect that by the last remark I wish to imply that the Entropy Law is nothing but a mere verbal convention.
4. One need only cite the low entropy of poisonous mushrooms to show that the condition is not also sufficient. Of course, even poisonous mushrooms might be indirectly useful to us through a divine order, *die göttliche Ordnung* of Johann Süssmilch. But that does not concern our problem.
5. P. W. Bridgman, *Reflections of a Physicist* (2nd edn., New York, 1955), p. 244.

Let us take the history of a copper sheet as a basis for discussion. What goes into the making of such a sheet is common knowledge: copper ore, certain other materials, and mechanical work (performed by machine or man). But all these items ultimately resolve into either free energy or some orderly structures of primary materials, in short, to *environmental* low entropy and nothing else. To be sure, the degree of order represented by a copper sheet is appreciably higher than that of the ore from which we have obtained the finished product. But, as should be clear from our previous discussions, we have not thereby bootlegged any entropy. Like a Maxwell demon, we have merely sorted the copper molecules from all others, but in order to achieve this result we have *used up irrevocably a greater amount of low entropy than the difference between the entropy of the finished product and that of the copper ore.* The free energy used in production to deliver mechanical work — by humans or machines — is irrevocably lost.

It would be a gross error, therefore, to compare the copper sheet with the copper ore and conclude: Lo! Man can create low from high entropy. In my lay opinion, the analysis of the preceding paragraph proves that, on the contrary, production represents a deficit in entropy terms: it increases total entropy by a greater amount than that which would result from the automatic shuffling in the absence of any productive activity. Indeed, it seems unreasonable to admit that our burning a piece of coal does not mean a speedier diffusion of its free energy than if the same coal were left to its own fate.[6] Only in consumption proper is there no entropy deficit in this sense. After the copper sheet has entered into the consumption sector the automatic shuffling takes over the job of gradually spreading its molecules to the four winds.

But, one may ask, why do we not sort out again the same molecules to reconstitute the copper sheet? The operation is not inconceivable, but in entropy terms no other project could be as fantastically unprofitable. This is what the promoters of entropy bootlegging fail to understand. To be sure, one can cite numberless scrap campaigns aimed at saving low entropy by sorting waste. If such campaigns have been successful it is only because in the *given circumstances* the sorting of, say, scrap copper required a smaller consumption of low entropy than any alternative way of obtaining the same amount of metal. It is equally true that the advance of technological knowledge may change the balance sheet of any scrap campaign, although history shows that past progress has benefited ordinary production rather than scrap saving. However, to sort out the copper molecules scattered all over land and the bottom of the seas would

6. To recall (Chapter Three, "Some Object Lessons from Physics," sections 9 and 12, above), the entire free energy incorporated in the coal-in-the-ground will ultimately dissipate into useless energy even if left in the ground.

require such a long time that the entire low entropy of our environment would not suffice to keep alive the numberless generations of Maxwell's demons needed for the completion of the project. This may be a new way of pinpointing the economic implications of the Entropy Law. But common sense caught the essence of the idea in the parable of the needle in the haystack long before thermodynamics came to the scene of the accident.

Economists' vision has reacted to the discovery of the first law of thermodynamics, i.e., the principle of conservation of matter-energy. Some earlier writers even emphasized the point that man can create neither matter nor energy.[7] But — a fact hard to explain — loud though the noise caused by the Entropy Law has been in physics and the philosophy of science, economists have failed to pay attention to this law, the most economic of all physical laws. Actually, modern economic thought has gradually moved away even from William Petty's old tenet that labor is the father and nature the mother of value, and nowadays a student learns of this tenet only as a museum piece. The literature on economic development proves beyond doubt that most economists profess a belief tantamount to thinking that even entropy bootlegging is unnecessary: the economic process can go on, even grow, without being continuously fed low entropy.

The symptoms are plainly conspicuous in policy proposals as well as in analytical writings. For only such a belief can lead to the negation of the phenomenon of overpopulation, to the recent fad that mere technical education of the masses is a cure-all, or to the argument that all a country — say, Somaliland — has to do to boost its economy is to shift its economic activity to more profitable lines. One cannot help wondering then why Spain takes the trouble to train skilled workers only to export them to other West European countries, or what stops us from curing the economic ills of West Virginia by shifting its activity to more profitable lines.

The corresponding symptoms in analytical studies are even more definite. First, there is the general practice of representing the material side of the economic process by a *closed system*, that is, by a mathematical model in which the continuous inflow of low entropy from the environment is completely ignored.[8] But even this symptom of modern econometrics was preceded by a more common one: the notion that the economic process is wholly *circular*. Special terms such as roundabout

7. E.g., A. Marshall, *Principles of Economics* (8th edn., New York, 1924), p. 63.
8. To my knowledge, the only mention of this flow in an analytical study is by T. C. Koopmans, "Analysis of Production as an Efficient Combination of Activities," in *Activity Analysis of Production and Allocation*, ed. T. C. Koopmans (New York, 1951), p. 37. Pathbreaking though Koopmans' paper was in other directions, it did not go beyond mentioning the existence of flows from nature.

process or circular flow have been coined in order to adapt the economic jargon to this view. One need only thumb through an ordinary textbook to come across the typical diagram by which its author seeks to impress upon the mind of the student the circularity of the economic process.

The mechanistic epistemology, to which analytical economics has clung ever since its birth, is solely responsible for the conception of the economic process as a closed system or circular flow. As I hope to have shown by the argument developed in this essay, no other conception could be farther from a correct interpretation of facts. Even if only the physical facet of the economic process is taken into consideration, this process is not circular, but *unidirectional*. As far as this facet alone is concerned, the economic process consists of a continuous transformation of low entropy into high entropy, that is, into *irrevocable waste*. The identity of this formula with that proposed by Schrödinger for the biological process of a living cell or organism vindicates those economists who, like Marshall, have been fond of biological analogies and have even contended that economics "is a branch of biology broadly interpreted." [9]

The conclusion is that, from the purely physical viewpoint, the economic process is entropic: it neither creates nor consumes matter or energy, but only transforms low into high entropy. But the whole physical process of the material environment is entropic too. What distinguishes then the first process from the second? The differences are two in number and by now they should not be difficult to determine.

To begin with, the entropic process of the material environment is *automatic* in the sense that it goes on by itself. The economic process, on the contrary, is dependent on the *activity* of human individuals who, like the demon of Maxwell, sort and direct environmental low entropy according to some definite rules — although these rules may vary with time and place. The first difference, therefore, is that while in the material environment there is only shuffling, in the economic process there is also sorting, or rather, a sorting activity.

But, since sorting is not a law of elementary matter, the sorting activity must feed on low entropy. Hence, the economic process actually is more efficient than automatic shuffling in producing higher entropy, i.e., waste. What could then be the *raison d'être* of such a process? The answer is that the true "output" of the economic process is not a physical outflow of waste, but the *enjoyment of life*. And this point represents the second difference between this process and the entropic march of the material environment. Without recognizing this fact and without introducing the concept of enjoyment of life into our analytical armamentarium we are not in the economic world. Nor can we discover the real source of economic value which is the value that life has for every life-bearing individual.

9. Marshall, *Principles*, p. 772.

It is thus seen that we cannot arrive at a completely intelligible description of the economic process as long as we limit ourselves to purely physical concepts. Without the concepts of *purposive activity* and *enjoyment of life* we cannot be in the economic world. And neither of these concepts corresponds to an attribute of elementary matter or is expressible in terms of physical variables.[10]

But even these new coordinates do not suffice to characterize completely the economic process. All living beings display a purposive activity and, from what we can judge, they too seem to enjoy being alive. So what distinguishes the economic from the biological process? The answer came — curiously enough — from biologists.

All living beings, in their role as Maxwellian demons sorting low entropy for the purpose of enjoying and preserving their lives, use their biological organs. These organs vary from species to species, their form even from variety to variety, but they are characterized by the fact that *each individual is born with them.* Alfred Lotka calls them *endosomatic instruments.*[11] If a few marginal exceptions are ignored, man is the only living being that uses in his activity also "organs" which are not part of his *biological constitution.* We economists call them capital equipment, but Lotka's term, *exosomatic instruments,* is more enlightening. Indeed, this terminology emphasizes the fact that broadly interpreted the economic process is a continuation of the biological one. At the same time it pinpoints the *differentia specifica* between the two kinds of instruments which together form one genus. Broadly speaking, endosomatic evolution can be described as a progress of the entropic efficiency of life-bearing structures. The same applies to the exosomatic evolution of mankind. Exosomatic instruments enable man to obtain the same amount of low entropy with less expenditure of his own free energy than if he used only his endosomatic organs.[12]

As already explained, the struggle for life which we observe over the entire biological domain is a natural consequence of the Entropy Law. It goes on between species as well as between the individuals of the same species, but only in the case of the human species has the struggle taken also the form of a social conflict. To observe that social conflict is an

10. I may point out that the enjoyment of life, though *caused* by a material flow, is not itself a flow. The only feature it has in common with flow is that its dimension too contains the factor time. The intensity of life enjoyment may thus be likened to the instantaneous rate of flow, but the parallelism stops here. For lack of a better choice, in the forthcoming monograph *Process, Value, and Development,* I have proposed to describe the enjoyment of life by the term "flux."

11. P. B. Medawar, *The Uniqueness of the Individual* (New York, 1958), p. 139.

12. The question why the expenditure of man's own free energy, even if continuously replaced, should be accompanied by a feeling of unpleasantness is, I think, a moot question. But without this feeling, man probably would not have come to invent exosomatic instruments, to enslave other men, or to domesticate animals of burden.

outgrowth of the struggle of man with his environment is to recognize a fairly obvious fact, but not to explain it. And since the explanation is of particular import for any social scientist, I shall attempt to sketch one here.

A bird, to take a common illustration, flies after an insect with *its own* wings and catches it with *its own* bill, i.e., with endosomatic instruments which by nature are the bird's individual property. The same is certainly true of the primitive exosomatic instruments used during the earliest phase of human organization, the primitive communism as Marx calls it. Then each familial clan lived by what *its own* bow and arrow could kill or *its own* fishing net could catch, and nothing stood in the way of all clan members' sharing the product more or less according to their basic needs.

But man's instincts, of workmanship and of idle curiosity, gradually devised exosomatic instruments capable of producing more than a familial clan needed. In addition, these new instruments, say, a large fishing boat or a flour mill, required more hands both for being constructed and for being operated than a single familial clan could provide.[13] It was at that time that production took the form of a *social* instead of a *clannish* activity.

Still more important is to observe that only then did the difference between exosomatic and endosomatic instruments become operative. Exosomatic instruments not being a natural, indissoluble property of the individual person, the advantage derived from their perfection became the basis of inequality between the various members of the human species as well as between different communities. Distribution of the communal income — income being understood as a composite coordinate of real income and leisure time[14] — thus turned into a social problem, the importance of which has never ceased to grow. And, if I may be allowed to venture an opinion, it will last as a center of social conflict as long as there will be any human society.

The preceding explanation seems *prima facie* to vindicate the basic creed of all strains of Marxism, namely, that the socialization of the means of production would mean the end of the social side of the economic conflict because mankind would thus return to a state essentially analogous to primitive communism where the evil inherent in the exosomatic instruments was not operative. Actually, my analysis points in another direction.

The perennial root of the social conflict over the distribution of income lies in the fact that our exosomatic evolution has turned production into

13. Cf. Karl Kautsky, *The Economic Doctrines of Karl Marx* (New York, 1936), pp. 8 ff.

14. Cf. section IV. 1 of my paper reprinted below, "Economic Theory and Agrarian Economics" (1960).

a social undertaking. Socialization of the means of production, clearly, could not change this fact. Only if mankind returned to the situation where every family (or clan) is a self-sufficient economic unit would men cease to struggle over their anonymous share of the total income. But mankind could never reverse its exosomatic any more than its endosomatic evolution.

Nor does socialization of the means of production implicitly warrant — as Marxists assert — a rational solution of the distributive conflict. Our habitual views on the matter may find it hard to accept, but the fact is that *communal ownership of the means of production is, in all probability, the only regime compatible with any distributive pattern.* A most glaring example is provided by feudalism, for we must not forget that land passed into private ownership only with the dissolution of the feudal estates, when not only the serfs but also the former lords legally became private owners of land. Besides, it is becoming increasingly obvious that social ownership of means of production is compatible even with some individuals' having an income which for all practical purposes is limitless in some, if not all, directions.

There is, however, another reason why the conflict between individuals over their share of the social income inevitably precipitates a class conflict in any society save primitive communism. Social production and its corollary, social organization, require a specific category of services without which they cannot possibly function. This category comprises the services of supervisors, coordinators, decision makers, legislators, preachers, teachers, newsmen, and so on. What distinguishes these services from those of a bricklayer, a weaver, or a mailman is that they do not possess an objective measure as the latter do. Labeling the former *unproductive* and the latter *productive* — as is the tradition in socialist literature — is, however, a misleading way of differentiating between the two categories: production needs both.

Now, even if the entire social product were obtained only with the aid of services having an objective measure, the problem of the income distribution would be sufficiently baffling. But the fact that society needs also services which have no objective measure adds a new dimensional freedom to the patterns of distribution. Economists know this from their lack of success in finding a measure for entrepreneurship. Even more telling is the fact that, as far as I know, no criterion has been suggested for determining the salary of, say, a senator or of a nation's president.

It is then obvious why in every society the members performing "unproductive" services constitute a kept-up class, to use Veblen's poignant expression. They, clearly, cannot claim a share according to the quantum of their services, for these services are not cardinally measurable. But this very fact is at the same time their everlasting weapon. The privileged elite that is found in every society known to history consists only of individuals who perform unproductive services. Elites — as Pareto argued — circulate, that is, one elite replaces another: only the name and the rationalization of the privileges are changed.[15] The preceding analysis, however, explains why this circulation is an inevitable consequence of the exosomatic evolution of mankind and also why an elite necessarily consists of the unproductive members of the society.

The class conflict, therefore, will not be choked forever if one of its phases — say, that where the captains of industry, commerce, and banking claim their income in the name of private property — is dissolved.[16] Nor is there any reason to justify the belief that social and political evolution will come to an end with the next system, whatever this system may be.

2. *The Boundaries of the Economic Process.* Controversy has been, with a varying degree of relative importance, a continuous stimulus in all spheres of intellectual endeavor, from literary criticism to pure physics. The development of economic thought, in particular, has been dependent on controversy to an extent that may seem exasperating to the uninitiated. It is nevertheless true that the doctrinaire spirit in which some fundamental issues have been approached has harmed the progress of our science. The most eloquent example of this drawback is the controversy over the boundaries of economics or, what comes to the same thing, over the boundaries of the economic process.

The problem was implicitly raised first by the German historical school, but it caused practically no stir until Marx and Engels set forth their doctrine of historical materialism. From that moment on, the proposition that constitutes the first pillar of that doctrine has been the subject of a sustained and misdirected controversy. This proposition is that the economic process is not an isolated system. The non-Marxist economists apparently believe that by proving the existence of some natural boundaries for the economic process they will implicitly expose the absurdity of historical materialism and, hence, its corollary: scientific socialism. However, whatever one may have to say about the other pillars of Marxism, one can hardly think of a plainer truth than that the economic process is not an isolated system. On the other hand, equally plain is the necessity of delimiting this process in some way: otherwise, there would be no sense at all in speaking of the economic process.

The problem is related to a point which I have endeavored to establish

15. Vilfredo Pareto, *Les systèmes socialistes* (Paris, 1926), I, 30 ff.
16. A single casual remark indicates that Marx may have seen, however dimly, that the class conflict transcends the problem of private ownership: "The whole economical history of society is summed up in the movement of this antithesis [the division of labor between town and country]." Karl Marx, *Capital* (Chicago, 1932), I, 387.

in the course of this essay, namely, that the boundaries of actual objects and, especially, events are dialectical penumbras. Precisely because it is impossible to say, for example, where the chemical process ends and where the biological one begins, even natural sciences — as I have argued earlier — do not have rigidly fixed and sharply drawn frontiers. There is no reason for economics to constitute an exception in this respect. On the contrary, everything tends to show that the economic domain is surrounded by a dialectical penumbra far wider than that of any natural science.

Within this wide penumbra — as every sophomore knows from the famous riddle of what happens to the national income if a bachelor marries his housekeeper — the economic intertwines with the social and the political. Malthus alone argued that there is also an interconnection between the biological growth of the human species and the economic process. Economists in general have rejected his doctrine because to this day they have failed to see that, in spite of his unfortunate choice of expressing it, Malthus was in essence right. This can be immediately seen from our entropic analysis of the economic process. The fact that biological and economic factors may overlap in some surprising ways, though well established, is little known among economists.

It was Francis Galton who in a celebrated study, *Hereditary Genius* (1869), showed how the desire for wealth — certainly, an economic factor — contributed to the biological extinction of twelve out of the thirty-one original English peerages. He found that peers more often than not married wealthy heiresses and thus introduced the gene of low fertility in their blood lines. Some forty years after Galton's discovery, J. A. Cobb pointed out that the phenomenon is far more general. In a society where personal wealth and social rank are highly correlated — as is the case under the regime of private ownership — the gene of low fertility tends to spread among the rich, and that of high fertility among the poor. On the whole, the family with very few children climbs up the social ladder, that with more than the average number of offspring descends it. Besides, since the rich usually marry the rich, the poor cannot marry but the poor. The rich thus become richer and the poor poorer because of a little-suspected interplay of economic and biological factors.[17]

The problem of delimiting the sphere of economics, even in a rough way, is therefore full of thorns. In any case, it is not as simple as Pareto urges us to believe through his argument that just as geometry ignores

17. For a masterly discussion of this category of problems — highly instructive for any student of economics — see R. A. Fisher, *The Genetical Theory of Natural Selection* (Oxford, 1930), chaps. x and xi. Also J. B. S. Haldane, *Heredity and Politics* (New York, 1938), pp. 118 ff.

chemistry so can economics ignore by abstraction *homo ethicus, homo religiosus,* and all other *homines.*[18] But Pareto is not alone in maintaining that the economic process has definite natural limits. The same position characterizes the school of thought which has followed the attractive paths opened by the early mathematical marginalists and which has come to be commonly referred to as standard economics. A more recent formulation of this position is that the scope of economics is confined to the study of how *given* means are applied to satisfy *given* ends.[19] In more specific terms: at any given instant of time the means at the disposal of every individual as well as his ends over the future are given; given also are the ways (technical and social) in which these means can be used directly or indirectly for the satisfaction of the given ends jointly or severally; the essential object of economics is to determine the allocation of the given means towards the optimal satisfaction of the given ends. It is thus that economics is reduced to "the mechanics of utility and self-interest." Indeed, any system that involves a conservation principle (given means) and a maximization rule (optimal satisfaction) is a mechanical analogue.[20]

Now, the economic nature of allocating given means for the optimal satisfaction of given ends cannot possibly be denied. In its abstract form, such allocation reflects a permanent preoccupation of every individual. Nor can one deny that frequently the problem presents itself in concrete terms and is susceptible of a numerical solution because all necessary data are actually *given.* The recent results achieved in this direction following the pioneering work of T. C. Koopmans deserve the highest praise. Yet, highly valuable though these results are, the new field of engineering (or managerial) economics does not cover the whole economic process, any more than husbandry exhausts all that is relevant in the biological domain.

Let me hasten to add that the usual denunciation of standard economics on the sole ground that it treats of "imaginary individuals coming to imaginary markets with ready-made scales of bid and offer prices" [21] is

18. Vilfredo Pareto, *Manuel d'économie politique* (Paris, 1927), p. 18.

19. By far the most articulate defense of this restrictive viewpoint is due to Lionel Robbins, *An Essay on the Nature and Significance of Economic Science* (2nd edn., London, 1948), p. 46 and *passim.*

20. Cf. Henri Poincaré, *The Foundations of Science* (Lancaster, Pa., 1946), p. 180. For a detailed examination of the strict analogy between the Pareto-Walras system and the Lagrange equations see V. Pareto, "Considerazioni sui principii fondamentali dell' economia politica pura," *Giornale degli economisti,* IV (1892), 409 ff. Irving Fisher, in *Mathematical Investigations in the Theory of Value and Prices* (New Haven, 1925), pp. 38–39 and *passim,* even constructed some mechanical analogues of the consumer. Even recently Frank H. Knight, in *The Ethics of Competition* (New York, 1935), p. 85, referred to mechanics as the "sister science" of economics.

21. Wesley C. Mitchell, "Quantitative Analysis and Economic Theory," *American Economic Review,* XV (1925), 5.

patently inept. Abstraction, even if it ignores Change, is "no exclusive *privilegium odiosum*" of the economic science,[22] for abstraction is the most valuable ladder of any science. In social sciences, as Marx forcefully argued, it is all the more indispensable since there "the force of abstraction" must compensate for the impossibility of using microscopes or chemical reactions.[23] However, the task of science is not to climb up the easiest ladder and remain there forever distilling and redistilling the same pure stuff. Standard economics, by opposing any suggestion that the economic process may consist of something more than a jigsaw puzzle with all its elements given, has identified itself with dogmatism. And that is a *privilegium odiosum* which has dwarfed the understanding of the economic process wherever it has been exercised.

From time indefinite, the natural sciences have cherished a positivist epistemology according to which scientific knowledge covers only those phenomena that go on irrespective of whether they are observed or not. Objectivity, as this criterion is often called, requires then that a proper scientific description should not include man in any capacity whatsoever. This is how some came to hold that even man's thinking is not a phenomenon.[24] True, the ideal of a man-less science is gradually losing ground even in physics — ever since the discovery of quantum of action and Heisenberg's indeterminacy.[25] However, for a science of man to exclude altogether man from the picture is a patent incongruity. Nevertheless, standard economics takes special pride in operating with a man-less picture. As Pareto overtly claimed, once we have determined the means at the disposal of the individual and obtained "a photograph of his tastes . . . the individual may disappear."[26] The individual is thus reduced to a mere subscript of the ophelimity function $\phi_i(X)$. The logic is perfect: man is not an economic agent simply because there is no economic process. There is only a jigsaw puzzle of fitting given means to given ends, which requires a computer not an agent.

If standard economics has not completely banished the individual from its discourse it is because a weakening assumption has been added to those outlined above. This assumption is that although every individual knows his own means and ends, no one knows the means and ends of others. "A farmer can easily calculate whether at the market prices it is

22. Joseph A. Schumpeter, *Essays*, ed. R. V. Clemence (Cambridge, Mass., 1951), p. 87.

23. Preface to the first edition of *Capital*, p. 12.

24. Cf. A. J. Ayer, *Language, Truth and Logic* (2nd edn., New York, 1946), pp. 57–58. But see E. Schrödinger, *Nature and the Greeks* (Cambridge, Eng., 1954), pp. 90 ff.

25. Cf. Niels Bohr, *Atomic Physics and Human Knowledge* (New York, 1958), p. 98; Werner Heisenberg, *Physics and Philosophy: The Revolution in Modern Science* (New York, 1958), pp. 52–53.

26. Pareto, *Manuel*, p. 170; V. Pareto, "Mathematical Economics," *International Economic Papers*, no. 5, 1955, p. 61.

more advantageous for him to use a horse or a tractor; but neither he nor anyone in the world can determine the effect [of the farmer's decision] on the prices of horses and tractors." [27] The puzzle can then be solved only by groping — *tâtonnement*. This is how the individual came to be endowed with some economic activity, that and only that of shifting resources by trial and error between various employments, contemporaneous or not. And since the founders of standard economics — like most economists — aspired to provide an analysis of the economic reality in which they actually lived, the rules of the *tâtonnement* as well as the nature of the ends were molded upon attitudes and practices prevalent in a capitalist society. One may therefore understand why Rosa Luxemburg defined economics as the study of how an uncoordinated, chaotic system such as capitalism can nevertheless function. Natural also is her conclusion that the economic science will die of starvation with the advent of the socialist society where scientific planning will replace the *tâtonnement*. [28]

So it is for its dogmatism, not for its use of abstraction, that standard economics is open to valid criticism. Casual observation of what happens in the sphere of economic organizations, or between these organizations and individuals, suffices to reveal phenomena that do not consist of *tâtonnement* with given means towards given ends according to given rules. They show beyond any doubt that in all societies the typical individual continually pursues also an end ignored by the standard framework: the increase of what he can claim as his income according to his current position and distributive norms. It is the pursuit of this end that makes the individual a true agent of the economic process.

Two are the methods by which he can pursue this particular end. First, he may seek ways by which to improve *qualitatively* the means he already possesses. Secondly, he may seek to increase his personal share of the stock or flow of social means, which is tantamount to changing the prevailing distributive relations. It is because even in a socialist society the individual activity is in the long run directed also towards these aims that new means are continually invented, new economic wants created, and new distributive rules introduced. [29]

The question is why a science interested in *economic* means, ends, and distribution should dogmatically refuse to study also the process by which new *economic* means, new *economic* ends, and new *economic* relations are created. Perhaps one would answer that what is to be included in the scope of any special science is a matter of convention or of division of

27. Pareto, *Manuel*, p. 335. My translation.
28. Rosa Luxemburg, "What Is Economics?" (mimeo., New York, 1954), pp. 46, 49.
29. The preceding remarks should be compared with those of Knight, *Ethics of Competition*, pp. 58 ff. However, I am not sure whether the particular activity described above coincides with what Knight calls "the institution of sport."

labor. To return to an earlier parallel, is it not true that husbandry constitutes a proper scientific endeavor and a very useful discipline despite the fact that it does not concern itself with biological evolution? There is, however, a very important reason why economics cannot follow the example of husbandry.

The reason is that the evolutionary pace of the economic "species" — that is, of means, ends, and relations — is far more rapid than that of the biological species. The economic "species" are too short-lived for an economic husbandry to offer a relevant picture of the economic reality. Evolutionary elements predominate in every concrete economic phenomenon of some significance — to a greater extent than even in biology.[30] If our scientific net lets these elements slip through it, we are left only with a shadow of the concrete phenomenon. No doubt, a navigator does not need to know the evolution of the seas; actual geography, as Pareto argued, suffices him.[31] But my point is that Pareto's illustration would be of no avail if the earth's geography evolved as rapidly as that of the economic world. It is beyond dispute, therefore, that the sin of standard economics is the fallacy of misplaced concreteness, by which Whitehead understands "neglecting the degree of abstraction involved when an actual entity is considered merely so far as it exemplifies certain [pre-selected] categories of thought." [32]

In retrospect, it appears natural that denunciations of the sterility of the standard armamentarium should have come from men such as Marx and Veblen, who were more interested in distributive relations than in the efficient allocations of means: the fallacy of misplaced concreteness is more conspicuous in the former than in the latter problem. However, although the disciples of Marx or Veblen like to claim the entire glory for their own master,[33] the shortcomings of the static analysis originated by Ricardo were pointed out long before Marx. J. B. Say, for example, in an 1821 letter warned Ricardo's contemporaries that future generations would laugh at the terror with which, because of the Ricardian analysis, they were viewing the effect of the technical progress upon the fate of the industrial workers.[34] It is nevertheless true that lessons, perhaps the only substantial ones, on how to transcend the

30. This is not the same thing as saying that the economic material is exposed to numerous disturbances, as Joseph A. Schumpeter says in *Business Cycles* (New York, 1939), I, 33. From a mechanistic viewpoint, every concrete phenomenon appears subject to innumerable disturbances.

31. Pareto, *Manuel*, p. 101.

32. Alfred North Whitehead, *Process and Reality: An Essay in Cosmology* (New York, 1929), p. 11.

33. E.g., Karl Korsch, *Karl Marx* (London, 1938), p. 156; John S. Gambs, *Beyond Supply and Demand* (New York, 1946), p. 10.

34. *Say's Letters to Malthus on Political Economy and Commerce* (London, 1936), p. 64.

static framework effectively have come from Marx, Schumpeter, and Veblen.[35]

One should not fail, however, to recognize also the unique endeavor of Marshall to instill some life into the analytical skeleton of standard economics. Schumpeter, with his tongue in cheek — as it often was — said that Marshall "wanted — strange ambition! — to be 'read by businessmen.'"[36] No doubt it *was* a strange ambition after all for Marshall to insist upon respect for relevance instead of succumbing to the temper of his age. To cite only one from the many eloquent examples: it was Marshall who showed in the most incontrovertible way that even such a basic concept as the supply schedule of an "increasing returns" industry slips through the analytical mesh because "increasing returns" is an essentially evolutionary phenomenon, necessarily irreversible and perhaps irrevocable as well.[37] Marshall expressed his respect for analysis in so many words, but his "thought ran in terms of evolutionary change — in terms of an organic, irreversible process."[38] But Schumpeter went on to say that Marshall's "vision of the economic process, his methods, and his results, are no longer ours."[39] Coming from an economist in whose work evolution occupied a prominent position, this last remark cannot be taken for anything but a veiled lament. Great minds — such as Lionel Robbins — who ultimately awake from "dogmatic slumber,"[40] are, unfortunately, rare exceptions.

The accusations that standard economics has deserted its main duty, namely, that of being a guide for practical politics, are commonplace. But, as I have endeavored to prove in this section, it is the very nature of the economic process, more precisely its evolutionary pace, that precludes a grasping of its relevant aspects by a static scheme, even if one studies economics only for the love of the art. As to where the boundary of the economic process should be appropriately set, I know of no better answer than Marshall's definition of economics as "the study of mankind in the ordinary business of life,"[41] provided one does not insist on an arithmomorphic interpretation of every term.

35. As all economists know, only Schumpeter formed no school. I wish to observe, however, that the American Institutionalists, though hailing Veblen as their prophet, have inherited little from him besides an aggressive scorn for "theory." Be this as it may, Paul T. Homan, in "An Appraisal of Institutional Economics," *American Economic Review*, XXII (1932), 10–17, has completely missed the issue raised by Veblen.

36. Joseph A. Schumpeter, *Ten Great Economists* (New York, 1951), p. 97.

37. Marshall, *Principles*, p. 808. See also Schumpeter, *Essays*, p. 53n2, and Knight, *Ethics of Competition*, pp. 166–167.

38. Schumpeter, *Ten Great Economists*, p. 101.

39. *Ibid.*, p. 92.

40. As Lionel Robbins admits for himself in his *The Economic Problem in Peace and War* (London, 1947), pp. 67–68.

41. Marshall, *Principles*, p. 1.

3. *Why Is Economics Not a Theoretical Science?* Everyone uses "theory" in multifarious senses. To wit, in one place Schumpeter uses it to mean "a box of analytical tools." [42] But in discriminating usage, the term generally denotes a logical edifice. Or, as I have explicitly put it (Chapter One, section 4, above), theory means a logical filing of *all* extant knowledge in some particular domain such that every known proposition be either contained in the logical foundation or deducible from it. That such a filing has the unique merit of affording *comprehensibility* is a leitmotif inherited from Aristotle. However, hardly any attention has been paid to the fact that there can be no comprehensibility without the *compressibility* of extant knowledge in only a relatively few ω-propositions. If our knowledge of a certain domain is not compressible, i.e., if its logical filing results in a very great number of ω-propositions, Aristotelian comprehensibility does not obtain. I have illustrated this point in connection with chemistry where, because of the frequency of novelty by combination, any logical foundation must contain far more numerous propositions than the β-class. For the same reason a logical foundation of chemistry would have to be continuously "under construction." A chemical theory, clearly, would serve no purpose whatsoever. The same applies with even greater force to any science concerned with evolution, for the scene of evolution is dominated by novelty.

After what has been said about the scope of economics the answer to the question at the head of this section is so obvious that to dwell further on it might seem superfluous. But since the view that the propositions about the economic process can be arranged into a theory is widely shared, it appears instructive to analyze briefly the most salient arguments advanced in its support.

The oldest, and also the most commonly held, argument is that economics must necessarily be a theoretical science because every economic phenomenon follows logically from a handful of elementary principles. The idea goes back to the Classical school, which taught that all economic phenomena are grounded in "the desire for wealth" which characterizes any "sane individual," and are governed by only two general laws. The first is that "a greater gain is preferred to a smaller"; the second is the propensity to obtain "the greatest quantity of wealth with the least labor and self-denial." [43] To these general laws the marginalists added two principles of more substantial content, the principles of decreasing utility and decreasing returns. But economists have continued to argue that the fundamentals of economics are known to us immediately by intuition, and hence their truth can be trusted "more confidently and certainly

42. Schumpeter, *Essays*, p. 227.
43. John Stuart Mill, *A System of Logic* (8th edn., New York, 1874), pp. 623 ff; Knight, *Ethics of Competition*, pp. 135 ff.

than . . . any statement about any concrete physical fact or event." [44] Still more important is the claim that because of this special property of its fundamental laws economics is the deductive science *par excellence*. Consequently, all economic propositions are valid in any institutional setting.[45]

No doubt, one can hardly think of a more obvious truth than the postulate *"Each individual acts as he desires."* [46] Or as the same idea is expressed in modern jargon, everybody acts so as to maximize his satisfaction in every given set of circumstances. Clearly, it is as absurd to think of an individual who prefers a smaller to a greater gain as to imagine a quadrangle with five sides. On the other hand, to compare the principle of maximum satisfaction with "any statement about any *concrete* physical fact" is an idle proposal, unless "satisfaction" too is more concretely described.

The last requirement is essential. Even standard theory could not ignore it: its theoretical edifice was not built upon a general and vague concept of satisfaction, but on the specific proposition that *only those goods and services an individual can enjoy personally influence his satisfaction*. Accordingly, in standard theory ophelimity is a function only of the quantities of such goods and services.

This particular formula — as I argued in a paper reprinted in this volume[47] — reflects an institutional trait proper (and, perhaps, specific as well) to the large urban communities of industrialized societies. The same is true of another cornerstone of standard theory, namely, the proposition that, for a seller, "gain" is measured solely by money-profit. But — to recall Marx's protest — "the bourgeois reason is not the normal human reason." [48] As Marshall carefully pointed out, it is not the general reason even in the bourgeois society.[49] Still less can we expect it to be valid in all institutional settings. Actually, in *peasant* communities the satisfaction of the individual depends not only on the quantities of goods and services at his disposal but also on other economic variables, and gain depends on other factors besides money-profit.

The statement that the fundamental principles of economics are universally valid, therefore, may be true only as their *form* is concerned. Their *content*, however, is determined by the institutional setting. And without this institutional content, the principles are nothing but "empty

44. Frank H. Knight, *On the History and Method of Economics* (Chicago, 1956), p. 164. Also, W. Stanley Jevons, *The Theory of Political Economy* (4th edn., London, 1924), p. 18.

45. Cf. Jevons, *Theory*, p. 19; Knight, *Ethics of Competition*, pp. 137–138 and *passim*.

46. Irving Fisher (note 20, above), p. 11; Pareto, *Manuel*, p. 62.

47. "Economic Theory and Agrarian Economics" (1960), section III. 2.

48. Karl Marx, *A Contribution to the Critique of Political Economy* (Chicago, 1904), p. 93.

49. Marshall, *Principles*, pp. 762 ff.

boxes," from which we can obtain only empty generalities. This is not to say that standard theory operates with "empty boxes." On the contrary, as we have seen, those boxes are filled with an institutional content distilled from the cultural patterns of a capitalist society. They may be only partly filled — as is certainly the case. Indeed, many traits of such societies have been left out because they were not quite ripe at the time the foundations of standard theory were laid, others because they cannot be fitted into the arithmomorphic structure a theory necessarily has.[50]

Let me repeat here a point made in the paper entitled "Economic Theory and Agrarian Economics" (1960), reprinted below. It is precisely because the boxes of standard theory were already filled with a specific institutional content that this theory was unceremoniously rejected by the students of the economic process in noncapitalist settings. The most salient examples are those of the historical school in Germany and of *Narodnikism* in Russia. Significant though the point is, it has received not more than casual attention. Marshall seems to be alone in reproaching the standard economists for having worked out "their theories on the tacit supposition that the world was made up of city men." [51] Yet, even Marshall's censure does not aim at the real issue.

No economist, even a Ricardo or a Walras, can be blamed for not having constructed a theory both *relevant and valid* for all institutional settings. Society is not an immutable entity, but evolves continuously in endless forms that differ with time and place as well. It is normal, therefore, that every great economist should have filled his analytical boxes with an institutional content inspired by the cultural patterns of the society he knew best: that in which he lived.

The economic profession should accept with immense pride Bridgman's accusation of practical opportunism.[52] Indeed, it would have been most regrettable had no Quesnay been interested in the specific economic problems of eighteenth-century France, had no Keynes studied the economic problems of modern state organizations, or had no contemporary economist been attracted by the problem of how to develop backward economies — which is *the* problem of our age. The standard economist, therefore, cannot be indicted, any more than Marx, for constructing his theory after the model of capitalist society. The egregious sin of the standard economist is of another kind. Because he denies the necessity of

50. The reader should have no difficulty in finding the reason why the preceding conclusions differ fundamentally from those of some well-known discussions of the same topic, such as that of Knight, *Ethics of Competition*, pp. 135 ff, or J. H. Clapham, "Of Empty Economic Boxes," *Economic Journal*, XXXII (1922), 305–314. These authors use "content" in its Paretoan sense, meaning the ensemble of all "standard" ophelimity and production functions.
51. Marshall, *Principles*, p. 762.
52. Bridgman, *Reflections*, pp. 443–444.

paying any attention to the evolutionary aspects of the economic process, he is perforce obliged to preach and practice the dogma that his theory is valid in *all* societies.[53]

The celebrated *Methodenstreit* apparently centered upon methodology. But, as should be clear from the preceding analysis, at bottom the *Streit* (i.e., the fight) was about the claim that it is possible to construct a universally valid economic theory. The adversaries of the Ricardians maintained that there is a Great Antinomy between this claim and the evolutionary nature of the economic process. Standard economists, as we have just seen, entrenched themselves behind the position of the *directly intuitive* basis of the fundamental economic laws. But another signal attempt at resolving the Great Antinomy proceeds from an *objective* basis. In essence, it is a chemical doctrine of society.[54]

A chemical doctrine claims, first, that all forms of societies can be objectively analyzed into a finite number of immutable elements, and second, that a society can possess no other property than those inherent in its elementary components. The Golden Horde, the medieval city of Florence, twentieth-century Switzerland, therefore, would not be different "animals" each with its specific behavior, but only stronger or weaker cocktails obtainable from a finite list of ingredients.

We owe to Walter Eucken the most cogent elaboration of a chemical doctrine of the economic process. He argues that the perennial ingredients of any economic system fall into three categories: the control (central or plural), the market (with its standard forms), and the monetary conventions (commodity-money, commodity-credit, money-credit).[55] Any economy is nothing but some combination of these ingredients, one from each category. All we need to know is the particular combinative formula in each case under consideration.

To clarify this epistemological position, Eucken resorts to an analogy: the works of composers, though different, have been created "by combining together a limited number of tones which they all used." [56] The choice is, however, most unfortunate, for through this analogy Eucken unwittingly lays bare the basic weakness of all chemical doctrines of society.

53. In justice to Marx I should note that he never endorsed this position. On the contrary, Marx repeatedly emphasized that his analysis pertains only to the capitalist system: e.g., Marx, *A Contribution*, p. 269. He also was aware of the fact that the differences between French and German economic schools were reflections of the institutional differences between the respective countries. *Ibid.*, p. 56n. Yet, in the end, Marx committed the great error of indiscriminately extending the laws of a capitalist society to the economy of a rural, agricultural society. See section I. 2 of my paper reprinted below, "Economic Theory and Agrarian Economics" (1960).

54. See above, Chapter Three, "Some Object Lessons from Physics," section 7. Actually, the term "chemical" is misappropriated, as will presently appear.

55. Walter Eucken, *The Foundations of Economics* (London, 1950), chap. ii.

56. *Ibid.*, pp. 226–227.

Musical scales have evolved and new ones are still in store for us. Besides, music requires instruments; new ones have been invented even during our generation. It is, therefore, patently false to say that *all* music is analyzable into a *given* set of tones and a *given* set of instruments. But that is not the major fault of a chemical doctrine.

From all we know, activity without a controlling agent is inconceivable; the existence of markets goes back to the dawn of history; some forms of capitalistic enterprises and money are found even in ancient societies. The obviousness of the general proposition that every economy consists of control, market, and monetary conventions, however, may be dangerously alluring. For, at least to anyone uncommitted to the fallacy of misplaced concreteness, it is equally obvious that this mixing formula fails to describe even partially the essential aspects of an extant economy.

As I had occasion to observe earlier, every chemical compound has some properties not possessed by any of its elements; moreover, there is no general principle by which to deduce every property of a compound from its chemical formula. If this were not so, it would be child's play — as C. P. Green remarked in a different connection — for the modern scientist who can count the protons in the whole universe to find by calculation the color spots on a bird in New Guinea.[57] Given that the "chemical" doctrine fails to work in the chemical domain, it would be foolhardy to count on its success in social sciences, where the number of compounds is almost limitless and quality dominates the scene to an incomparably greater degree than in the domain of elementary matter.

It is highly significant that a modern mathematician, not a medieval mystic, raised the devastating question: how can a naturalist who has studied only the chemical composition of the elephant know anything about the behavior of that animal?[58] But biology, despite the increasing tribute it pays to chemical knowledge, did not wait for the intervention of an outsider to reject the chemical doctrine. As a Nobel laureate instructs us, for modern biology "a gene is known by its performance and not by its substantive properties." [59] This simple statement epitomizes the new biological conception, which has come to be known as the organismic epistemology.[60] It is a belated recognition of the existence of novelty by combination, but free from any vitalist overtones.

The same conception did not fare as well in the social sciences, still less in economics. The job of the economist being that of studying a process which often evolves faster than he can complete his professional training,

57. C. P. Green, "Time, Space, Reality," *Philosophy*, IX (1934), 463.
58. Poincaré, *Foundations of Science*, p. 217.
59. P. B. Medawar, *The Future of Man* (New York, 1960), p. 119.
60. The essence of this idea, however, is much older than its propounders seem to realize. See Plato, *Philebus*, 14 ff.

it is normal for him to thirst more than anyone else for the objectivity of Classical physics. To be sure, such a thirst becomes even more pressing when fed propositions which defy any algebra, such as the tenet that "society is not a sum of individuals." Let us observe, however, that this is a rather unfortunate way of saying that society has properties the individual *by himself* cannot have. It may seem superfluous to some, futile to others, to point out that although every inch of the devastation left by a mob could be traced back to an act of some particular individual, an individual by himself can never display the peculiar properties of a mob. Nor can a single individual have all the manifestations of a religious sect, nor those we observe at religious revivals. Marx was completely right in ridiculing the economics of Robinson Crusoe,[61] where there are no monopolists, no labor unions, no conflict over the distribution of sacrifice and reward.

On the other hand, we may as well recognize that the reluctance of most of us to part with the tenet that society is a sum of individuals, is rooted in a historical condition: the only instance where the tenet is *roughly* true is the bourgeois society in which we have been reared and which is the nearest approximation to Hegel's Civil Society.[62] However, even the bourgeois society evolves, and nowadays it probably no longer fits Hegel's bill of requirements.[63]

Viewed as a theoretical reduction of a phenomenal domain, any chemical doctrine is fallacious from the start — save in the case of those physical phenomena which are indifferent to scale. At most, it can be accepted as a procedural code for morphological analysis. In this role it has proved its usefulness in chemistry, in nuclear physics, and to a lesser extent in the biology of the organism. In all probability that is the limit, considering that as keen an economist as Eucken could reap only a few vague generalities of little value even for morphological analysis. His doctrine leaves the economist as enlightened as a naturalist told only that the common denominator of all organisms is nutrition, defense, and reproduction.

The import of the conclusion that economics cannot be a theoretical and at the same time a pertinent science may seem purely academic. Unfortunately, this is not so. For the tenacity with which we cling to the tenet that standard theory is valid in all institutional settings — either because its principles are universally valid or because all economic systems are mere mixtures of some invariable elements — has far-reaching consequences for the world's efforts to develop the economy of nations

---

61. Marx, *Critique*, p. 266.
62. *Hegel's Philosophy of Right*, tr. T. M. Knox (Oxford, 1953), pp. 124 ff, 267.
63. For a few brief remarks on this point see section III. 2 of my "Economic Theory and Agrarian Economics" (1960), below.

which differ in their institutions from the capitalist countries. These consequences may go down in history as the greatest monument to the arrogant self-assurance of some of science's servants.

For example, most of us now swear by the axiom — which actually goes back to Marx — that industrial development is the only road to general economic development, that is, to the development of the agricultural sector as well. As factual evidence we invoke the incontrovertible fact that industrialization did result in the over-all development of the South of the United States. But the ingrained outlook of the standard economist prevents us from noting first, that the South is part and parcel of the most advanced capitalist economy, and second, that the American farmer is not institutionally identical (or even comparable) to the Indian or any other *peasant*. In fact, the greater the industrial development achieved by an underdeveloped nation plagued by a predominant, over-populated, and disorganized agricultural sector, the stronger the evidence such a nation offers of the fallacy of the industrialization axiom. There the peasantry is still as poverty-stricken as ever — a passive gloomy onlooker at the increasing well-being of the exclusive circle that delights in the Square Dance of Effective Demand, which alone moves faster and faster with each day. But, for one who believes that distributive relations form the core of the economic process, even this situation has its simple explanation. It is one phase in the evolution of that process.

4. *Arithmomorphic Models and Qualitative Analysis.* Economics is not a theoretical science because, as Kipling said in an often-quoted passage:

There are nine and sixty ways of constructing tribal lays
And-every-single-one-of-them-is-right!

What he forgot to add is that they could not all be right for the same tribe. The laws of the capitalist society, for instance, could not be right for an agrarian overpopulated economy. But in order to prove such a point or to answer any question concerning either economy we must possess a description of each as exhaustive as possible.

One may think then that the first task of economics is to establish some general criteria for classifying all known economic systems into genera, species, and varieties. Unfortunately, our economic knowledge in this direction is so little that even an economic Linnaeus would not be able to design a system of classification. The most we can do at this stage is to observe each economic reality by itself without necessarily looking for taxonomic characteristics. Our aim should be to construct an ideal-type that would make "pragmatically *clear* and *understandable*" the specific features of that particular reality.[64] But without a classi-

64. Max Weber, *The Methodology of the Social Sciences* (Glencoe, Ill., 1949), p. 90.

ficatory code — one may argue — even this lesser task cannot be accomplished. Too many of us hold today that classificatory systems, abstract analytical concepts, and, according to K. Popper, even theories "are prior to observations" [65] as if all these things were found by science ready-made. We seem to forget not only that science emerged from undirected observation but that some pre-scientific thought always precedes the scientific one. [66]

The absence of a classifying code did not prevent the Classical economists — to cite a single example — from discovering the significant features of the capitalist economy. There are some tasks in every science, not only in economics, which require an appreciable dose of "delicacy and sensitiveness of touch." [67]

Once we have arrived at a workable body of descriptive propositions for a given reality, to construct an arithmomorphic model is a relatively simple task. Each economic reality should be provided with such a model as soon as feasible. [68] All the harder to understand is the position that even in the case of a capitalist system "it is premature to theorize." [69] Actually, judging from the immense difficulties encountered by Werner Sombart and other inspired economists we should rather agree with Marshall in saying that economics is not yet ripe for historizing. [70]

Arithmomorphic models, whether in physics or any other science, subserve legitimate needs of Understanding and, in my opinion, of Didactics even more. The scientist who would deny that his mind, at least, does not grasp a diagrammatic representation and, if he had some training, a mathematical model more firmly and faster than a verbal analysis of the same situation, is free to step forward any time, if he so wishes. Besides, of all men of science, economists should not let their slip show by opposing the use of the mathematical tool in economic analysis, for this amounts to running counter to the principle of maximum efficiency. But on the same principle we must deplore the exaggerated fondness for mathematics which causes many to use that tool even when a simple diagram would suffice.

65. Karl R. Popper, *The Poverty of Historicism* (Boston, 1957), p. 98; Jevons, *Theory*, p. 22.
66. Albert Einstein, *Ideas and Opinions* (New York, 1954), p. 276.
67. Marshall, *Principles*, p. 769.
68. For the loss incurred by not doing so, see section I. 4 of my "Economic Theory and Agrarian Economics," reprinted below. No doubt, the analytical tools developed by standard economics could prove themselves handy in many other situations. That is no reason to say with Schumpeter, *Essays*, p. 274*n*, that a model in which factor prices are not proportional to their marginal productivities is "still marginal-productivity theory."
69. Gambs, *Beyond Supply and Demand*, p. 64.
70. *Memorials of Alfred Marshall*, ed. A. C. Pigou (London, 1925), p. 489.

Let me add that the position taken by many of my colleagues that "mathematics *is* language" [71] tends rather to obscure the fact that, whenever the mathematical tool can be used, the analytical process can be accomplished faster than if carried out by ordinary logic alone. No doubt the mathematical armamentarium, if traced back to its genesis, is the product of ordinary logic, just as capital equipment resolves phylogenetically into labor, and living organisms into elementary matter. However, once these forms emerged from their *causa materialis* they displayed novel qualities that have ever since differentiated them from ordinary logic, labor, and inert matter, respectively. To obtain, say, a horse we do not go back and retrace the evolutionary process by which the horse gradually emerged from lifeless substance. Nor do we produce steel hammers by using stone hammers found accidentally in nature. It is more efficient to take advantage of the fact that we can obtain a horse from a horse and capital equipment with the aid of capital equipment. By the same token, it would be utterly absurd to rely on ordinary logic alone whenever a mathematical tool can be used or every time we want to prove a mathematical proposition. If we do teach mathematics from ABC in schools, it is only because in this way we aim to develop the mathematical skill of future generations. It is ghastly to imagine the destruction of all present capital equipment, still ghastlier to think of all men suddenly forgetting all mathematics. But such thought may make us see that qualitatively mathematics is not just language, and though man-made it is not an arbitrary game of signs and rules, like, say, chess.

But the immense satisfaction which Understanding derives from arithmomorphic models should not mislead us into believing that their other roles too are the same in both social and natural sciences. In physics a model is also "a *calculating* device, from which we may compute the answer to *any* question regarding the physical behavior of the corresponding physical *system.*" [72] The same is true for the models of engineering economics. The specific role of a physical model is better described by remarking that such a model represents an *accurate blueprint* of a particular sector of physical reality. But the point, which I made in the paper reprinted below, "Economic Theory and Agrarian Economics," and which I propose to explain now in greater detail, is that an economic model is not an accurate blueprint but an *analytical simile.*

Economists are fond of arguing that since no model, whether in physics or economics, is accurate in an absolute sense we can only choose between a more and a less accurate model. Some point out also that after all how

71. P. A. Samuelson, "Economic Theory and Mathematics," *Papers and Proceedings, American Economic Review,* XLII (1952), 56.
72. P. W. Bridgman, *The Nature of Physical Theory* (Princeton, 1936), p. 93. Italics mine.

accurate we need to be depends on our immediate purpose: at times the less accurate model may be the more rational one to use.[73] All this is perfectly true, but it does not support the further contention — explicitly stated by Pareto — that it is irrelevant to point out the inaccuracy of economic models. Such a position ignores an important detail, namely, that in physics a model must be accurate in relation to the sharpest measuring instrument existing at the time. If it is not, the model is discarded. Hence, there is an *objective* sense in which we can say that a physical model is accurate, and this is the sense in which the word is used in "accurate blueprint." In social sciences, however, there is no such objective standard of accuracy. Consequently, there is no acid test for the validity of an economic model. And it is of no avail to echo Aristotle, who taught that a model is "adequate if it achieves that degree of accuracy which belongs to its subject matter." [74] One may always proclaim that his model has the proper degree of accuracy. Besides, the factors responsible for the absence of an objective standard of accuracy also render the *comparison* of accuracy a thorny problem.

To illustrate now the difference between blueprint and simile, let me observe that one does not need to know electronics in order to assemble a radio apparatus he has purchased in kit form. All he needs to do is follow automatically the accompanying blueprint, which constitutes an *operational* representation by symbols of the corresponding mechanism. The fact that no economic model proper can serve as a guide to *automatic action* for the uninitiated, or even for a consummate economist, necessitates no special demonstration. Everyone is familiar with the dissatisfaction the average board member voices after each conference where some economic consultant has presented his "silly theory." Many graduate students too feel greatly frustrated to discover that, in spite of all they have heard, economics cannot supply them with a manual of banking, planning, taxation, and so forth. An economic model, being only a simile, can be a guide only for the initiated who has acquired an analytical insight through some laborious training. Economic excellence cannot dispense with "delicacy and sensitivity of touch" — call it art, if you wish. And it is only too bad if at times the economist lets himself be surpassed in this respect by the layman. The widespread view that the economist's role is to analyze alternative policies, whereas their adoption is the art of statesmanship,[75] is no excuse. An artless analysis cannot subserve an art.

Jevons' hope that economics will ultimately become an exact science

73. Pareto, *Manuel*, pp. 11, 23, and *passim;* also Milton Friedman, *Essays in Positive Economics* (Chicago, 1953), pp. 3–43.

74. Aristotle, *Ethica Nicomachea*, 1094[b] 12–14.

75. Cf. Homan (note 35, above), p. 15.

has filled the hearts of many great economists. Irving Fisher still nourished it at eighty.[76] And since by exact or genuine science they all understood a science of calculating devices — a definition that goes back to the Enlightenment era[77] — they all endeavored to point out the quantitative nature of the economic domain. Schumpeter even argued that economics is "the most quantitative . . . of *all* sciences" because its observables are "made numerical by life itself" [78] — an argument far more impressive than Jevons'. Some, also like Jevons, went further and argued that even pleasure can be submitted to accurate calculation.[79]

Others, however, gradually came to weakening the classical definition of exact science by distinguishing between *quantitative* and *numerical* devices.[80] An economic model is still exact even if it does not serve as a calculating device, provided that it constitutes a paper-and-pencil representation of reality.

To recall, Pareto argued with his characteristic aggressiveness that Walras had already transformed economics into an exact science. But while firmly holding that we can determine the value of any parameter we choose, he explicitly stated that, in opposition to Walras, he did not believe in the possibility of effectively solving a concrete Walrasian system.[81] Pareto, like Cournot before him, saw in the immensity of equations the only obstacle to economics' being a numerical science, like astronomy.[82]

Many still share the idea that the Walrasian system would be an accurate calculating device for a Laplacian demon. But let us imagine a new demon, which with the speed of thought can make all the necessary observations for determining all ophelimity and production functions, solve the system, and communicate the solution to everyone concerned. Pareto's position is that everyone will be perfectly happy with the solution and that the economy will remain in equilibrium, if not forever, at least until disturbed by new forces from the *outside*.

This logic ignores a most crucial phenomenon: the very fact that an

76. Ragnar Frisch, "Irving Fisher at Eighty," *Econometrica*, XV (1947), 74.
77. Cf. *The Logic of Hegel*, tr. W. Wallace (2nd edn., London, 1904), p. 186.
78. Schumpeter, *Essays*, pp. 100–101.
79. Surprising though it may seem, this very idea is found in Plato: "If you had no power of calculation you would not be able to calculate on future pleasure, and your life would be the life, not of a man, but of an oyster or *pulmo marinus*." *Philebus*, 21.
80. Robbins, *An Essay* (note 19, above), p. 66; Joseph A. Schumpeter, *History of Economic Analysis* (New York, 1954), p. 950.
81. V. Pareto, "Teoria matematica dei scambi foresteri," *Giornale degli economisti*, VI (1894), 162. I need to add that this source shows that G. Demaria is wrong in saying that Pareto thought that his system would enable economists to make the same kind of predictions as astronomers. See V. Pareto, *Scritti teorici*, ed. G. Demaria (Milan, 1952), p. xix.
82. A. Cournot, *Researches into the Mathematical Theory of Wealth* (New York, 1897), p. 127.

individual who comes to experience a new economic situation may alter his preferences. *Ex post* he may discover that the answer he gave to our demon was not right. The equilibrium computed by our demon is thus immediately defeated not by the intervention of exogenous factors but by endogenous causes. Consequently, our demon will have to keep on recomputing running-away equilibria, unless by chance he possesses a divine mind capable of writing the whole history of the world before it actually happens. But then it would no longer be a "scientific" demon. Pareto, first among many, would have nothing to do with clairvoyance.

There is, however, at least one additional difficulty into which our demon would certainly run with the Walrasian system. It is the Oedipus effect, which boils down to this: the announcement of an action to be taken changes the evidence upon which each individual bases his expectations and, hence, causes him to revise his previous plans. Preferences too may be subject to an Oedipus effect. One may prefer a Rolls-Royce to a Cadillac but perhaps not if he is told that his neighbor will buy a Rolls-Royce.

Edgeworth once said that "to treat *variables* as *constants* is the characteristic vice of the unmathematical economist." [83] But an economist who sticks only to mathematical models is burdened with an even greater vice, that of ignoring altogether the qualitative factors that make for endogenous variability. Bridgman was thus right in reproaching the social scientist for failing to pick up the significant factors in describing social reality.[84]

Time and again, we can see the drawback of importing a gospel from physics into economics and interpreting it in a more catholic way than the consistory of physicists. It is all right for physics to trust only what is amenable to sense-perception, i.e., only observables, because that is the sole contact we have with the outside world. It is equally understandable for physics to treat as fiction and view with mistrust the unobservables it had to invent in order to unify into one picture disparate observables and thus simplify its logical foundation. But there is absolutely no reason for economics to treat as fiction the very springs of economic action — wants, beliefs, expectations, institutional attitudes, etc. For these elements are known to us by immediate acquaintance, that is, more intimately than any of the economic "observables" — prices, sales, production, and so forth.

No doubt, many mathematical economists must have been aware of the fact that in an arithmomorphic model there is no room for human propensities. Jevons started them searching for a cardinal measure of utility. More recently, others tried to establish such a measure for un-

83. F. Y. Edgeworth, *Mathematical Physics* (London, 1932), p. 127n.
84. Bridgman, *Reflections*, pp. 447–448.

certainty. All these painstaking endeavors should be viewed with pride because science should leave no stone unturned. However, through these very endeavors we gradually came to realize that measurability, whether ordinal or cardinal, requires very stringent conditions. Some of these conditions were brought to light for the first time in my 1936 article reprinted below, "The Pure Theory of Consumer's Behavior." By pursuing this line of thought in several other papers, included in Part II of this volume, I was able to show — convincingly, I hope — that neither wants nor expectations fulfill the conditions of measurability. The apparent solidity of all demonstrations of how to establish a measure for wants or expectations derives from "the ordinalist fallacy" — as I proposed to call the idea that a structure where we find "more" and "less" is necessarily a *linear* continuum.

But our thirst for measure is so great that some have tried to dispose of all evidence and logical arguments against the measurability of human propensities by arguing that if mental attitudes are "inaccessible to science and measurement, the game is lost before the first move is made." [85] Clearly, the game to which the statement applies cannot be other than the game of "science is measurement." But why should this be the only game a scientist can play? It is precisely because of this question that I have tried to present here all the evidence I could muster — however technical or tedious this evidence may seem at first — in order to prove that no science can completely avoid dialectical concepts. The reason, as I have explained, is that no science can ignore Change forever. The idea that human propensities, which are the main vehicle of economic Change, are not arithmomorphic concepts, therefore, is not a fancy of some unscientific school of thought.

The obvious conclusion is that if economics is to be a science not only of "observable" quantities but also of man, it must rely extensively on dialectical reasoning.[86] Perhaps this is what Marshall meant by "delicacy and sensitiveness of touch." But in the same breath he added that the economic science "should not be invertebrate . . . [but] have a firm backbone of careful reasoning and analysis." [87] It is highly significant that Marshall did not say "exact reasoning." For dialectical reasoning cannot be exact. But as I argued earlier (in Chapter Two, "Concepts, Numbers, and Quality," section 6), dialectical reasoning can be correct and ought to be so. There are two known methods for testing the correctness of dialectical reasoning: Socratic analysis and analytical simile. Surprisingly enough, we owe them to Plato, who used them freely

85. S. S. Stevens, "Measurement and Man," *Science*, CXXVII (1958), 386.
86. Let me remind the reader that my meaning of dialectical reasoning differs from that of Hegel and, hence, of Marx. Cf. Chapter Two, above ("Concepts, Numbers, and Quality"), note 24; also below, note 92.
87. Marshall, *Principles*, p. 769.

throughout the *Dialogues*.[88] Two thousand years later, in 1690, William Petty surprised political scientists by proposing to apply one of Plato's methods to economic reasoning: "The method I take to do this, is not yet very usual; for instead of using only comparative and superlative Words, and intellectual Arguments, I have taken the course . . . to express myself in Terms of *Number, Weight,* or *Measure,* [which] at worst are sufficient as Suppositions to shew the way to that Knowledge I aim at." [89]

Perhaps the most obvious merit of an arithmomorphic model is the one that is acknowledged by almost every criticism of mathematical economics: the merit of bringing to light important errors in the works of literary economists who reasoned dialectically. In this respect, the role of a mathematical model in economics as well as in many other sciences is analogous to that of the rule of casting out nines in arithmetic. Both are expedient ways of detecting errors in some mental operations. Both work negatively: if they reveal no error, it does not mean that the dialectical argument or the arithmetical calculation is wholly correct. Important though this last point is, only F. H. Knight, it seems, saw that economic theory shows "what is 'wrong' rather than what is 'right.' " [90]

The second role of an arithmomorphic model is that of illustrating certain points of a dialectical argument in order to make them more understandable. One may use, for instance, an ophelimity function containing a special parameter in order to discuss didactically the problem of change in tastes or a probability distribution to illustrate the situation of an individual confronted with uncertainty.[91] Or, like Walras or Leontief, we may construct a system of indefinite dimensions in order to illustrate some important aspects of a whole economy.[92]

These two roles of the mathematical model circumscribe the *raison d'être* of what currently passes for "economic theory," which is to supply our dialectical reasoning with a "firm backbone." An analytical simile,

88. "The higher ideas . . . can hardly be set forth except through the medium of examples" (*Statesman*, 277), suffices as an illustrative quotation.
89. *The Economic Writings of Sir William Petty*, ed. C. H. Hull (Cambridge, Eng., 1899), I, 244–245.
90. Knight, *On the History* (note 44, above), p. 177.
91. I used precisely this Platonic method in analyzing the hysteresis and novelty effects in consumer's choice. Cf. "The Theory of Choice and the Constancy of Economic Laws" (1950), reprinted below. The conclusion at which I arrived — symmetrical to Marshall's observation concerning long-run supply schedules — is that demand curves too are irreversible. The same analytical simile also enabled me to pinpoint the delusion that experiments with an individual leave him as he was at the outset.
92. Let me add that an analytical simile would not work in case the epistemological approach to the economic process follows exactly the Hegelian Dialectics, as was Marx's case. See my essay below on "Mathematical Proofs of the Breakdown of Capitalism" (1960), and see especially my "Some Thoughts on Growth Models: A Reply," *Econometrica*, XXXI (1963), 230 ff.

therefore, must be formulated with the utmost rigor without any regard for its factual applications. That is why there is no room in "pure theory" even for pseudo-arithmomorphic concepts, such as price index, cost of living, aggregate production, and the like. They have been denounced by almost every theoretical authority[93] — and rightly as far as pure theory is concerned.

In spite of all the denunciations these pseudo-arithmomorphic concepts fared increasingly well. Macro-economics by now has smothered micro-economics almost completely. The phenomenon, far from being perplexing, has a very simple reason. Coordinates such as standard of living, national real income, aggregate production, etc., are far more significant for the analysis of the economic process than Mr. X's tastes or the pricing rule of entrepreneur Y. Like all other vital coordinates of the same process, they are dialectical notions. They differ from others only because if abstractly reduced to an individual and to an instant they can be represented by a number. From this number we can then construct a pseudo measure, which is always some sort of average. The fact that we can never tell which formula we should choose in order to compute this average, nor why a greater or a smaller number than the one obtained by some formula would also do, shows that a pseudo measure is in essence a dialectical concept.

As is often the case, the same reason why pseudo measures are poison to "theory" accounts for their success in the description and analysis of concrete facts. In proper use, an index or an aggregate is not a fine bullet, but a piece of putty which covers a dialectical target, such as "the standard of living" or "the national product," better than a bullet. That is why an increasing number of economists share the view that macro-analysis, though it is only vaguely clear, is far more productive than the traditional micro-economics with its Ockham's razor. But, perhaps, the real reason is that they have come to realize that the more significant variables pertain to society, not to the individual.

The preceding observations should not be interpreted as a motion to place the mathematical macro-model on a high pedestal in the gallery of blueprints. Actually, as a blueprint a macro-model is vulnerable from more sides than a micro-model.

To begin with, a macro-model, in contrast with that of Walras-Pareto, is admittedly incomplete because, we are told, the significant macro-coordinates are too numerous for our power of calculation. The excuse is

93. E.g., N. G. Pierson, "Further Considerations on Index-Numbers," *Economic Journal*, VI (1896), 127 ff; Lionel Robbins, *An Essay*, p. 66; W. W. Leontief, "Implicit Theorizing: A Methodological Criticism of the Neo-Cambridge School," *Quarterly Journal of Economics*, LI (1937), 350.

familiar. The truth, however, is that their number exceeds our analytical power and, hence, we are unable even to say which are the significant coordinates. To recall the earlier discussion of objective accuracy, we understand why it is not very clarifying to explain *ex post* that a model is not a blueprint because some significant variables were left out. Yet that is what we are compelled to explain most of the time.

Secondly, macro-economic models generally consist of a system of equations which has a quite special structure: they involve only analytical functions. Now, the peculiar property of an analytical function, $f(x)$, is that its value for *any* x is completely determined by the values $f(x)$ has in *any interval, however small*. The reason why we use only such functions is obvious. Without analytical functions we would not be able to extrapolate the model beyond the range of past observations.[94] But why should economic laws, or any other laws for that matter, be expressed by analytical functions? Undoubtedly, we are inclined to attribute to reality a far greater degree of orderliness than the facts justify. That is particularly true for the linear macro-models — save perhaps the case of models such as Leontief's which deal only with material flows. Yet even linear macro-models are usually hailed for having run successfully the most terrific gantlet of statistical analysis. But we often forget to ask whether the gantlet was not a mere farce. The validity of statistical tests, even the nonparametric ones, requires conditions which a rapidly changing structure such as the economic process may fulfill only by sheer accident. Besides, if one formula does not pass the test, we can always add another variable, deflate by another, and so on. By clearly choosing one's chisels, one can prove that inside any log there is a beautiful Madonna.

Thirdly, the very idea of a mathematical (read arithmomorphic) relation between pseudo measures, like those used in economics, is a manifest contradiction in terms. For, in contrast with the conditions prevailing in other domains, in economics there is no basis for the average income, for instance, to be represented by the same average formula at all times or in all places. Though a statement such as "average real income increases with the proportion of industrial production in the gross national product" is not sharply clear, it raises far fewer questions than if it were replaced by some mathematical formula. It is also a less deceptive guide for action. That is why many economists interested in the problems of economic development have shifted from mathematical macro-models to a less exact but more valuable analysis of the sort professed, especially, by S. Kuznets. Such analysis may not seem sophisticated enough. But sophistication is not an end in itself. For, as more than one physicist or

94. Let me add a thought that seems important: without analytical functions we would be unable also to argue that a law changes with the scale.

economist has observed, "if you cannot — in the long run — tell everyone what you have been doing, your doing has been worthless." [95]

From whatever angle we may look at arithmomorphic models, we see that their role is "to facilitate the argument, clarify the results, and so guard against possible faults of reasoning — *that is all.*" [96] This role is not only useful, as everyone admits, but also indispensable — a point some tend or want to ignore. Unfortunately, we are apt, it seems, to be fascinated by the merits of arithmomorphic models to the point of thinking only of the scalpel and forgetting the patient. That is why we should keep reminding ourselves that an arithmomorphic model has no value unless there is a dialectical reasoning to be tested. To return to an earlier analogy, the rule of casting out nines is of no use if we have no arithmetic calculation to check. If we forget this point we run the great risk of becoming not "mathematicians first and economists afterwards" — as Knight once said [97] — but formula spinners and nothing else.

5. *Economics and Man.* Arithmomorphic models, to repeat, are indispensable in economics, no less than in other scientific domains. That does not mean also that they can do all there is to be done in economics. For, as Schrödinger argued in the case of biological life, the difficulty of the subject of economics does not lie in the mathematics it needs, but in the fact that the subject itself is "much too involved to be fully accessible to mathematics." [98] And what makes this subject not fully amenable to mathematics is the role that cultural propensities play in the economic process. Indeed, if man's economic actions were independent of his cultural propensities, there would be no way to account for the immense variability of the economic pattern with time and locality.

The well-known conflict between standard economics and all other schools of economic thought is a striking illustration in point. The conflict stemmed from the cultural differences between the economic process known to one school and that known to another. Nothing is more natural than the inability of the standard economists to understand their German colleagues who insisted on bringing such "obscurantist" ideas as *Geist* or *Weltanschauung* into the economic science. On the other hand, it was equally normal for the German school to reject an idea which reduces the economic process to a mechanical analogue.

The much better faring of standard economics notwithstanding, it is the position of the historical school that is fundamentally the correct one.

95. E. Schrödinger, *Science and Humanism* (Cambridge, Eng., 1951), pp. 8–9. The same opinion is held by Werner Heisenberg, *Physics and Philosophy: The Revolution in Modern Science* (New York, 1958), p. 168; J. K. Galbraith, *Economics and the Art of Controversy* (New Brunswick, N.J., 1955), p. 43.

96. Knut Wicksell, *Value, Capital and Rent* (London, 1954), p. 53. Italics mine.

97. Knight, *Ethics of Competition*, p. 49.

98. E. Schrödinger, *What Is Life?* (Cambridge, Eng., 1944), p. 1.

The point seems to be winning the consent, however tacit, of an increasing number of economists. And perhaps it is not too involved after all.

That in all societies man's economic actions consist of choosing is beyond question. It is equally indisputable that the outcome of the economic choice is expressible as a vector $X(x_1, x_2, \ldots, x_n)$, the coordinates of which are quantities of some commodities. Now, some economic choices are *free choices*, that is, the individual is as free to choose one of the alternatives as if he had to choose a card out of a deck or a point on a line. But the most important choices usually are not free in this sense. They imply a certain action by the agent. It follows then that in its general form the economic choice is not between two commodity vectors, Y and Z, but between two complexes (Y, B) and (Z, C), where B and C stand for the actions by which Y or Z is attainable. Ordinarily, there exist several actions, $B_1, B_2, \ldots, B_k$ by which, say, Y may be attained. One may beg for a dollar, or pinch the cash register, or ask his employer to give him one for keeps. What on the average one will do depends on the cultural matrix of the society to which he belongs. The point is that whether the outcome of choice is Y or Z depends also upon the *value* the actions B and C have according to the cultural matrix of the economic agent. To leave an employer with whom one has been for some long years only because another employer pays better, certainly is not an action compatible with every cultural tradition. The same can be said about the action of an employer who lets his workers go as soon as business becomes slack.

Cultures differ also in another important respect. In some societies, most actions have either a great positive or a great negative value according to the prevailing cultural matrix. These values then count heavily in the choice of the individual. At the other extreme, there is the Civil Society, where, except for actions specifically barred by the *written* laws, the choice is determined only by the commodity vectors Y and Z. We can now see clearly why standard economics has fared so well in spite of its *homo oeconomicus*. For this *homo oeconomicus* chooses freely, that is, according to a choice-function involving only the commodity vector.

It is customary to refer to the societies where choice is determined also by the action factor as "traditional societies." But the term is, obviously, a pleonasm: every society has its own tradition. That of the Civil Society is that only the written law, sometimes only the opinion of the court, tells one whether an action is allowed or forbidden.

The opinion that the choice-function of the *homo oeconomicus*, that is, the utility index, adequately represents the economic behavior in any society is still going strong. I can foresee the argument that after all one can include the actions in the commodity vector by distinguishing, say, between x_k obtainable through action B and the same x_k obtainable through action C. That this suggestion only covers a difficulty by a paper-and-pencil artifact needs no elaboration. More familiar, however, is the position epitomized by Schumpeter's argument that "the peasant sells his calf just as cunningly and egotistically as the stock exchange member his portfolio of shares." [99] The intended implication is that the standard utility function suffices to describe economic behavior even in a peasant community. But Schumpeter, obviously, referred to a peasant selling his calf in an urban market to buyers whom he scarcely knows. In his own community, however, a peasant can hardly behave as a stock exchange broker. As an increasing number of students of peasant societies tell us, for the peasant it does matter whether he can buy cheap only because a widow, for example, must sell under the pressure of necessity. The stock broker does not care why the seller sold cheap: he has no means of knowing from whom he buys.

In recent years, a great number of economists have been engaged in the study of the peasant economies in various underdeveloped countries. Their attachment to the utility and the profit functions as "rational choice-functions," has led many to proclaim that the peasant — or in general, any member of a "traditional" society — behaves *irrationally*. In fact, a substantial amount of work has been done on how to make the peasant behave *rationally*. But most of these writers do not seem to realize that what they propose to do is to make the peasant communities choose as the Civil Society does, according to a utility and a profit function. Whether such a pattern of economic behavior is the rational one, is actually a pseudo problem. To be sure, one could very well argue that at this stage all peasant societies are already under the unavoidable impact of urban institutions, and that ultimately new rural societies will be born. But the voluntary midwives do not always appear to realize the most important difficulties of the task: the inertia any tradition has because of the superb *internal* logic of many of its articulations. Were it not for this internal logic mankind would not have come to build lasting traditions.

Traditions often seem to be an obstacle to progress, and perhaps they actually are. But their role in the evolution of mankind is not superfluous. The economic process, to recall, does not go on by itself. Like any non-automatic process, it consists of sorting. Sorting, in turn, requires an agent of the kind illustrated by Maxwell's fable. Moreover, it is the sorting agent that constitutes the most important factor in any such process. For, as explained earlier, low entropy will turn into high entropy in any case. But it depends upon the type of sorting activity whether a greater or a smaller amount of environmental low entropy is absorbed

99. Joseph A. Schumpeter, *The Theory of Economic Development* (Cambridge, Mass., 1949), p. 80.

into or retained by the process. In other words, it depends upon what sort of Maxwellian demon keeps the process going. It suffices to compare two different varieties of the same species living within the same environment in order to convince ourselves that not all Maxwellian demons are identical. Not even two specimens of the same race are always identical Maxwell demons.

In the case of a single cell, the corresponding Maxwellian activity seems to be determined only by the physico-chemical structure inherited by the cell; in the case of a higher organism, it is a function of its innate instincts as well. An eagle can fly because it is born both with wings and with the instinct to fly. But man too can fly nowadays even though he has neither a biological constitution fit for flying nor an innate instinct to do so. The upshot is obvious: the Maxwellian activity of man depends also on what goes on in his mind, perhaps more on this than on anything else. And it is the role of tradition to transmit knowledge as well as propensities from one generation to another. In other words, tradition is a substitute for heredity; actually, it has all the fundamental attributes of heredity.

The intense interest in the problem of the economic development of the "underdeveloped" countries has brought an increasing number of scholars and students in direct contact with numerous "traditional societies." Numerous are those who have thus come to realize the importance that cultural propensities have in the economic process and also for the strategy of inducing economic development. Unfortunately, however, most policies of economic development still rest on the old fallacy bred by the mechanistic philosophy, the fallacy that *it is the machines that develop man, not man that develops machines*. Highly surprising though it may seem, the most frank and pinpointed recognition of the fallacy has come from a Soviet author — which means from the Communist "brain trust" itself: "It is not the machine created by man, but man himself who is the highest manifestation of culture, for the thoughts and dreams, the loves and aspirations of man, *the creator*, are both complex and great." [100]

Anthropologists and historians have long since thought that the introduction of any economic innovation in a community is successful only if the community can adapt itself culturally to it, i.e., only if the innovation becomes socially approved and understood.[101] Among the Anglo-American economists at one time only a rebel such as Veblen argued that it is dangerous to place modern machines in the hands of people still

100. S. T. Konenkov, "Communism and Culture," *Kommunist*, no. 7, 1959. English translation in *Soviet Highlights*, no. 3, I (1959), 3–5. Italics mine.

101. G. Sorel in the "Introduction" to G. Gatti, *Le socialisme et l'agriculture* (Paris, 1902), p. 8; Richard Thurnwald, *Economics in Primitive Communities* (London, 1932), p. 34; V. Gordon Childe, *Social Evolution* (New York, 1951), p. 34.

having a feudal economic *Anschauung*.[102] No doubt, "dangerous" is hardly the proper term here, but probably Veblen wanted to emphasize the immense economic loss as well as the great social evils resulting from a forced introduction of modern industries into a community deprived of the corresponding propensities.[103]

The point has obvious implications for any policy aimed at speeding up the growth rate of an economy. They have been recognized sporadically and mainly by "unorthodox" economists. Leonard Doob, for instance, insisted that no planning can succeed unless it is based on a knowledge of the social environment, that is, of the people who will be affected by it. An even stronger thesis is put forward by J. J. Spengler, who argues that the rate of economic growth depends upon the degree of compatibility between the economic and noneconomic components of the respective culture.[104] These observations should not be dismissed easily, for all analyses of why the results of our economic foreign aid often have not been proportional to its substance converge on one explanation: local mores.

Actually, there are a few facts which suggest that the influence of the economic *Anschauung* upon the economic process is far more profound than the authors quoted above suspected. I shall mention only the most convincing ones. Soviet Russia, at a time when she had hardly introduced any innovation besides central planning, felt the need to act upon the economic *Anschauung* of the masses: "The purpose of politically educative work [in the forced-labor camps] is to eradicate from convicted workers the old habits and traditions born of the conditions prevailing in the pattern of life of former times." [105] Strong though the pressure exercised through numerous similar educative works has been on the people of the USSR, the result was such that, at the Twenty-First Congress of the CPSU, Nikita Khrushchev still had to announce: "To reach communism . . . we must rear the man of the future right now." [106]

A far more familiar case in point is the great economic miracle of Japan. There is no doubt in my mind that only the peculiar economic *Anschauung*

102. Gambs, *Beyond Supply and Demand*, p. 25.

103. Or as P. N. Rodenstein-Rodan was to put it in "Problems of Industrialization of Eastern and South-Eastern Europe," *Economic Journal*, LIII (1943), 204, "An institutional framework different from the present one is clearly necessary for the successful carrying out of industrialization in internationally depressed areas."

104. Leonard Doob, *The Plans of Men* (New Haven, 1940), pp. 6–7; J. J. Spengler, "Theories of Socio-Economic Growth," in *Problems in the Study of Economic Growth*, National Bureau of Economic Research (New York, 1949), p. 93. See also K. Mannheim, "Present Trends in the Building of Society," in *Human Affairs*, ed. R. B. Cattell *et al.* (London, 1937), pp. 278–300.

105. Resolution of the 1931 All-Russian Congress of Workers of the Judiciary in *Report of the Ad Hoc Committee on Forced Labor*, United Nations, ILO, Geneva, 1953, pp. 475–476.

106. Quoted in Konenkov, "Communism and Culture."

of the average Japanese can explain that miracle. For, I am sure, no expert on planning could draw up an economic plan for bringing an economy from the conditions of Japan in 1880 to those existing today. And if he could, he must have known beforehand that the people were the Japanese and also realized that the complete data in any economic problem must include the cultural propensities as well.

Nothing is further from my thought than to deny the difficulties of how to study the economic *Anschauung* of a society in which one has not been culturally reared. Nor am I prepared to write down a set of instructions on how to go about it mechanically. But if we deny man's faculty of empathy, then there really is no game we can play at all, whether in philosophy, literature, science, or family. Actually, we must come to recognize that the game is not the same in physical sciences as in sciences of man; that, contrary to what Pareto and numberless others preached, there is not only one method by which to know the truth.[107]

In physics we can trust only the pointer-reading instrument because we are not inside matter. And yet there must be a man at the other end of the instrument to read it, to compare readings and to analyze them. The idea that man cannot be trusted as an instrument in the process of knowing is, therefore, all the more incomprehensible. Man may not be as accurate an instrument as a microscope, but he is the only one who can observe what all the physical instruments together cannot. For if it were not so, we should send some politoscopes to reveal what other people think, feel, and might do next — not ambassadors, counselors, journalists, and other kinds of observers; and as we have yet no politoscopes, we should then send nothing.

But perhaps one day we will all come to realize that man too is an instrument, the only one to study man's propensities. That day there will be no more forgotten men, forgotten because today we allegedly do not know how to study them and report on what they think, feel, and want.

107. Pareto, *Manuel*, p. 27.

PART II

Choice: Utility and Expectation

1

The Pure Theory
of Consumer's Behavior

I. INTRODUCTION

All essential differences between static and dynamic economics center upon the fundamentally distinct ways in which their mathematical treatments are elaborated. Most, if not all, of the functional relations used in the set-up of static problems are likely, because of the rationale of the problem itself, to be resolved into some simple type of general function in the sense given to this concept by Dirichlet.[1] The use of such functions in dynamic problems has not yet been justified by any analysis. The functions used in many attempts at a mathematical treatment of dynamic economics, besides involving the time element introduced in a way that seems to imply a certain causal relation between *facts* and *time*, have been limited also in their structural properties for convenience in carrying out the necessary transformations.

For the sake of this rationale of the general set-up, the introduction of the concept of marginal utility was regarded as an advance over Cournot's approach to the static problem which uses only the tools of demand and supply curves. The demand and supply laws appear today to be derived concepts, and their justification is sought in terms of the ultimate considerations that find their place within the frame of economic science, i.e., the reasons that induce individuals to produce and exchange goods.

For the same reason, the introduction of indifference varieties or of the marginal rate of substitution was regarded as new refinements of the set-up. And thus the theory of static equilibrium, in so far as it is concerned with the position of consumers, which is beyond doubt the most delicate side of the problem, has reached a degree of rigorousness and

NOTE: This paper is reprinted from *Quarterly Journal of Economics*, L (1936), 545–593.

1. In this sense, u is said to be a function of $x, y, z, \ldots$ if to any set of values of the latter there corresponds unequivocally one value of u. Another type of function may be defined so as to make the value of u depend not only upon the values $x, y, z, \ldots$ but also upon the path by which any particular set of these values is reached.

exactness that was considered, not very long ago, as the appanage of the natural sciences only. One need but read the admirable two-part paper of R. G. D. Allen and J. R. Hicks[2] in order to be convinced of the vast possibilities that are open to the mathematical treatment of static economics.

The method of economics remains — and it seems that it will remain despite many attempts in the opposite direction — that of the mental experiment aided by introspection. There are well-known attacks directed against this procedure for supporting scientific laws. Nevertheless, we may defend our position by arguing that, so far as we deal with the consumer's position, introspection is justified by the problem itself.

At the same time we may seek a safer line of approach. This might be reached, for instance, by formulating our mental experiment in such a way as to suggest, and direct step by step, the pattern of an actual experiment which may be carried out in the future, subject to technical possibilities in the matter.

Such a scheme will implicitly require, if we want to proceed in a scientific way, that all the results of the mental experiment that are not "evident" a priori be stated as additional assumptions. We can arrive in this way at the formulation of a necessary and sufficient set of assumptions for handling the problem, and thus obtain a kind of measure of the extent to which our mental experiment may diverge from a similar actual investigation. For maintaining further this parallelism between the mental and actual experiment, the formulation of our postulates in such a way as to outline in a straightforward fashion the corresponding physical investigation is undoubtedly the most advisable procedure.[3]

In so far as our analysis is of the type just mentioned, care must be taken not to postulate the very thing we want to explain. A phenomenon may be regarded as justified merely by its existence, if this can be experimentally established. If this is not possible, we have to seek for another kind of explanation. It is true that the meaning of explanation is in this case formal only. Nevertheless, the advantage of an explanation, even formal, of the exchange equilibrium in terms of concepts outside economics cannot be denied.

Pareto seems to have realized the advantage of such a position, for we find in his latest writings an attempt to describe how the existence of

2. R. G. D. Allen and J. R. Hicks, "A Reconsideration of the Theory of Value," *Economica*, I (1934), 52–76, 196–219.

3. It is far from my belief that all the points connected with the nature of indifference elements can be elucidated by an actual experiment. For the subtlety of some of these — perhaps of the most important ones — is beyond the usual degree of accuracy of our measurements. But, as is clearly seen from what precedes, my advocacy of a parallelism between the two kinds of experiments has entirely a different aim.

indifference directions might possibly be tested by an actual experiment.[4] But he failed to state explicitly the essential assumption thus introduced, and because of this the whole problem appeared to have a unique issue.[5]

The failure to state explicitly one's assumptions in connection with the individual's behavior accounts also for the apparent paradox to which the nonintegrability case led.

It is the purpose of the present paper to undertake the task of such an analysis as that outlined above. I shall take this opportunity for correcting some errors that have slipped into the most recent papers on the subject and, finally, for developing an alternative theory of the nature of indifference curves.

II. THE MAIN POSTULATES OF THE THEORY OF CHOICE:
CLASSICAL SCHEME; STABILITY CONDITIONS

I shall not go back as far as Jevons' concept of total and marginal utility, for this way of presenting a theory of exchange has been proved to constitute a minor view in comparison with the latest developments, which, however, it must not be forgotten, had Jevons' construction as a starting point. Consequently, I shall confine myself to the consideration of the most recent presentations, as they are found in the writings of Edgeworth, Irving Fisher, and Pareto, namely:

1. The theory of indifference varieties.
2. The theory of choice.

Since the second way of approaching the exchange problem has lately been regarded as an improvement over the first, I shall begin by considering the theory of choice. The theory of indifference varieties will not be taken into account until the point of view just mentioned has been analyzed.

I shall assume perfect continuity of the entire field under consideration and make also the further assumption that the mathematical functions which are introduced have derivatives up to the order desired. The first assumption is only formal, for the discontinuous fields may, by a suitable interpolation, be converted into continuous ones. The second assumption is more delicate; it may, however, be looked upon as the mathematical interpretation of the "regularity" of human behavior.

4. Vilfredo Pareto, *Manuel d'économie politique* (edn. of Paris, 1927), p. 542; also his "Economie mathématique," *Encyclopédie des Sciences Mathématiques*, tome I, vol. 4, fasc. 4, pp. 596–598.

5. See Sections II and VI of this essay.

Let S be an ordinal and continuous set of combinations, i.e., a set such that any combination (C_r) belonging to the set may be completely characterized by its rank r, and vice versa. Let us assume further that the combination (C_α) is always preferred to (C_β) if $\alpha > \beta$. When this last condition is fulfilled, the set (S) will be called preferential. If the individual is already in possession of a third combination (T) which is preferred to (C_β) if $\beta < r$, and (C_α) is preferred to (T) if $\alpha > R > r$, we can distinguish in the set (S) at least two classes of combinations: those that are preferred to (T) and those to which (T) is preferred. Evidently these two classes cannot overlap each other. There must consequently be a certain boundary separating the two classes. Two cases may occur. For the time being I shall suppose that:

A. *There is a unique combination* (C_t) *that separates the nonpreferred combinations from the preferred ones and is indifferent in respect to* (T).

The essential implication of this postulate is that the mental comparison within a preferential set is as accurate as any other objective physical measurement can theoretically be. A perfect similarity with regard to the possibility of discerning differences in a monotonic series is thus assumed between a mental and a physical experiment. Because of this fact, the postulate A may justifiably be very much questioned. Later on (Section VI), I shall consider the second alternative, where there is no (C_t).

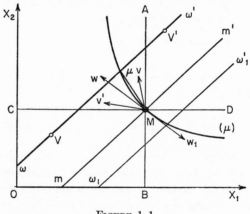

FIGURE 1–1

For the geometrical representation, I shall choose, as support of the developments to follow, the case where the individual is faced with only two commodities, X_1 and X_2. Any differences that may arise in the argument from the consideration of more than two commodities will be mentioned as they come up.

Let us assume that the individual's income per unit of time is (x_1, x_2), where x_1, x_2 are physical amounts of the two goods X_1, X_2, and let us represent his initial position by the point $M(x_1, x_2)$ in the plane X_1OX_2 (Fig. 1–1).

From the beginning we have to distinguish two cases according to whether there is or is not a saturation region. Let us analyze first the problem under the assumption:

B. *There is no saturation point.*

This means that the individual will prefer to M any other position within the right angle AMD, and that he will take the trouble to move from M to $M'(x_1 + \Delta x_1, x_2 + \Delta x_2)$, if he can do so without further conditions, no matter how small the positive increments Δx_1 and Δx_2 are. On the contrary, M will be preferred to any combination within the angle CMB.

Let us consider all combinations lying on a positively inclined straight line $\omega\omega'$. These combinations form a preferential set. As there are positions such as V', that are preferred to M, and others, such as V, that are nonpreferred to M, there will be a unique point μ on the line $\omega\omega'$ which will represent an indifferent combination in respect to M. As the line $\omega\omega'$ moves parallel to itself towards mm', the direction $\overrightarrow{M\mu}$ will tend towards a definite position $\overrightarrow{Mw}$. This position will be the tangent at M to the locus of μ.

By repeating the same argument for the line $\omega_1\omega_1'$, we shall obtain another limiting direction $\overrightarrow{Mw_1}$.

Any direction within the angle w_1Mw, such as $\overrightarrow{Mv}$, will be a preference direction; in other words the individual will move away from M on any such direction if he has the opportunity of doing so. The directions within the angle wMw_1, such as $\overrightarrow{Mv'}$, will be nonpreference directions.

We shall further assume that:

C. *The limiting directions $\overrightarrow{Mw}$, $\overrightarrow{Mw_1}$ are directly opposite.*

In this case the element defined at the point M by the slope of ww_1 is the indifference element. The last assumption expresses the fact that the individual will exchange either X_1 for X_2 or X_2 for X_1, at any given rate of exchange, with the exception of that rate which equals the slope of the corresponding indifference element. This may be looked upon as an interpretation of the fact that the converse of what we prefer is always nonpreferred.

Nothing has yet been said about how the indifference direction varies with the change of the slope (ω) of the line $\omega\omega'$.[6] Clearly, two alternatives are possible. However, in order to follow the classical point of view, I shall introduce the assumption:

D. *The indifference direction at any point is uniquely determined.*

Subject to assumptions A, B, C, D, the individual's tastes can be described by the differential equation

(1) $$\varphi_1(x_1, x_2)dx_1 + \varphi_2(x_1, x_2)dx_2 = 0,$$

where φ_1, φ_2 are so far subject only to the conditions

(2) $$\varphi_1 \geq 0, \qquad \varphi_2 \geq 0,$$

the $=$ signs not being taken simultaneously.

Proceeding to the case of three commodities, and applying the same kind of reasoning to the case where the choice of the individual is limited to the combinations represented by a plane which passes through a preference direction positively inclined with respect to all coordinate axes, we reach the result that the indifference element is represented by the total differential equation

(3) $$\varphi_1 dx_1 + \varphi_2 dx_2 + \varphi_3 dx_3 = 0.$$

It is clear that we could not possibly have any other form for this element that would be consistent with the choice of the individual in a two-dimensional space as constructed above. The linearity of the differential equation defining the indifference element, once established for two dimensions, must be extended to the case of n dimensions. Therefore, the indifference element will be in general defined by

(4) $$\Sigma_i \varphi_i dx_i = 0.$$

The meaning of this is that a direction defined by the increments Δx_1, $\Delta x_2, \ldots, \Delta x_n$ will be one of preference, indifference, or nonpreference, according to whether:

(5) $$\Sigma_i \varphi_i \Delta x_i >, =, < 0.$$

As to the variation of the indifference element from point to point some further assumptions are necessary.

We may, as Pareto suggested,[7] assume that in most cases the amounts $x_2, x_3, \ldots, x_n$ remaining constant, the individual will be willing to accept a higher rate of exchange (x_1/x_i) if x_1 increases, and vice versa.

---

6. I speak here only of the slope of a straight line, although the locus of the combinations forming a preferential set is not necessarily a straight line. But on such a locus, only the infinitesimal element around the indifferent combination μ matters; and this can be always assimilated to a linear element.

7. Pareto, *Manuel*, pp. 573–574.

But this way of considering the matter cannot explain some particular aspects of the exchange problem; as for instance that usually referred to as the Giffen Paradox.[8]

For the case of two commodities, the general condition expressed in a mathematical form was worked out for the first time by W. E. Johnson.[9] His psychological interpretation of it is somewhat heavy and difficult to grasp at once.

R. G. D. Allen was the first to state the necessary and sufficient condition for more than two commodities. Unfortunately an error has slipped into his mathematical treatment of it.[10]

Let M' be a point infinitesimally near to M, such that $\overrightarrow{MM'}$ be an indifference direction for M. The stability of equilibrium of exchange *with constant prices* will be secured if the direction $\overrightarrow{MM'}$ is a nonpreference direction for M'. It is clear that this condition is necessary and sufficient.

Let us write $B_2, B_3, \ldots, B_n$ instead of $\varphi_2/\varphi_1, \varphi_3/\varphi_1, \ldots, \varphi_n/\varphi_1$; the indifference element corresponding to the point $M'(x_i + \Delta x_i)$ may be written, abstracting from infinitesimals of higher order,

$$(6) \qquad dx_1 + \sum_{2} {}_i(B_i + \Delta B_i)dx_i = 0,$$

if Δx_i are sufficiently small. And, because $\overrightarrow{MM'}$ is an indifference direction for M, we have (5),

$$(7) \qquad L = \Delta x_1 + \sum_{2} {}_i B_i \Delta x_i = 0.$$

According to (5), the condition of stability is

$$(8) \qquad \Delta x_1 + \sum_{2} {}_i(B_i + \Delta B_i)\Delta x_i < 0.$$

Owing to (7), this last relation becomes

$$(9) \qquad \sum_{2} {}_i \sum_{1} {}_k B_{i,k} \Delta x_i \Delta x_k < 0,$$

where

$$B_{i,k} = \frac{\partial B_i}{\partial x_k}.$$

The system of (7) and (9) is satisfied when the principal minors, including always the elements of the first row and first column, of the determinant

8. Alfred Marshall, *Principles of Economics* (8th edn., New York, 1949), p. 132.
9. W. E. Johnson, "The Pure Theory of Utility Curves," *Economic Journal*, XXIII (1913), 483–513.
10. R. G. D. Allen, "The Foundations of a Mathematical Theory of Exchange," *Economica*, XII (1932), 197.

$$(10) \quad D = \begin{vmatrix} 0 & 1 & B_2 & B_3 & \cdots & B_n \\ 1 & 0 & B_{2,1} & B_{3,1} & \cdots & B_{n,1} \\ B_2 & B_{2,1} & 2B_{2,2} & B_{2,3} + B_{3,2} & \cdots & B_{2,n} + B_{n,2} \\ B_3 & B_{3,1} & B_{3,2} + B_{2,3} & 2B_{3,3} & \cdots & B_{3,n} + B_{n,3} \\ \cdot & \cdot & \cdot & \cdot & \cdots & \cdot \\ B_n & B_{n,1} & B_{n,2} + B_{2,n} & B_{n,3} + B_{3,n} & \cdots & 2B_{n,n} \end{vmatrix}$$

are, starting with the third order, alternatively positive and negative;[11] i.e.,

$$(11) \quad \begin{vmatrix} 0 & 1 & B_2 \\ 1 & 0 & B_{2,1} \\ B_2 & B_{2,1} & 2B_{2,2} \end{vmatrix} > 0; \quad \begin{vmatrix} 0 & 1 & B_2 & B_3 \\ 1 & 0 & B_{2,1} & B_{3,1} \\ B_2 & B_{2,1} & 2B_{2,2} & B_{2,3} + B_{3,2} \\ B_3 & B_{3,1} & B_{3,2} + B_{2,3} & 2B_{3,3} \end{vmatrix} < 0; \cdots$$

This is the very point on which I do not agree with Allen's way of stating the same condition, which he formulates as:[12]

$$(12) \quad \begin{vmatrix} 1 & B_2 \\ B_{2,1} & B_{2,2} \end{vmatrix} < 0; \quad \begin{vmatrix} 1 & B_2 & B_3 \\ B_{2,1} & B_{2,2} & B_{2,3} \\ B_{3,1} & B_{3,2} & B_{3,3} \end{vmatrix} > 0; \quad \cdots$$

The conditions (12) secure the uniqueness of equilibrium, but not necessarily its stability — unless the total differential equation (4) is integrable.[13] Indeed, from (12), it follows that the system

$$(13) \quad \begin{aligned} & B_2 = B_2^0, \ B_3 = B_3^0, \ \cdots, \ B_n = B_n^0, \\ & x_1 + \sum_2 {}_i B_i x_i = x_1^0 + \sum_2 {}_i B_i^0 x_i^0, \end{aligned}$$

where $B^0 = B(x_1^0, x_2^0, \ldots, x_n^0)$, can be satisfied only by $x_1 = x_1^0, x_2 = x_2^0, \ldots, x_n = x_n^0$. And since (13) is equivalent to

$$(14) \quad \begin{aligned} & B_2 = B_2^0, \ B_3 = B_3^0, \ \cdots, \ B_n = B_n^0, \\ & x_1 + \sum_2 {}_i B_i^0 x_i = x_1^0 + \sum_2 {}_i B_i^0 x_i^0, \end{aligned}$$

we conclude that in a given plane

$$(15) \quad x_1 - x_1^0 + \sum_2 {}_i B_i^0 (x_i - x_i^0) = 0$$

(i.e., for a given budget equation) there is only one point whose indifference element is contained in this plane.

It is, however, true that from (12) it follows that the stability condition is fulfilled for *some* indifference directions. For instance, in the case of

11. For the analytical proof, see "Mathematical Note" at end of this paper.
12. Allen, "The Foundations," p. 220.
13. See my "Mathematical Note" at end of this paper.

three commodities, from (12) it follows that the marginal rate of substitution between X_1 and X_2 decreases along the corresponding indifference direction. A simple numerical verification is sufficient to prove that from the above conditions it does not necessarily follow that either of the inequalities

$$\begin{vmatrix} 1 & B_3 \\ B_{3,1} & B_{3,3} \end{vmatrix} < 0, \qquad \begin{vmatrix} B_2 & B_3 \\ \dfrac{\partial}{\partial x_2}\left(\dfrac{B_3}{B_2}\right) & \dfrac{\partial}{\partial x_3}\left(\dfrac{B_3}{B_2}\right) \end{vmatrix} < 0,$$

which express the same property for the marginal rates of substitution of X_1 for X_3 and of X_2 for X_3 are true.[14]

Only in the case where the differential equation of the indifference elements is integrable are the two sets (11) and (12) equivalent. It is only in this last case that the condition of stability resolves completely into the principle of decreasing marginal rate of substitution in any direction.

Now, from (11) we deduce that

$$(16) \quad \Phi = \Delta x_1 + \sum_2{}_i (B_i + \Delta B_i)\Delta x_i \equiv L + \alpha^2 L^2 - L_1^2 - L_2^2 - \cdots - L_{n-1}^2,$$

where $L, L_1, \ldots, L_{n-1}$ are homogeneous and linear functions of $\Delta x_1, \Delta x_2, \ldots, \Delta x_n$. For obvious reasons, these functions are linearly independent.

In the space $(\Delta x_1, \Delta x_2, \ldots, \Delta x_n)$, $\Phi = 0$ represents a hyperquadric; the plane tangent to this hyperquadric in the origin is $L = 0$. Therefore, in a small domain round the origin, we have

$$(17) \qquad\qquad \Phi < 0 \quad \text{if} \quad L \leq 0,$$

and the condition (8) can be stated in a more comprehensive form

$$(18) \quad \Delta x_1 + \sum_2{}_i(B_i + \Delta B_i)\Delta x_i < 0 \quad \text{if} \quad \Delta x_1 + \sum_2{}_i B_i \Delta x_i \leq 0,$$

although we must still regard the former as the necessary and sufficient condition for stability.

The interpretation of (18) is that any direction $(\Delta x_1, \Delta x_2, \ldots, \Delta x_n)$ which constitutes for $M(x_1, x_2, \ldots, x_n)$ either an indifference or a nonpreference direction, will be a nonpreference direction for the infinitesimally near point $M'(x_i + \Delta x_i)$. Pursuing further the same direction, and applying the result (18) as we pass from M' to M'', from M'' to M''' and so on, we conclude that no indifference or nonpreference direction can become, if prolonged, a preference or indifference direction for any of its points.

14. In Allen and Hicks's more recent paper, "A Reconsideration of the Theory of Value," part II, p. 203 (see my note 2, above) these last conditions are added, without further argument, to the former ones. Even this would not do. Along an indifference direction, other than those obtained by leaving constant one of the quantities x_i, the stability condition might still not be fulfilled. Besides, we increase unnecessarily the number of conditions and do not arrive at a set of necessary and sufficient conditions.

Therefore, on any straight line there is only one point of equilibrium. This fact is of great importance for the understanding of how the equilibrium is reached, the movements of an individual being always described by straight lines in the case of constant prices.

III. SATURATION REGION: TWO COMMODITIES

The next step of the present analysis will be to consider the alternative cases derived by adopting only some of the assumptions A, B, C, D. Let us then assume that A, C, and D are true, but that

B_1. *There is a saturation region.*

This case has been considered prior to the present paper.[15] Thus, it would no longer seem necessary to deal with the subject again, if it were not for the sake of a more general treatment of the problem. It is true that opinions have been expressed that this case presents no interest. I should entirely agree with such views if they meant only that the absolute saturation region is very unlikely to be reached. But I believe that it is very difficult to disprove the existence of such a region. In any case, the consideration of a saturation region has, as will be shown, a theoretical importance, and some of the results connected with such a case will help us towards a better understanding of other patterns.

At this point of the argument the difficulty arises in defining the saturation region. The usual definition of it is formulated in terms of maximum total utility or zero marginal utility; but these concepts cannot be used in a rigorous analysis which aims exactly at an elucidation of their standing. Thus I propose to define a saturation point $M_s(s_i)$ by the fact that the direction $\overrightarrow{MM_s}$ will be for any non-saturation point M a preference direction. This definition seems to correspond entirely with our idea of a saturation point in the sense that reaching such a point is the ultimate motive of the individual's behavior. Besides, it accords with the very spirit of the theory of choice.[16]

15. See, for instance, R. G. D. Allen, "The Nature of Indifference Curves," *Review of Economic Studies*, I (1934), 110–121.

16. The case considered in the preceding section appears as a particular instance of the case where a saturation region exists. For, saying that independently of what the indifference element might be, any direction within the angle AMD (Fig. 1–1) is a preference direction, is simply interpreting the fact that all points at infinity within the quadrant X_1OX_2 are saturation points. We see that, as the position of saturation points puts a certain limit to the possible slopes of the indifference elements, the former are more intricately connected with human behavior than the term "saturation point" suggests at first.

Referring to equation (1),

$$\varphi_1 dx_1 + \varphi_2 dx_2 = 0,$$

we can easily assume that the coefficients φ_1, φ_2 will be finite and not simultaneously zero at any regular point, i.e., at a non-saturation point.

The preference side of the indifference direction will be that which will have the same sign as

(19) $$S = \varphi_1(s_1 - x_1) + \varphi_2(s_2 - x_2).$$

We may suppose that this sign is positive for all regular points.

If $M'_s(s'_i)$ is also a saturation point, it is the same for any point of the segment $M_s M'_s$. Indeed, from

$$\varphi_1(s_1 - x_1) + \varphi_2(s_2 - x_2) > 0,$$
$$\varphi_1(s'_1 - x_1) + \varphi_2(s'_2 - x_2) > 0,$$

it follows that

$$\varphi_1\left(\frac{s_1 + \lambda s'_1}{1 + \lambda} - x_1\right) + \varphi_2\left(\frac{s_2 + \lambda s'_2}{1 + \lambda} - x_2\right) > 0$$

for any positive value of λ.

Consequently, the saturation region must consist either of a convex closed area or of the degenerated forms of it, i.e., a segment of straight line or a single point.[17] The angle under which the saturation region is viewed from a regular point M is the fundamental preference angle in the sense that all directions within it are known as preference directions independently of what the indifference element may be. Under the assumption B, the fundamental angle is a right angle whose sides are parallel to the coordinate axes.

In treating the problem of a saturation region, I shall confine myself to the case where there is only one saturation point, this case being susceptible of simpler and more complete analytical treatment. It is, however, possible to justify, to a certain extent, the elimination of the other cases by the fact that the uniqueness of the indifference combinations in an ordinal set (assumption B) will harmonize better with the uniqueness of the saturation point. Besides, if we want to maintain a certain parallelism between the two cases considered so far, it is necessary to assume the uniqueness of the saturation point in order to reach the fundamental result obtained above, namely that on any straight line there is only one point of equilibrium. But these are only formal considerations, and consequently cannot be regarded as sufficient logical reasons for disallowing the possible existence of a saturation segment or a saturation area.

17. No distinction is necessary between the points at infinity and the others. Thus, a second-degree parabola will be regarded as enclosing a convex area.

For simplicity in the calculations to follow, let us choose the saturation point as the origin of coordinates. The only conditions imposed then upon φ_1, φ_2 are the condition of stability

$$(20) \qquad \begin{vmatrix} \varphi_1 & \varphi_2 \\ \dfrac{\partial}{\partial x_1}\left(\dfrac{\varphi_2}{\varphi_1}\right) & \dfrac{\partial}{\partial x_2}\left(\dfrac{\varphi_2}{\varphi_1}\right) \end{vmatrix} < 0$$

and

$$(21) \qquad x_1\varphi_1 + x_2\varphi_2 < 0.$$

According to the definition adopted, the indifference direction corresponding to the saturation point must be indeterminate.

This might happen in two cases:

1. φ_1 or φ_2 are indeterminate at the origin.
2. φ_1 and φ_2 are both zero at the origin.

From (21) we deduce that

$$(22) \qquad x_1\varphi_1(x_1, 0) < 0,$$

or

$$\varphi_1(-\alpha, 0) > 0, \qquad \varphi_1(+\alpha, 0) < 0,$$

for any positive value of α. Hence, either $\varphi_1(0, 0) = 0$ or $\varphi_1(0, 0)$ is indeterminate. In any case, on any path avoiding the origin and joining the two points $(+\alpha, 0)$, $(-\alpha, 0)$ there must be at least one zero of φ_1. Consequently the curve $\varphi_1 = 0$ approaches infinitesimally close to the origin, but we cannot decide if it passes through it or not. This shows that the only singular point of $\varphi_1 = 0$ might be the origin. If the origin is a singular point then φ_1 will be indeterminate for $x_1 = x_2 = 0$. If not, φ_1 will vanish for the same values.

Trying to derive the form of the integral curves, i.e., those defined by (1),[18] without introducing further assumptions about the behavior of the functions φ_1, φ_2 around the origin, would be an almost insurmountable task. It is necessary to realize that practically all mathematical transformations are valid only for functions subject to important restrictions. Consequently let us assume that φ_1, φ_2 are regular functions in a domain around the origin.

From (21) it follows that

$$(23) \qquad \varphi_{11}^0 dx_1^2 + (\varphi_{12}^0 + \varphi_{21}^0)dx_1 dx_2 + \varphi_{22}^0 dx_2^2 < 0,$$

where

$$\varphi_{ik} = \frac{\partial \varphi_i}{\partial x_k},$$

and the superscript means that the values are taken at the origin.

18. The reason why the term indifference curves is not used here, as has always been done, will appear later (Section V).

For a small domain round the origin the equation (1) can be replaced by

(24) $(x_1\varphi_{11}^0 + x_2\varphi_{12}^0)dx_1 + (x_1\varphi_{21}^0 + x_2\varphi_{22}^0)dx_2 = 0,$

and by putting $x_2 = y_2x_1$, the latter becomes

(25) $$2\frac{dx_1}{x_1} + \left[\frac{T'(y_2)}{T(y_2)} + \frac{\varphi_{21}^0 - \varphi_{12}^0}{T(y_2)}\right]dy_2 = 0,$$

where

(26) $$T(y_2) = \varphi_{11}^0 + (\varphi_{12}^0 + \varphi_{21}^0)y_2 + \varphi_{22}^0 y_2^2.$$

Owing to (23), the trinomial T has imaginary roots, and therefore there are only two cases to be considered.

1. $\varphi_{12}^0 \neq \varphi_{21}^0$. The integral of (24) is

(27) $$\log\left[x_2^2 T\left(\frac{x_2}{x_1}\right)\right] + \frac{2(\varphi_{21}^0 - \varphi_{12}^0)}{\delta}\tan^{-1}\left[\frac{1}{\delta}T'\left(\frac{x_2}{x_1}\right)\right] = \text{const.}$$

where $\delta^2 = 4\varphi_{11}^0\varphi_{22}^0 - (\varphi_{12}^0 + \varphi_{21}^0)^2$. The integral curves envelop the saturation point in the same way as logarithmic spirals turn round their asymptotic point, as in Fig. 1–2(a).

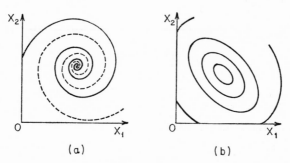

FIGURE 1-2

2. $\varphi_{12}^0 = \varphi_{21}^0$. The term involving $\tan^{-1}$ in (27) disappears and thus the equation of integral curves near the origin is

(28) $$\varphi_{11}^0 x_1^2 + (\varphi_{12}^0 + \varphi_{21}^0)x_1x_2 + \varphi_{22}^0 x_2^2 = \text{const.}$$

They are convex closed curves enveloping the saturation point as a center, as in Fig. 1–2(b).[19]

The second case has been considered prior to the present paper, but

19. The preceding argument is only an intuitive sketch of proof. The replacement of (1) by (24) in a small domain around the origin for finding out the shape of the integral curves calls for further restrictions to be imposed upon the functions φ_1, φ_2. These restrictions could be considered as one of the aspects of the "regularity" of human behavior. For a rigorous and complete discussion of the problem under consideration the reader is invited to refer to H. Poincaré, *Oeuvres*, tome I (Paris, 1951), p. 17, and to M. H. Dulac, *Points singuliers des équations différentielles*, Mémorial des Sciences Mathématiques, fasc. LXI, Paris, 1934.

its treatment has always been approached through the utility concept. The first one is most curious. As we shall presently see, it has great theoretical importance. Suffice it to point out now that in the case where the integral curves are ellipse-like, an index of ophelimity can be built *analytically*. This is no longer possible in the case where the integral curves are spiral-like. For if it were, then by keeping the amount of one commodity, say x_2, constant, we would obtain an index of ophelimity (I), the variation of which with respect to x_1 is indicated by Fig. 1–3.

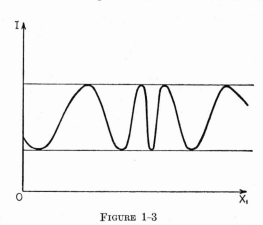

FIGURE 1-3

IV. THE NONINTEGRABILITY CASE

Let us consider the plane

$$(29) \qquad x_1 + B_2^0 x_2 + B_3^0 x_3 = x_1^0 + B_2^0 x_2^0 + B_3^0 x_3^0$$

which contains the indifference element corresponding to the point $M_0(x_1^0, x_2^0, x_3^0)$. The intersection of this plane with the indifference elements

$$(30) \qquad dx_1 + B_2 dx_2 + B_3 dx_3 = 0$$

determines in each point of (29) an indifference direction. *The theory of choice can be applied to the movements of the individual in this plane exactly as in the case of two commodities.* We must therefore expect the indifference directions thus obtained to fulfill the conditions established in the preceding section. An analysis in this direction may be regarded as one way of checking the conclusions already obtained.

If we project the indifference directions on the plane $X_2 O X_3$, their equation is

$$(31) \qquad (\overline{B}_2 - B_2^0) dx_2 + (\overline{B}_3 - B_3^0) dx_3 = 0,$$

where for brevity

(32) $\overline{B}_i = B_i[x_1^0 + B_2^0(x_2^0 - x_2) + B_3^0(x_3^0 - x_3), x_2, x_3].$

Any point $M(x_1, x_2, x_3)$ in the plane (29) satisfies the inequality

(33) $x_1 - x_1^0 + \overline{B}_2(x_2 - x_2^0) + \overline{B}_3(x_3 - x_3^0) < 0,$

for the direction $(x_1 - x_1^0, x_2 - x_2^0, x_3 - x_3^0)$ being an indifference direction for M_0, is a nonpreference direction for M, (18). Owing to (29), the last inequality yields

(34) $(\overline{B}_2 - B_2^0)(x_2^0 - x_2) + (\overline{B}_3 - B_3^0)(x_3^0 - x_3) > 0,$

which shows that $m_0(x_2^0, x_3^0)$ is a saturation point for (31). This is the only saturation point. Indeed, the corresponding conditions

$$\overline{B}_2 - B_2^0 = 0, \qquad \overline{B}_3 - B_3^0 = 0,$$

are satisfied only when $x_2 = x_2^0$, $x_3 = x_3^0$ (13). Since, on the other hand, $M_0(x_1^0, x_2^0, x_3^0)$ is the equilibrium point corresponding to the budget equation (29), we see that the equilibrium position is one of *relative saturation*. This is true of any equilibrium position.

From the condition of stability,

$$(B_{2,1}dx_1 + B_{2,2}dx_2 + B_{2,3}dx_3)dx_2 + (B_{3,1}dx_1 + B_{3,2}dx_2 + B_{3,3}dx_3)dx_3 < 0$$

when

$$dx_1 + B_2dx_2 + B_3dx_3 = 0,$$

joined to

$$dx_1 + B_2^0dx_2 + B_3^0dx_3 = 0,$$

we obtain the other necessary condition (20):

(35) $\begin{vmatrix} \overline{B}_2 - B_2^0 & \overline{B}_3 - B_3^0 \\ \dfrac{\partial}{\partial x_2}\left(\dfrac{\overline{B}_3 - B_3^0}{\overline{B}_2 - B_2^0}\right) & \dfrac{\partial}{\partial x_3}\left(\dfrac{\overline{B}_3 - B_3^0}{\overline{B}_2 - B_2^0}\right) \end{vmatrix} < 0.$

As to the form of the integral curves round the saturation point m_0, we have to distinguish the two cases set forth in the preceding section:

1. If

(36) $\left(\dfrac{\partial \overline{B}_2}{\partial x_3}\right)^0 = \left(\dfrac{\partial \overline{B}_3}{\partial x_2}\right)^0,$

the superscript meaning that the values are taken at $m_0(x_2^0, x_3^0)$, the integral curves will be ellipse-like.

2. If, on the contrary,

(37) $\left(\dfrac{\partial \overline{B}_2}{\partial x_3}\right)^0 \neq \left(\dfrac{\partial \overline{B}_3}{\partial x_2}\right)^0,$

the integral curves will be spiral-like.

Relation (36) yields

(38) $(B_{2,3} - B_3B_{2,1} + B_{3,2} - B_2B_{3,1})^0 = 0$

and if this is satisfied identically, it means that the total differential equation (30) is integrable, and vice versa.[20]

A perfect correspondence is thus established between the integrable and nonintegrable cases on the one hand and the ellipse-like and spiral-like integral curves on the other. Because of this correspondence a more intuitive treatment of the nonintegrable case is made possible. At the same time, it is seen that the difference between the case involving two commodities only and that of more than two was illusory because the problem concerning a saturation point had not been worked out to a sufficient extent. The impossibility of an analytical construction of an ophelimity index is a feature that belongs to both cases.

V. INDIFFERENCE VS. INTEGRAL VARIETIES:
TRANSITIVITY AND INTEGRABILITY

Thus far, we have mainly been concerned with a "geometric" development of the four postulates set out near the beginning of the paper (Section II), and to this extent the whole analysis has refrained completely from any interpretation of the results obtained. We may now proceed further and analyze the implications of the theory of choice and its standing as a satisfactory explanation of the uniqueness and stability of exchange equilibrium.

According to the postulate A, we can, given a direction (ω) and a point $M_0(x_i^0)$, construct a surface,

$$(39) \qquad F_\omega(x_i; x_i^0) = 0,$$

the locus of all combinations $\mu(x_i)$ that are indifferent in comparison to M_0. The subscript ω is used in order to emphasize the dependence of the indifference surface thus constructed upon the slope of the line $\omega\omega'$ (Section II, Fig. 1–1). The last relation can be written in the form

$$(40) \qquad f_\omega(x_i; x_i^0) = f_\omega(x_i^0; x_i^0)$$

which explicitly shows that M belongs to (39).

So far we have succeeded in constructing a preferential field. In this field, F_ω the result of the comparison of $N(y_i)$ with $M(x_i)$ is always known. Symbolically we may write

$$(41) \qquad \begin{aligned} (N = M)_\omega &\quad \text{if} &\quad f_\omega(y_i; x_i) &= f_\omega(x_i; x_i), \\ (N > M)_\omega &\quad \text{if} &\quad f_\omega(y_i; x_i) &> f_\omega(x_i; x_i), \\ (N < M)_\omega &\quad \text{if} &\quad f_\omega(y_i; x_i) &< f_\omega(x_i; x_i). \end{aligned}$$

20. See, for instance, G. C. Evans, *Mathematical Introduction to Economics* (New York, 1930), pp. 119–120.

In the case of two commodities the curves (μ) are all tangent to the integral curves (I) of

$$dx_1 + B_2 dx_2 = 0,$$

the point of contact being precisely the corresponding point M_0 (Fig. 1–4). A similar picture can be drawn for the case of three commodities, with the difference that the integral surfaces (I) may not exist.

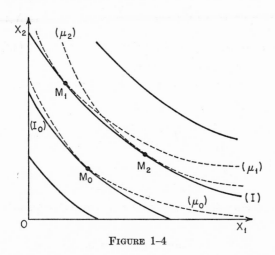

FIGURE 1–4

The preceding considerations show that even if the indifference elements are integrable, no meaning can be attached to the integral varieties without introducing further assumptions. Two points belonging to the same integral surface do not necessarily represent equivalent combinations. The integrability of indifference elements alone does not imply that the individual will be able to tell, independently of any restriction, which combinations give him the same satisfaction.[21] For the field (ω) is in general inconsistent. Indeed, we may find a path going always in a preference direction from M to P, and still have $M < P$, or vice versa. The movement along such a path will lead to an "illusion."

It is true that along any indifference direction, for infinitesimally near points $M(x_i)$, $M'(x_i' = x_i + dx_i)$, $M''(x_i' + dx_i')$, we have $(M = M')$ and $(M' = M'')$. But from this we cannot advance that $(M = M'')$.

Such "illusions" will no longer be possible in a consistent field; this will be so if the equivalence defined above, (41), is transitive: i.e., if from

$$(P = N)_\omega \qquad \text{and} \qquad (N = M)_\omega$$

it follows that

$$(P = M)_\omega.$$

21. N. Georgescu-Roegen, "Note on a Proposition of Pareto," *Quarterly Journal of Economics*, XLIX (1935), 706–714.

In this case, the relations

(42) $$F_\omega(z_i; y_i) = 0, \qquad F_\omega(y_i; x_i) = 0,$$

must yield

(43) $$F_\omega(z_i; x_i) = 0.$$

Let us introduce the following notations

(44) $$\frac{\partial F_\omega(X_i; Y_i)}{\partial X_k} = u_k(X_i; Y_i), \qquad \frac{\partial F_\omega(X_i; Y_i)}{\partial Y_k} = v_k(X_i; Y_i).$$

Leaving x_i constant and differentiating in (42) and (43), we obtain

(45) $$\Sigma_k u_k(z_i; y_i)dz_k + \Sigma_k v_k(z_i; y_i)dy_k = 0,$$
$$\Sigma_k u_k(y_i; x_i)dy_k = 0, \qquad \Sigma_k u_k(z_i; x_i)dz_k = 0.$$

As dy, dz are arbitrary, it follows that

(46) $$u_k(y_i; x_i) = \lambda v_k(z_i; y_i),$$
$$u_k(z_i; x_i) = \mu u_k(z_i; y_i).$$

Since (43) is satisfied for $z_i = y_i$ we obtain

(47) $$u_k(y_i; x_i) \equiv \mu u_k(y_i; y_i).$$

The differential equation of $F(y_i; x_i) = 0$ can consequently be written in the form

(48) $$\Sigma_k u_k(y_i; y_i)dy_k = 0.$$

This shows that the tangential element corresponding to any point of the variety (μ) is independent of x_i. In other words, (μ) and (I) coincide; the existence of an integral-indifference variety is thus proved. The equation (40) becomes

(49) $$f(x_i) = f(x_i^0).$$

This equation, according to the postulate D, is independent of ω. It follows also that the transitivity is valid with respect to the preference and nonpreference changes and that the comparisons are reversible. In other words, if $M = N$, then $N = M$.

The preceding analysis shows clearly that without the transitivity postulate the integral varieties and the indifference varieties are two distinct things. If a point of saturation exists, the indifference varieties will be always concave closed surfaces (postulate A), while the integral ones need not necessarily be so. The much-discussed paradox of the nonintegrable case is due to the confusion of these two concepts.

The existence of indifference elements satisfying certain conditions expressed by inequalities is sufficient for an explanation of the uniqueness and stability of equilibrium in an exchange with constant prices. As R. G. D. Allen has pointed out, the integrability condition is too severe.[22]

22. Allen, "The Foundations," p. 223.

I should add "and without any meaning outside the transitivity condition."

VI. THRESHOLD IN CHOICE

The main argument leveled against the existence of integral-indifference varieties consists in denying the possibility of a mental comparison at a finite distance. This argument was logically questioned because of its failure in the case of two commodities. But despite this fact, it was generally accepted that a mere introspection is enough to prove our hesitation if faced with a problem of choice. As shown in the preceding section, this failure was only apparent, and the paradox was due to a confusion between the integral varieties and integral-indifference varieties. This difficulty being eliminated, the standing of the argument mentioned is very much improved.

Hesitation, however, does not necessarily mean impossibility of choice. We may, for instance, be in doubt about which one out of two masses is heavier, if we have to judge their weight by lifting them only. From this it does not follow that we are unable to distinguish with certainty any differences in weight whatever.

Let us suppose that we have recorded the number of times an individual decided that $A > B$ in a series of mental comparisons so conducted as to avoid as much as possible any hereditary influence from test to test. It is natural to assume that the relative frequence p of the decision $A > B$ is perfectly stable and that, *ceteris paribus*, it depends only on A and B. What appears to be haphazard in a single case is no longer so in a series of similar attempts. If we denote by q the complementary frequency, i.e., that of the cases in which the decision is $A < B$, we have

$$p + q = 1.$$

The graphical representation of such a law is indicated in Fig. 1–5(a). The function $p(y, x)$ represents the probability that the stimulus x be considered greater than y. The basis of this probability is explained in terms of errors of perceptions.[23] If $f(y, x)dx$ is the probability that y be perceived as being comprised between x and $x + dx$ — Fig. 1–5(b) — then the probability that $y < x$ is

(50) $$p(x, y) = \int_{y_0}^{x} f(x, y)dx,$$

hence

23. Cf. G. S. Fullerton and J. M. Cattell, *On the Perception of Small Differences*, Publications of the University of Pennsylvania, 1892.

(51) $$f(x, y) = \frac{\partial p(x, y)}{\partial x}.$$

A natural assumption to make is that the errors of perceptions are equally likely to be negative or positive. In this case

(52) $$\int_{y_0}^{y} f(y, x)dx = \int_{y}^{y_1} f(y, x)dx$$

and

(53) $$p(y, y) = q(y, y) = \frac{1}{2}.$$

The last relation constitutes the definition of physical equality. It is the basis of all means of detecting the equality between two quantities.

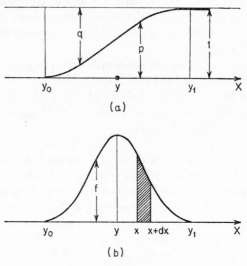

(a)

(b)

FIGURE 1-5

The interval (y_0, y_1) is known as the psychological threshold. Outside it, there is no doubt as to which stimulus is greater. The length of the interval (y_0, y_1) depends upon many causes, but the most important factor in this respect is the interval of time the individual is allowed to perceive the two stimuli before formulating his judgment. The greater this interval of time, the smaller will be the psychological threshold. At the limit when the time of experimenting is infinite, the threshold is zero. The latter condition is essentially the basis of postulate A. Its use in a static analysis is justified by the very characteristic of such a system, which is that of being independent of time. A dynamic system could not be analyzed on the same basis. The individual's doubt as to which of two combinations he will finally choose would have to be taken into account and postulate A modified accordingly.

Let us adopt now the following postulate:

A_1. *If the individual has already experimented during a certain interval of time with two combinations of commodities $M(x_i)$, $N(y_i)$ there is a probability expressed by a function $\omega(x_i; y_i)$ that he will consider the combination $M(x_i)$ preferable to $N(y_i)$.*

For obvious reasons ω will satisfy the following relation:

$$(54) \qquad\qquad \omega(x_i; y_i) + \omega(y_i; x_i) = 1.$$

The only condition to be imposed for the time being upon ω is that

$$(55) \qquad\qquad \frac{\partial \omega}{\partial x_i} > 0 \quad \text{if} \quad \omega < 1.$$

*   *   *

In the analytical treatment of the present scheme I shall confine myself to the case of two commodities.

Let $N(y_i)$ be the individual's initial position. If he is faced with a given rate of exchange p, between X_1 and X_2, there will be the probability π_1 that he will exchange X_1 against X_2, and the probability π_2 that he will do the reverse. For convenience, I shall speak of each price line as having two directions, each of them being distinguished by the corresponding probability π_1 or π_2.

The probabilities π_1, π_2 might both vary from 0 to 1, subject to the condition $\pi_1 + \pi_2 = 1$. There will be consequently in each point of the plane X_1OX_2 an angle of indifference (II', JJ'), meaning by this that for all price lines within this angle, the individual might be a seller as well as a buyer of either of the two commodities involved (Fig. 1–6).

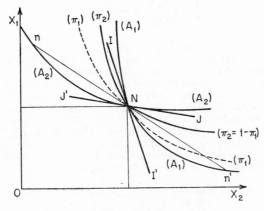

FIGURE 1–6

As the price line rotates from the position II' to JJ', π_1 will decrease monotonically from 1 to 0, while π_2 will increase in the same manner from 0 to 1.[24] It follows that all the curves

$$(56) \qquad \omega(x_i; y_i) = \pi, \quad (0 \leq \pi \leq 1),$$

x_i varying, pass through $N(y_i)$, and consequently $\omega(x_i; y_i)$ is indeterminate, for $x_i = y_i$. The value of $\omega(y_i; y_i)$ depends upon the path the point $M(x_i)$ follows toward $N(y_i)$.

If we keep x_2 constant and let only x_1 vary, the variation of $\omega(x_i; y_i)$ will be represented by a graph similar to that of Fig. 1–5(a); i.e., ω will be a monotonically increasing function from 0 to 1. The probability that, trying to find on the straight line $x_2 = $ const. an equivalent combination to $N(y_i)$, the individual will choose a point between x_1 and $x_1 + dx_1$, is (51)

$$(57) \qquad \frac{\partial \omega(x_1, x_2; y_1, y_2)}{\partial x_1} \, dx_1 = f(x_1, x_2; y_1, y_2) dx_1.$$

If we now perform the same operation, but keeping x_1 constant and letting x_2 vary, we shall have instead (*mutatis mutandis*) the probability

$$(58) \qquad \frac{\partial \omega(x_1, x_2; y_1, y_2)}{dx_2} \, dx_2 = g(x_1, x_2; y_1, y_2) dx_2.$$

The same considerations apply to any path along which ω is monotonically increasing. If (γ) is such a path, then the probability analogous to (57) or (58) is expressed in terms of a directional derivative,[25] i.e.,

$$(59) \qquad \left(\frac{\partial \omega}{\partial s}\right)_\gamma ds = (f \cos \alpha + g \sin \alpha) ds.$$

The probability that the position equivalent to N is on the path (γ) below[26] the point $M(x_1, x_2)$ of this curve is, according to (50),

$$(60) \qquad \int_{(\gamma)}^{M} \left(\frac{\partial \omega}{\partial s}\right) ds = \omega(x_i; y_i).$$

24. The question of measuring the physical quantities of goods is left, in the present scheme, to some outside scales to which the individual refers. He knows that $x \gtrless x'$ by using the scales and not as a consequence of his feelings of the equal or different degrees of utility he can derive from the consumption of x or x'. The individual's preference for $x + dx$ in comparison to x is justified on the ground that, although the individual will not feel the addition of $\frac{1}{100}$ apple daily, he will certainly appreciate an increment of three apples yearly. The individual's behavior appears therefore as a resultant of two different types of measurement: a physical one, which is supposed to tell him the exact amounts of commodities, and a psychological one, which represents his capability of comparing satisfactions. The fact that these two kinds of measurements are both involved in the present scheme constitutes an important point in the problem.

25. See W. F. Osgood, *Advanced Calculus* (New York, 1925), p. 143.

26. By a point "below" M is meant here any point for which the corresponding value of ω is smaller than $\omega(x_i; y_i)$.

If the path is a straight line nNn' passing thru N (Fig. 1–6),

(61) $$x_1 - y_1 + p(x_2 - y_2) = 0,$$

the formula (59) is valid for the separate directions $\overrightarrow{nN}$ and $\overrightarrow{n'N}$. In this case, the probabilities that from N the individual will move to any point on $\overrightarrow{Nn}$ or $\overrightarrow{Nn'}$ are respectively

(62)
$$\Pi = \lim_{ds=0} \omega(y_1 - ds \cos \alpha, y_2 - ds \sin \alpha; y_1, y_2),$$
$$\Pi' = \lim_{ds=0} \omega(y_1 + ds \cos \alpha, y_2 + ds \sin \alpha; y_1, y_2).$$

The directions $\overrightarrow{Nn}$, $\overrightarrow{Nn'}$ are, consequently, tangent respectively to the curves $\omega(x_i; y_i) = \Pi$, $\omega(x_i; y_i) = \Pi'$. The probabilities Π and Π' then are identical with π_1 and π_2 defined above in connection with the price line nn'. The curves (56) present a cusp point in N, except in the case where $\pi_1 = \pi_2 = \frac{1}{2}$.

By joining two branches of the curves

(63) $$\omega(x_i; y_i) = \pi, \qquad \omega(x_i; y_i) = 1 - \pi,$$

we form two curves that have no longer any irregularity in N (Fig. 1–6). We shall refer to such curves as (π) curves, where (π) is the greater of π and $1 - \pi$. We shall also introduce the term "limiting curves" for designating those for which $(\pi) = 1$.

The angle of indifference (II', JJ') can be represented by a quadratic differential equation,

(64) $$\Phi = \varphi_{11}dx_1^2 + 2\varphi_{12}dx_1dx_2 + \varphi_{22}dx_2^2 = 0,$$

$\varphi_{11}, \varphi_{12}, \varphi_{22}$ being given functions of x_1, x_2. This equation may be regarded at the same time as the differential equation of the limiting curves.

By a suitable choice of signs in (64), the conditions imposed upon the functions φ, can be written

(65)
$$\Delta = \varphi_{11}\varphi_{12} - \varphi_{12}^2 < 0,$$
$$\varphi_{11}, \varphi_{12}, \varphi_{22} > 0.$$

In order to determine the nature of a given direction it is convenient to decompose the form Φ into linear factors

(66) $$\Phi = \varphi_{11}(dx_1 + a_1dx_2)(dx_1 + a_2dx_2),$$

where $a_1, a_2 > 0$.[27]

The direction $\Delta x_1, \Delta x_2$ is an indifference, preference, or nonpreference direction according to whether

27. For more than two commodities this decomposition is no longer possible. In general, for n commodities:

$$\varphi = A_1^2 - A_2^2 - \cdots - A_n^2,$$

$n - 1$ terms being negative.

$$\Delta x_1 + a_1 \Delta x_2 \gtreqless 0, \qquad \Delta x_1 + a_2 \Delta x_2 \lesseqgtr 0: \qquad \text{indifference;}$$
$$\Delta x_1 + a_1 \Delta x_2 > 0, \qquad \Delta x_1 + a_2 \Delta x_2 > 0: \qquad \text{preference;}$$
$$\Delta x_1 + a_1 \Delta x_2 < 0, \qquad \Delta x_1 + a_2 \Delta x_2 < 0: \qquad \text{nonpreference.}$$

Further conditions to be imposed upon the functions φ are those pertaining to the stability of equilibrium of exchange. These conditions are that the limiting directions $\overrightarrow{NI}$, $\overrightarrow{NJ}$ must become indifference directions as we move away from N along them; similarly $\overrightarrow{NI'}$, $\overrightarrow{NJ'}$ must become

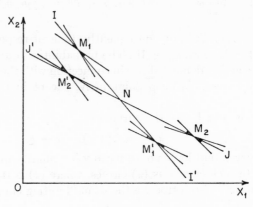

FIGURE 1–7

nonpreference directions (Fig. 1–7). The analytical formulation of these restrictions is

$$d\Phi \gtreqless 0 \qquad \text{when } dx_1 + a_1 dx_2 = 0 \qquad \text{if } dx_1 + a_2 dx_2 \lesseqgtr 0,$$
$$d\Phi \gtreqless 0 \qquad \text{when } dx_1 + a_2 dx_2 = 0 \qquad \text{if } dx_1 + a_1 dx_2 \lesseqgtr 0;$$

or,

$$(dx_1 + a_2 dx_2)d\Phi < 0 \qquad \text{when } dx_1 + a_1 dx_2 = 0,$$
$$(dx_1 + a_1 dx_2)d\Phi < 0 \qquad \text{when } dx_1 + a_2 dx_2 = 0.$$

Hence,

$$\varphi_{11}(dx_1 + a_2 dx_2)^2 d(dx_1 + a_1 dx_2) < 0 \qquad \text{when } dx_1 + a_1 dx_2 = 0,$$

and as

$$dx_1 + a_2 dx_2 \neq 0 \qquad \text{when } dx_1 + a_1 dx_2 = 0,$$

(for according to (65), $a_1 \neq a_2$ and $\varphi_{11} > 0$) we obtain the condition

$$d(dx_1 + a_1 dx_2) < 0 \qquad \text{when } dx_1 + a_1 dx_2 = 0.$$

Therefore

$$(67) \qquad E_1 = \frac{\partial a_1}{\partial x_2} - a_1 \frac{\partial a_1}{\partial x_1} < 0, \qquad E_2 = \frac{\partial a_2}{\partial x_2} - a_2 \frac{\partial a_2}{\partial x_1} < 0.$$

From (66) we deduce

$$(68) \qquad 2\frac{\varphi_{12}}{\varphi_{11}} = a_1 + a_2, \qquad 2\frac{\varphi_{12}}{\varphi_{22}} = \frac{1}{a_1} + \frac{1}{a_2}, \qquad \frac{\varphi_{22}}{\varphi_{11}} = a_1 a_2.$$

Hence:

$$2\partial\left(\frac{\varphi_{12}}{\varphi_{11}}\right) = \partial a_1 + \partial a_2,$$

$$(69) \qquad 2\partial\left(\frac{\varphi_{12}}{\varphi_{22}}\right) = -\frac{\partial a_1}{a_1^2} - \frac{\partial a_2}{a_2^2},$$

$$\partial\left(\log\frac{\varphi_{22}}{\varphi_{11}}\right) = \frac{\partial a_1}{a_1} + \frac{\partial a_2}{a_2}.$$

This yields

$$U_0 = E_1 + E_2 = 2\frac{\partial}{\partial x_2}\left(\frac{\varphi_{12}}{\varphi_{11}}\right) - 4\frac{\varphi_{12}}{\varphi_{11}}\frac{\partial}{\partial x_1}\left(\frac{\varphi_{12}}{\varphi_{11}}\right) + 2\frac{\partial}{\partial x_1}\left(\frac{\varphi_{22}}{\varphi_{11}}\right),$$

$$U_1 = \frac{E_1}{a_2} + \frac{E_2}{a_2} = \frac{\partial}{\partial x_2}\left(\log\frac{\varphi_{22}}{\varphi_{11}}\right) - 2\frac{\partial}{\partial x_1}\left(\frac{\varphi_{12}}{\varphi_{11}}\right),$$

$$(70)$$

$$U_2 = \frac{E_1}{a_1^2} + \frac{E_2}{a_2^2} = \frac{\partial}{\partial x_1}\left(\log\frac{\varphi_{11}}{\varphi_{22}}\right) - 2\frac{\partial}{\partial x_2}\left(\frac{\varphi_{12}}{\varphi_{22}}\right),$$

and further

$$(71) \qquad \begin{vmatrix} U_0 & U_1 \\ U_1 & U_2 \end{vmatrix} = \begin{vmatrix} 1 & 1 \\ \dfrac{1}{a_1} & \dfrac{1}{a_2} \end{vmatrix}^2 \cdot E_1 E_2.$$

But

$$(72) \qquad U_0 = 2\frac{\varphi_{12}}{\varphi_{11}} U_1 - \frac{\varphi_{22}}{\varphi_{11}} U_2;$$

and introducing this last relation in (71) we get

$$(73) \qquad \varphi_{11} U_1^2 - 2\varphi_{12} U_1 U_2 + \varphi_{22} U_2^2 = -\begin{vmatrix} 1 & 1 \\ \dfrac{1}{a_1} & \dfrac{1}{a_2} \end{vmatrix}^2 \cdot \varphi_{11} E_1 E_2.$$

It is easily seen now that the conditions (67) may be replaced by

$$(74) \qquad U_1 > 0, \qquad \varphi_{11} U_1^2 - 2\varphi_{12} U_1 U_2 + \varphi_{22} U_2^2 < 0.$$

In the case under consideration because of a_1, $a_2 > 0$, we have also

$$U_0 < 0, \qquad U_2 < 0,$$

but these are not independent of (74).

The equation of the limiting curves is obtained by integrating (64) or, what comes to the same thing,

$$(75) \qquad dx_1 + a_1 dx_2 = 0, \qquad dx_1 + a_2 dx_2 = 0.$$

Thus we obtain in the plane X_1OX_2 two families of curves

(76) $$A_1(x_1, x_2) = \alpha_1, \qquad A_2(x_1, x_2) = \alpha_2,$$

which taken together are identical with

(77) $$\omega(x_i; y_i) = 1, \qquad \omega(x_i; y_i) = 0.$$

The curves A_1 and A_2 are, according to the assumptions made, negatively inclined (65) and convex toward the coordinate axes (67).

Two curves belonging to different families cannot intersect more than once. Indeed, if they intersected in m and n, we should have

$$\left(\frac{dx_2}{dx_1}\right)_{\alpha_1} \geq \left(\frac{dx_2}{dx_1}\right)_{\alpha_2},$$

in one point and $\left(\frac{dx_2}{dx_1}\right)_{\alpha_1} \leq \left(\frac{dx_2}{dx_1}\right)_{\alpha_2}$ in the other. But this would mean that there are points for which

$$\left(\frac{dx_2}{dx_1}\right)_{\alpha_1} = \left(\frac{dx_2}{dx_1}\right)_{\alpha_2},$$

which is in contradiction with our main condition (65).

As on the other hand, through each point $N(y_i)$ passes only one curve of each family, such curves divide the plane into four regions, the meaning of which is evident (Fig. 1–6).

Owing to these properties of the curves A, a one-to-one correspondence can be established between the points of the plane X_1OX_2 and those of the plane $\alpha_1O\alpha_2$ (Fig. 1–8). To each point $M(x_i)$ in the former, there corresponds an image $m(\alpha_1, \alpha_2)$ in the latter. In this last plane, the straight lines $\alpha_1 = $ const. and $\alpha_2 = $ const. correspond to the curves A_1, A_2. Taking α_1, α_2 as new variables, (66) becomes

(78) $$\Phi = \mu(\alpha_1, \alpha_2)d\alpha_1 d\alpha_2, \qquad (\mu > 0).$$

The indifference directions being characterized by $\Phi < 0$, the corresponding indifference region consists in $\alpha_1O\alpha_2$ of two quadrants amc, bmd. To the preference and nonpreference regions correspond respectively the quadrants amd and cmb.

The present scheme of the behavior of an individual does not give rise to any "illusion." Indeed, the image of any continuous path going from M always in a preferred direction is in $\alpha_1O\alpha_2$ a curve along which $d\alpha_1$, $d\alpha_2 > 0$, and consequently the latter can lead only to some point within the quadrant amd. Similar reasoning can be applied to a path going always in a nonpreferred direction (Fig. 1–8). The inequality of satisfaction is transitive.

The relation between indifferent combinations is, however, not transitive. Thus from

$$m = n, \qquad n = p,$$

it follows that $m = p$ (Fig. 1-8); but, although $n = p'$, p' is preferred to m. In other words, the relation between indifferent combinations is transitive if they can be connected by a path going always in an indifference direction and such as along it the direction of the movement, $\dfrac{dx_2}{dx_1}$, be continuous in all points. If this is not true, we cannot tell whether the transitivity is valid or not.

Consequently from

$$M(x_i) = N(y_i), \qquad N(y_i) = P(z_i),$$

if $x_i - y_i$ and $y_i - z_i$ have the same sign, we can always conclude that $M = P$.

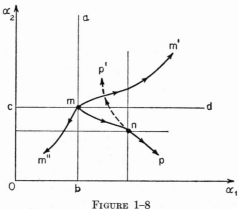

FIGURE 1-8

The lack of transitivity with respect to the indifference situation is easily explained. What we really mean by $M = N$ is that there is a probability ω for M to be considered preferable to N. The comparisons between M and N, N and P, and M and P are three independent phenomena. There is no way of establishing a relation between the probabilities connected with each of them.

* * *

An equation similar to (64) defines any curve (π)

(79) $$\psi_{11}dx_1^2 + 2\psi_{12}dx_1dx_2 + \psi_{22}dx_2^2 = 0,$$

where ψ_{11}, ψ_{12}, ψ_{22} are functions of x_1, x_2, π, satisfying the conditions

$$\psi_{11}\psi_{22} - \psi_{12}^2 < 0, \qquad \psi_{11}, \psi_{12}, \psi_{22} < 0,$$

with the exception that for $\pi = \frac{1}{2}$ we have

$$\psi_{11}\psi_{22} - \psi_{12}^2 \equiv 0.$$

For this particular value (79) becomes

(80) $$\varphi_1dx_1 + \varphi_2dx_2 = 0$$

which is the differential equation of the curves of complete indifference. The use of this term is justified by the fact that for any of two combinations belonging to such a curve there is a real doubt as to which one is preferable; i.e., both answers $A > B$, $B > A$ have the same probability equal to $\frac{1}{2}$.

According to the principle of decreasing marginal rate of substitution the curves (79) are convex toward the coordinate axes and consequently the functions ψ satisfy relations analogous to (76). Each curve (π) is divided by any curve of complete indifference into two parts, a superior and an inferior one. All the combinations situated on the (π) curve passing through N are preferred to N with the frequency π if they belong to the superior part, with the frequency $1 - \pi$ in the other case.

* * *

I shall proceed now to examine briefly the consequences which the scheme described in the present section has for the equilibrium of exchange with constant prices. Let us consider an individual whose income is x_1^0 and who can obtain one unit of X_2 in exchange for p units of X_1. The properties of the limiting curves as outlined above lead to the following pattern:

As the individual starts to move on the budget line

$$x_1^0 = x_1 + px_2$$

from A towards B (Fig. 1–9), he will — if $\overrightarrow{AB}$ is not an indifference direction for A — move for a certain interval in a preference direction.

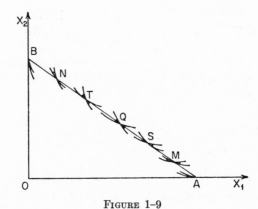

FIGURE 1–9

There exists on AB a point $S(\xi_1, \xi_2)$ for which $\overrightarrow{SB}$ is a limiting direction between the preference and indifference regions, and further — if $\overrightarrow{BA}$ is not an indifference direction for B — another point $T(\tau_1, \tau_2)$ for which

$\overrightarrow{TB}$ is the limiting direction between the indifference and nonpreference regions. Between S and T there are only points for which AB is in both senses an indifference direction; $\overrightarrow{AB}$ is a nonpreference direction for any point situated between B and T. This shows that any point between S and T might be an equilibrium point and that the system defining it is

$$(81) \qquad x_1^0 = x_1 + px_2, \qquad p^2\varphi_{11} - 2p\varphi_{12} + \varphi_{22} < 0.$$

In which one of these points the equilibrium would finally end cannot be determined in advance in the case of a single bargain. A multitude of causes related to the actual situation of the buyer and of the seller will influence the final position of equilibrium. This is no longer true in the case of a series of repeated bargains in which the price p and the income x_1^0 remain constant. There exists a definite probability for the bargain to end in any given point between S and T.

Let us suppose that the individual has gone on the straight line AB as far as $Q(y_1, y_2)$. This position accepted, there is a certain probability (59) that he move from Q to a point Q' on the same line and having the ordinate comprised between x_2 and $x_2 + dx_2$. And as in this case

$$\cos \alpha = \pm \frac{p}{\sqrt{1 + p^2}}, \qquad \sin \alpha = \mp \frac{1}{\sqrt{1 + p^2}},$$

and

$$ds = \frac{dx_1}{\cos \alpha} = \frac{dx_2}{\sin \alpha},$$

the probability (59) can be worked out into the form $k(y_2, x_2)dx_2$. This probability depends consequently on the previous position. The next step will be to express the probability of a change of position starting from Q'; this probability depends upon the position of Q', and so forth. We recognize here a set of random events subject to a chain relationship.

Let $P(x_2)dx_2$ be the probability that independently of what the previous position may have been the equilibrium be established in Q'. The function P is defined by the integral equation[28]

$$(82) \qquad P(x_2) = \int^{\tau_2} P(y_2)k(y_2, x_2)dy_2.$$

But this law of probability could not be used for a determination of the final equilibrium position in a single bargain any more than we can use a mortality table to determine how long a given individual will live.

In fact, $P(x_2)dx_2$ is the probability that under the given circumstances the quantity of X_2 demanded will lie between x_2 and $x_2 + dx_2$. The demand law can no longer be pictured as a curve relating prices and

28. G. Darmois, "Analyse et comparaison des séries statistiques qui se développent dans le temps," *Metron*, VIII (1929), 238–239.

quantities, but as a bivariate distribution between the same variables.

Let us denote by $P_i(x/p)dx$ the probability that the quantity demanded at the price p by the individual (i) lie between x and $x + dx$ and let $s_i(p)$ be the variance of the distribution thus defined. The probability connected with the collective demand of n individuals is easily worked out in terms of P_i. Let $P(X/p)dX$ be the probability that the market demand at the price p be comprised between X and $X + dX$. We have

$$P(X/p)dX = \int P_1(x_1/p)P_2(x_2/p)\cdots P_n(x_n/p)dx_1dx_2\cdots dx_n,$$

the integration being subject to the condition

$$x_1 + x_2 + x_3 + \cdots + x_n = X.$$

It is obvious that the range of indeterminateness of the market demand is equal to the sum of the corresponding ranges of the individual demands. The variance of X is, according to a well-known theorem,

$$S = s_1 + s_2 + \cdots + s_n,$$

and consequently the larger the market the greater will be the variance of its demand.

Only relative values, like the relative deviation $(X - \overline{X})/\overline{X}$, $\overline{X}$ being the mean value of X, or the quantity demanded per capita X/n lead to a smaller variance as the size of the market increases.[29]

The transformation of quantitative data into some kind of relative values as the preliminary step of a statistical treatment of such series is thus shown to be not only justified but also necessary.

VII. CONCLUSIONS

The aim of this last section will be both to summarize the main results reached thus far — presenting them at the same time in a more concise form — and to advance further considerations upon the bearing of some of these points on the exchange problem.

1. Four main points have been shown to affect any analytical construction of a theory of choice. Each of these point presenting two alternatives, there will be as many versions of such a theory as there are consistent combinations of the postulates to which these alternatives give rise. Among the consistent aspects of the theory we find the theory of utility varieties of Edgeworth as well as the theory of choice of Pareto.

29. The variance of $(X - \overline{X})/\overline{X}$ which is $\sigma = S/\overline{X}^2$ will, under very general assumptions, decrease as n increases. The variance of X/n is S/n^2 and will also decrease as n increases if s_i has an upper limit.

Pure analysis can do no more in this direction. Which form of the postulates A, B, C, and D we have to accept is a question that can be decided by actual experiment only. Some of these postulates, however, namely C and D, are very unlikely to lend themselves successfully to such a treatment. This fact is rather discomforting, for the uniqueness of the indifference elements — and therefore that of the equilibrium of exchange — depends precisely on these points. Yet not all the roads are completely barred. The question as to whether the indifference combinations are transitive or not plays an important role regarding the conclusions of such an analysis. It seems that this point could be easily submitted to an experimental verification. We should really lose all hope in this direction only if the answer to such an investigation should be negative. For, as has been shown (Section V), while the transitivity confirms the postulates C and D, from these it does not necessarily follow that the indifferent combinations have this special property.[30]

Nevertheless, many significant formal considerations could be advanced as to the standing of these postulates.

Once the invariance of the tangential element in M to the variety (μ) in respect to the variation of ω is accepted, I cannot think of any objection compatible with the spirit of the static pattern which could be leveled against the extension of this invariance to the entire variety. The postulates C and D are sufficient for an analytical explanation of the uniqueness of exchange equilibrium, but they fail to allow the construction of a consistent picture of human behavior. Can the "illusion" described above (Section V) be interpreted? We have been willing to consider — and later to embody in the theory — the impossibility of accurate mental comparison at a finite distance, on the mere grounds that anybody can verify this assumption by a simple introspection. But this type of argument may be reversed. Thus, why should we not reject a scheme which everybody finds paradoxical? One has to realize that what we aim at is a theory molded on a type of individual that really exists and not on a "necessary and sufficient" one.

30. So far as I know, there has been only one attempt at an experimental investigation of the nature of the indifference curves. Cf. L. L. Thurstone, "The Indifference Function," *Journal of Abnormal and Social Psychology*, II (1931), p. 139. Professor Thurstone's experiment is, however, very unlikely to help us in deciding anything about the forms of the postulates here analyzed. The investigation having been carried out by way of questions and answers, we cannot be sure whether the prices ruling on the market at the time of the experiment had or had not influenced the subjects in their answers. Some of the diagrams in Professor Thurstone's paper, namely 13 and 17, suggest on the contrary that they had. Besides, the result of mere visualization cannot be relevant to a theory concerned with an actual choice, unless the combinations used in the experiment are those with which the subject is familiar because of his latest experience. This last condition restricts the range of the experiment to a degree which simply makes the investigation useless. It seems that we cannot avoid the necessity of letting the subject experience the satisfaction before making his choice.

These considerations might, *in extremis*, be simply ignored in economics, were they confined to the explanation of the exchange process only. But they are encountered again and again under the same form in connection with items no less important, such as the marginal utility of money, price indices, individual and social welfare, etc. A consistent theory of choice forms a basic part of economic science.

2. The necessary and sufficient analytical conditions for the stability of exchange equilibrium have been established in their correct form, and the integrability of indifference elements has been shown to be necessary for their resolution into the principle of decreasing marginal rate of substitution.

The geometric interpretation of these conditions is that the curvature of the integral varieties — whenever they exist — is at any point greater than that of a linear variety, i.e., greater than zero. It is worth emphasizing here that the conditions thus obtained are valid only for an exchange with constant prices. Their form will be entirely different if the bargain is not perfected on a constant price basis, i.e., if the individual supply curve is not perfectly elastic. In the last case the budget equation no longer represents a linear variety. The stability conditions then require that the curvature of the integral varieties be greater than that of the variety represented by the budget equation. The last condition is the general one.

3. The analysis also showed the necessity of distinguishing between indifference and integral varieties. The confusion of these two concepts accounted for the paradox of the nonintegrability case. From the mere existence of the integral varieties we are not entitled to deduce that these represent the loci of constant ophelimity. The integrability of the indifference elements has less to do with the existence of constant ophelimity varieties than the condition of transitivity has. This argument is intuitively illustrated by the spiral-like integral curves (Section III).

4. Finally, a scheme derived from a different version of the postulates A and C was proposed and analyzed. The main characteristic of this scheme is the existence of a threshold in the comparison of satisfactions. As a direct consequence, a certain range of indeterminateness was shown to exist in connection with the equilibrium of exchange. The demand law appears, in this case, as a multivariate distribution between prices and quantities, or, to use a term introduced by Professor Taussig in connection with a similar problem, as a *penumbra.*[31]

From this scheme, by assuming the interval of time allowed for the mental comparison to be infinite, we obtain as a limiting case the classical static scheme formulated in terms of utility varieties (Edgeworth). In a

31. F. W. Taussig, "Is Market Price Determinate?" *Quarterly Journal of Economics,* XXXV (1921), 394.

dynamic analysis this last supposition can no longer be accepted and such an analysis will have to drop the assumption that a demand law is a rigid connection between prices and quantities.

The penumbra of demand makes way for an explanation of many facts that are observed in actual markets. In this way we may account for certain aspects of the imperfection of markets.

The general means used by the entrepreneurs for inducing people to buy more of one commodity — which necessarily implies less of some others — might be separated into two groups: those that aim at influencing the position of the buyer within the existing penumbra and those that seek to modify his tastes. The immediate task of the salesman, special sales, loss leaders, etc., belongs to the first group. The second group consists mainly of advertising on a large scale and for a long time; in other words, it consists of a real propaganda to convince buyers of the advantages of consuming more of the advertised commodity.

Staple commodities, like meat, bread, milk, etc., that are more regularly consumed and consequently experimented with longer, will naturally present a smaller threshold and thus a smaller penumbra of demand. This seems to be in complete agreement with the fact that the first group of actions on the part of the seller is practically absent in the marketing of these commodities. Such actions will be especially important and conspicuous in the case of those commodities whose use is less frequent.

There is thus a sense in which two different commodities, not necessarily competitive, can be considered as competing against each other in the short run as well as in the long.

MATHEMATICAL NOTE ON THE CONDITIONS OF STABILITY OF EQUILIBRIUM IN AN EXCHANGE WITH CONSTANT PRICES

1. The condition of stability is (9)

(i)
$$\sum_i \sum_k B_{i,k} \Delta x_i \Delta x_k < 0$$

subject to (7)

(ii)
$$\Delta x_1 + \sum_i B_i \Delta x_i = 0.$$

Eliminating Δx_1 from these two relations we get

(iii)
$$F = \sum_i \sum_k (B_{i,k} - B_{i,1} B_k) \Delta x_i \Delta x_k < 0.$$

F is a homogeneous quadratic form in $n - 1$ variables $\Delta x_2, \Delta x_3, \ldots, \Delta x_n$ which may be written under the form

(iv) $$F = \frac{1}{2}\sum_{i}{}_2\sum_k{}_2 f_{ik}\Delta x_i \Delta x_k,$$

where

(v) $$f_{ik} = f_{ki} = B_{i,k} + B_{k,i} - B_{i,1}B_k - B_{k,1}B_i.$$

Except for a constant factor of proportionality, the discriminant of the form F is

(vi) $$|F| = \begin{vmatrix} f_{22} & f_{23} & \cdots & f_{2n} \\ f_{32} & f_{33} & \cdots & f_{3n} \\ \cdot & \cdot & \cdot & \cdot \\ f_{n2} & f_{n3} & \cdots & f_{nn} \end{vmatrix}.$$

The necessary and sufficient condition for F to be definite and negative, (iii), is that the principal minors of $|F|$ should be alternatively negative and positive:[32]

(vii) $$|f_{22}| < 0; \qquad \begin{vmatrix} f_{22} & f_{23} \\ f_{32} & f_{33} \end{vmatrix} > 0; \qquad \cdots$$

By bordering the determinant (vi), F may be written

(viii) $$|F| = - \begin{vmatrix} 0 & 1 & B_2 & \cdots & B_n \\ 1 & 0 & 0 & \cdots & 0 \\ B_2 & 0 & & & \\ \cdot & \cdot & & f_{ik} & \\ \cdot & \cdot & & & \\ B_n & 0 & & & \end{vmatrix}.$$

By adding to the elements of third, fourth, $\ldots$, $(n + 1)$th row those of the first row multiplied respectively by $B_{2,1}, B_{3,1}, \ldots, B_{n,1}$, and by performing a similar operation in respect to the columns, we obtain, according to (v),

(ix) $$|F| = - \begin{vmatrix} 0 & 1 & B_2 & B_3 & \cdots & B_n \\ 1 & 0 & B_{2,1} & B_{3,1} & \cdots & B_{n,1} \\ B_2 & B_{2,1} & 2B_{2,2} & B_{2,3} + B_{3,2} & \cdots & B_{2,n} + B_{n,2} \\ \cdot & \cdot & \cdot & \cdot & \cdot & \cdot \\ B_n & B_{n,1} & B_{n,2} + B_{2,n} & B_{n,3} + B_{3,n} & \cdots & 2B_{n,n} \end{vmatrix}.$$

Since, thru the same procedure, we obtain

32. See, for instance, T. J. I'A. Bromwich, *Quadratic Forms and Their Classification by Means of Invariant Factors*, Cambridge Tracts, no. 3, 1906, p. 19.

$$|f_{22}| = -\begin{vmatrix} 0 & 1 & B_2 \\ 1 & 0 & B_{2,1} \\ B_2 & B_{2,1} & 2B_{2,2} \end{vmatrix},$$

(x)
$$\begin{vmatrix} f_{22} & f_{23} \\ f_{32} & f_{33} \end{vmatrix} = -\begin{vmatrix} 0 & 1 & B_2 & B_3 \\ 1 & 0 & B_{2,1} & B_{3,1} \\ B_2 & B_{2,1} & 2B_{2,2} & B_{2,3} + B_{3,2} \\ B_3 & B_{3,1} & B_{3,2} + B_{2,3} & 2B_{3,3} \end{vmatrix}$$

$$\cdot \quad \cdot \quad \cdot \quad \cdot \quad \cdot \quad \cdot \quad \cdot \quad \cdot \quad \cdot \quad \cdot \quad \cdot \quad \cdot \quad \cdot \quad ,$$

the conditions (vii) are shown to be equivalent with (11), i.e.,

(xi)
$$\begin{vmatrix} 0 & 1 & B_2 \\ 1 & 0 & B_{2,1} \\ B_2 & B_{2,1} & 2B_{2,2} \end{vmatrix} > 0; \quad \begin{vmatrix} 0 & 1 & B_2 & B_3 \\ 1 & 0 & B_{2,1} & B_{3,1} \\ B_2 & B_{2,1} & 2B_{2,2} & B_{2,3} + B_{3,2} \\ B_3 & B_{3,1} & B_{3,2} + B_{2,3} & 2B_{3,3} \end{vmatrix} < 0; \quad \cdots$$

2. Let

(xii)
$$\Delta x_i = \sum_{2} {}_k \alpha_{ik} y_k, \qquad (i = 2, 3, \ldots, n),$$

be a linear transformation that leads to

(xiii)
$$F = \frac{1}{2}\sum_{2} {}_k \varphi_k y_k^2.$$

Then

(xiv)
$$\sum_{2} {}_i \sum_{2} {}_k f_{ik} \alpha_{ir} \alpha_{ks} = \begin{cases} 0 & if & r \neq s, \\ \varphi_r & if & r = s. \end{cases}$$

Let us put

(xv)
$$\alpha'_{ik} = \alpha_{ki}$$

and perform the following multiplication of matrices:

(xvi)
$$[\alpha'_{ri}] \times [f_{ik} + g_{ik}] \times [\alpha_{ks}] = [\beta_{rs}],$$

where for the time being $[g_{ik}]$ is an arbitrary skew symmetrical matrix, i.e.,

(xvii)
$$g_{ik} + g_{ki} = 0.$$

According to the rules for the multiplication of matrices:[33]

(xviii)
$$\beta_{rs} = \sum_{2} {}_i \sum_{2} {}_k \alpha'_{ri} (f_{ik} + g_{ik}) \alpha_{ks}$$
$$= \sum_{2} {}_i \sum_{2} {}_k f_{ik} \alpha_{ir} \alpha_{ks} + \sum_{2} {}_i \sum_{2} {}_k g_{ik} \alpha_{ir} \alpha_{ks}.$$

33. See, for instance, H. W. Turnbull, *The Theory of Determinants, Matrices and Invariants* (London, 1928), p. 63.

But from (xvii) it follows that

(xix)
$$\sum_{2} {}_{i} \sum_{2} {}_{k} g_{ik} \alpha_{ir} \alpha_{ks} = \begin{cases} 0 & \text{if } r = s, \\ \psi_{rs} & \text{if } r \neq s, \end{cases}$$

with

(xx)
$$\psi_{rs} + \psi_{sr} = 0.$$

Owing to (xiv) and (xix), we obtain

(xxi)
$$\beta_{rs} = \begin{cases} \varphi_r & \text{if } r = s, \\ \psi_{rs} & \text{if } r \neq s, \end{cases}$$

and (xvi) yields

(xxii)
$$|f_{ik} + g_{ik}| \cdot |\alpha_{ik}|^2 = \begin{vmatrix} \varphi_2 & \psi_{23} & \cdots & \psi_{2n} \\ \psi_{32} & \varphi_3 & \cdots & \psi_{3n} \\ \cdot & \cdot & & \cdot \\ \psi_{n2} & \psi_{n3} & \cdots & \varphi_n \end{vmatrix}.$$

Let us develop the last determinant with respect to the elements of the principal diagonal. As all the principal minors of the determinant

(xxiii)
$$\begin{vmatrix} 0 & \psi_{23} & \cdots & \psi_{2n} \\ \psi_{32} & 0 & \cdots & \psi_{3n} \\ \cdot & \cdot & & \cdot \\ \psi_{n2} & \psi_{n3} & \cdots & 0 \end{vmatrix}$$

are skew symmetrical, and consequently those of odd order vanish identically, and since

(xxiv)
$$|f_{ik}| \cdot |\alpha_{ik}|^2 = \varphi_2 \varphi_3 \cdots \varphi_n,$$

the development leads to

(xxiv)
$$|f_{ik} + g_{ik}| \cdot |\alpha_{ik}|^2 - |f_{ik}| \cdot |\alpha_{ik}|^2$$
$$= S\varphi_4 \varphi_5 \cdots \varphi_n \begin{vmatrix} 0 & \psi_{23} \\ \psi_{32} & 0 \end{vmatrix} + S\varphi_6 \varphi_7 \cdots \varphi_n \begin{vmatrix} 0 & \psi_{23} & \cdots & \psi_{25} \\ \psi_{32} & 0 & \cdots & \psi_{35} \\ \cdot & \cdot & \cdot & \cdot \\ \psi_{52} & \psi_{53} & \cdots & 0 \end{vmatrix} + \cdots$$

All the determinants of the right-hand member, being skew symmetrical of even order, are perfect squares. On the other hand, the form F being definite and negative, all the products $\varphi_4 \varphi_5 \cdots \varphi_n$, $\varphi_6 \varphi_7 \cdots \varphi_n$, $\cdots$ will be positive or negative according to whether n is odd or even.

Therefore

(xxvi)
$$\begin{aligned} |f_{ik} + g_{ik}| &\geq |f_{ik}| \quad \text{for } n \text{ odd,} \\ |f_{ik} + g_{ik}| &\leq |f_{ik}| \quad \text{for } n \text{ even,} \end{aligned}$$

the equality being possible only in the case where all $\psi_{rs} = 0$.

Thus far, g_{ik} had arbitrary values. Let us now put

(xxvii)
$$g_{ik} = B_{i,k} - B_{k,i} + B_{k,1} B_i - B_{i,1} B_k.$$

Then (v) yields

(xxviii) $$f_{ik} + g_{ik} = 2(B_{i,k} - B_{i,1}B_k).$$

We have further

(xxix) $$|f_{ik} + g_{ik}| = \begin{vmatrix} 1 & B_2 & \cdots & B_n \\ 0 & & & \\ \cdot & & & \\ \cdot & & f_{ik} + g_{ik} & \\ \cdot & & & \\ 0 & & & \end{vmatrix}$$

and by adding the elements of the first row multiplied successively by $2B_{2,1}, 2B_{3,1}, \ldots, 2B_{n,1}$ to the remaining rows we obtain

(xxx) $$|f_{ik} + g_{ik}| = 2^{n-1} \begin{vmatrix} 1 & B_2 & \cdots & B_n \\ B_{2,1} & B_{2,2} & \cdots & B_{2,n} \\ \cdot & \cdot & \cdot & \cdot \\ B_{n,1} & B_{n,2} & \cdots & B_{n,n} \end{vmatrix}.$$

And according to (xxvi)

(xxxi) $$2^{n-1} \begin{vmatrix} 1 & B_2 & \cdots & B_n \\ B_{2,1} & B_{2,2} & \cdots & B_{2,n} \\ \cdot & \cdot & \cdot & \cdot \\ B_{n,1} & B_{n,2} & \cdots & B_{n,n} \end{vmatrix} \begin{matrix} \geq |f_{ik}|, \\ \leq |f_{ik}|. \end{matrix}$$

Returning now to the stability conditions (vii) or (xi), we see that if these are satisfied, from (xxxi) it follows that we shall also have

(xxxii) $$\begin{vmatrix} 1 & B_2 \\ B_{2,1} & B_{2,2} \end{vmatrix} < 0; \quad \begin{vmatrix} 1 & B_2 & B_3 \\ B_{2,1} & B_{2,2} & B_{2,3} \\ B_{3,1} & B_{3,2} & B_{3,3} \end{vmatrix} > 0; \quad \cdots$$

but that the reciprocal proposition is not true.

The two sets of conditions are equivalent if and only if all $\psi_{rs} = 0$. But this cannot happen unless $g_{ik} = 0$. Indeed, the quadratic form

(xxxiii) $$G = \sum_{2} {}_i \sum_{2} {}_k g_{ik} \Delta x_i \Delta x_k$$

is transformed by the linear substitution (xii) into

(xxxiv) $$G = \sum_{2} {}_r \sum_{2} {}_s \psi_{rs} y_r y_s,$$

and if the last one is identical null, so must be the former.

The conditions

(xxxv) $$g_{ik} = B_{i,k} - B_{k,i} + B_{k,i}B_i - B_{i,1}B_k = 0$$

are easily recognized to be those of the integrability of the total differential equation

(xxxvi) $dx_1 + B_2 dx_2 + B_3 dx_3 + \cdots + B_n dx_n = 0,$

and thus we reach the conclusion that the sets of conditions (xi) and (xxxii) are equivalent only when the differential equation of the indifference elements is integrable; otherwise the latter is a consequence of the former but not vice versa. This difference therefore does not exist in the case where $n = 2$.

2

The Theory of Choice and the Constancy of Economic Laws

I. INTRODUCTION

The first attempts of mathematical economists, such as Walras and Jevons, to introduce mathematical models for explaining consumer's behavior met with overwhelming criticism concerning the legitimacy of forcing human nature into the rigid frame of a mathematical structure. Later on, with the work of Edgeworth and Pareto, such criticism was partially met by discarding the concept of a measurable utility as a necessary element of a coherent theory of demand and by substituting the theory of choice in place of the old utility theory.

As the theory of choice was further improved, the controversy over the justification of a mathematical formulation of consumer's behavior, at least for the purposes of static theory, slowly faded away. In the meantime, modern mathematical economists carried the theory of choice to such a level that apparently very little could still be added to refine it further.[1] Today, it seems that even nonmathematical economists no longer question the results of this chapter of mathematical economics, and that the conclusion that the law of demand is a uniquely defined relationship between prices and quantities is a soundly enthroned economic law.

The adequacy of the mathematical instrument for a production theory was never challenged by the first critics of mathematical economics; for it was thought that since production consists of a series of physical and chemical processes, to analyze it mathematically meant simply to extend further the domain of natural science already shaped for such an approach.

NOTE: This paper is reprinted from *Quarterly Journal of Economics*, LXIV (1950), 125–138. In Section V of the present version, the argument illustrating the hysteresis effect by Fig. 2–2 has been recast in a more explicit form than in the original article of 1950.

1. For a complete and very instructive presentation of the present status of the theory of choice, see Paul A. Samuelson, *Foundations of Economic Analysis* (Cambridge, Mass., 1947), pp. 90 ff.

But while the theory of choice was gaining the acceptance of an increasing number of economists, some mathematical economists began, paradoxical as this may seem, to be apprehensive of previously unforeseen difficulties connected with the mathematical theory of production, such as defining the factors of production in a proper way, finding a unit measure for entrepreneurship, etc.

It is the purpose of the present paper, after submitting to a critical re-examination the postulates of the theory of choice as they stand in modern writings, to show that the paradox mentioned is only superficial, and that the difficulties related to the complexity of human nature were only avoided, but not solved, by such a theory. The bearing of the new point of view upon the constancy of static laws will also be discussed.[2]

II. THE THEORY OF CONSUMER'S CHOICE

The main postulate of the theory of choice is embodied in the proposition that, confronted with two different combinations of commodities $C^1(x_1^1, x_2^1, \ldots, x_n^1)$ and $C^2(x_1^2, x_2^2, \ldots, x_n^2)$, where x_j^i stands for a certain flow of the commodity X_j per unit of time, the consumer *is able* to decide upon their ophelimity ordering in a unique way, i.e., on one and only one of the following possibilities:

(1) He prefers C^1 to C^2;
(2) He prefers C^2 to C^1;
(3) The combinations are indifferent to him.

The postulate is further supplemented by some transitivity conditions which lead to the conclusion that, given C^1, all combinations indifferent to it will be represented in the space $(X_1, X_2, \ldots, X_n)$ by an $(n-1)$-dimensional variety.[3]

The usual attitude adopted by different writers on the subject, including the author, was to investigate, as a next step, the shape of the indifference varieties. But, on second thought, this appears to be rather a rash logical procedure. For the above postulate cannot be transformed into a useful theoretical tool without further qualifications.

In the first place, one needs to know whether this postulate is to be regarded as a rigid law or as a limit formulation of a stochastic relation-

2. The author wishes to acknowledge the helpful criticism of Professors W. W. Leontief and A. Smithies and especially the suggestions he has received from Professor E. H. Chamberlin, who has revised the manuscript. Naturally, the author is solely responsible for whatever faults the present paper may contain.
3. In the case of two commodities, this means that the combinations indifferent to C^1 form a curve passing through C^1 and therefore do not cover a plane area.

ship. In other words, under the assumption that prices $(p_1, p_2, \ldots, p_n)$ and income are given, does the postulate imply that, by mutually comparing all combinations compatible with the given data, the individual will hit *at once* upon the most preferred combination or that such a position will be reached after many trials and errors, if ever? If the latter alternative is accepted, the classical economic theory of consumer's equilibrium will be but little affected. The economic models alone, when confronted with the reality, would have to include some random components proper to the stochastic structure of consumer's choice and distinct from the random errors related to actual measurement.[4]

Another point, of a much more profound nature, to be raised in connection with the postulate of choice as formulated above is whether the mapping of the indifference varieties does or does not depend upon the economic experience of the individual. In other words, whether the indifference varieties of the *same* individual could or could not be treated as an invariant element by the static theory.

The evidence in support of assuming that they are not invariant could be found in the shift of the budget distribution of an individual who has experienced temporarily either a different income or a different price constellation. Such temporary experiences show their traces when the individual comes back to his previous budget limitations.[5]

Moreover, it is a currently observed fact that an induced increase in the consumption of some commodities brings about a change in the demand for them even if the factors that induced that increase have disappeared.[6] Were it not so, that part of commercial propaganda which seeks to increase the demand for a certain commodity by inducing people to become more accustomed to its use would lose all rationality.

From another point of view, neglecting the influence of new experiences

4. Such a point of view was presented in an earlier paper of the author. It leads to considering the demand law as a bivariate distribution instead of a unique curve relating prices and quantities. See "The Pure Theory of Consumer's Behavior" [the preceding essay in the present volume]. Stochastic aspects of the theory of choice related to a dynamic scheme based on subjective risk and uncertainty, entirely different from those explained in terms of the individual's inability to make a sure choice between two different present combinations of goods, have been treated by Gerhard Tintner in "The Theory of Choice under Subjective Risk and Uncertainty," *Econometrica*, IX (1941), 298–304.

5. An extremely interesting illustration of the influence of past experiences upon the individual's present behavior is to be found in the scheme presented by James S. Duesenberry, "Income-Consumption Relations and Their Implications," in the volume *Income, Employment and Public Policy* (New York, 1948), in honor of Alvin H. Hansen, and by Franco Modigliani, "Fluctuations in the Saving-Income Ratio: A Problem in Economic Forecasting," in *Studies in Income and Wealth*, vol. XI, National Bureau of Economic Research (New York, 1949). Both these authors have argued, and supported their argument by actual macro-economic data, that the individual's spending depends not only upon his current real income but also on his highest real income in the past.

6. Cf. Alfred Marshall, *Principles of Economics* (8th edn., New York, 1949), p. 807.

upon the indifference map would be equivalent to setting an arbitrary limit to an evident and essential characteristic of human nature, namely that of being subject to change as a result of education, travel, personal experiences, etc. To set a limit to the domain where such hereditary influences are a normal feature of human nature would be logically unjustifiable.

It should, however, be stressed that the hereditary mechanism just described must not be confused with the influence upon individual tastes due to exogenous (from the point of view of the individual) economic actions taken by others, such as advertising, conspiracy for creating false social distinction, etc. The proper hereditary influences are an inborn feature of the individual. They are present irrespective of whether exogenous forces are at work or not. The latter may be overlooked in a static theory, the former never.[7]

The modifications of the theory of choice necessary to allow for hereditary influences will affect both the economic theory and the econometric approach. To facilitate the analysis of the new pattern of consumer's behavior, a tentative set of postulates will be here presented. Though they will be formulated in a mathematical form, the discussion will be supplemented as much as possible by ordinary argumentation, for in the author's opinion the model under discussion offers an instructive example of a structural aspect of economic laws which requires the introduction in economic theory of a new type of operator. Mention of this structural aspect is to be found in most economic writings, although a mathematical framework for describing it has not yet been attempted. It will be seen that this new operator will help to clarify some blurred and controversial points related to the constancy of economic laws, a problem of equal interest for mathematical and nonmathematical economists.

III. FOUR QUALIFYING POSTULATES

Only those postulates dealing with the existence and the nature of the indifference varieties will be examined here. This does not mean that the shape of the indifference varieties is not worthy of investigation, but simply that the shape-properties are of little relevance for the object of this paper.

The first postulate affirms the latent capability of the individual to draw his own indifference map:

7. Unless by static theory we understand a theory dealing with a petrified economy, from which all change is eliminated from the start, and from which, to be consistent, all maps and curves should be eliminated also.

A. The Choice Postulate

1. The consumer *is able*, at any moment of his life, to choose between two combinations of flows of commodities C^1 and C^2, meaning by such a choice that there is a definite theoretical frequency (or probability), $p(C^1, C^2, t_1, t_2)$ connected with the choice of C^1 in preference of C^2. By t_1 and t_2 we represent respectively the length of time during which the individual has already experimented with C^1 and C^2.[8]

Let us introduce the notations

$C^1 \supset C^2$	if	$p(C^1, C^2, t_1, t_2) \equiv 1,$
$C^1 \subset C^2$	if	$p(C^1, C^2, t_1, t_2) \equiv 0,$
$C^1 \supset\subset C^2$ or $C^2 \supset\subset C^1$	if	$0 < p(C^1, C^2, t_1, t_2) < 1,$

where $C^1 \supset C^2$ means that C^1 is definitely preferred to C^2 and $C^1 \supset\subset C^2$, that, although the individual makes a choice, there exists some doubt in his mind about this being the right one.

B. The Transitivity and Sequence Postulate

1. From $C^1 \subset C^2$, $C^2 \subset C^3$, it follows that $C^1 \subset C^3$.
2. From $C^1 \subset C^2$, $C^2 \supset\subset C^3$, it follows that either $C^1 \subset C^3$ or $C^1 \supset\subset C^3$.
3. If C^2C^3 is a non-negative vector (i.e., if the combination C^3 is formed by adding certain positive quantities of some commodities to C^2), then, whatever be C^1,

$$p(C^1, C^2, t_1, t_2) \leq p(C^1, C^3, t_1, t_3),$$

the equality being possible only when both p's are equal to 1 or 0.

4. Whenever C^1C^2 is non-negative, $C^1 \subset C^2$.

Both 3 and 4 correspond to the fact that, irrespective of his past experience, the consumer reacts favorably to any increase in the flow of commodities.

Another convenient, yet not absolutely necessary, postulate will greatly simplify the stochastic structure of the model and will give a meaning to the indifference varieties:

C. The Limit Postulate

1. If t_1 and t_2 tend towards infinity, $p(C^1, C^2, t_1, t_2)$ tends towards one of the following limits: 0, $\frac{1}{2}$, 1.

If $\lim p(C^1, C^2, t_1, t_2) = 0$ (or $= 1$), we shall write $C^1 < C^2$ (or $C^1 > C^2$). If $\lim p = \frac{1}{2}$, C^1 and C^2 will be called indifferent and we shall write $C^1 = C^2$.

---

8. It is rather curious that this time aspect of the problem has never been considered by the pioneers of marginal utility. The case they apparently had in mind was that of a consumer with perfect knowledge, corresponding to t_1 and t_2 equal to infinity. This may be accepted only in the case of some essential commodities, such as water, bread, etc. One may find in this the reason why the earlier writers used only commodities of this type for illustrating the principle of marginal utility.

We may further assume that the indifference varieties constitute an invariant of the stochastic scheme:

2. For two indifferent combinations C^1 and C^2

$$p(C^1, C^2, t_1, t_2) \equiv \tfrac{1}{2}$$

for any values of t_1 and t_2.

The last postulate could be justified by the fact that, in a stochastic scheme, "true" values should be independent of the degree of precision and therefore be an invariant for the entire sampling field.[9] It will enable us to leave aside, from now on, the stochastic aspect of consumer's behavior and to limit ourselves to the indifference varieties. Their existence is guaranteed by $(B, 3)$ while their real significance follows from $(C, 2)$.

D. The Heredity Postulate

1. The indifference varieties depend upon the economic experience of the individual.

For a more precise formulation of this postulate, let us denote by S the set of all combinations which the individual has experienced in the past. At a given moment of the consumer's life, his indifference varieties are represented by the equation

(1) $\varphi(C; S) = \text{const.}$

where φ is a point function of C and a set function of S.

The structure of the set function φ is certainly far from being a simple one. There seems to be no evidence to support the introduction of simplifying assumptions in order to restrict the generality of this structure.[10]

In respect to a given economic experience of the individual, all combinations can be divided into three groups:

a. Those that have been experimented with in the past. A re-experimenting with such combinations will not alter the indifference map.

b. Those that have not been experimented with, but for which experimentation will bring about no change in the indifference map. Such combinations are, so to speak, implicitly covered by the given past experience.

c. Those that have not been experimented with, but for which experimentation will alter the indifference map. They represent *new relevant* experiences.

9. Such is the case for any physical quantity. Its "true" value should be independent of the degree of accuracy of the observations, if the latter are free from systematic errors.

10. Referring to the simplest types of set functions, φ is neither *additive* nor is it a *measure function*. Cf. Hans Hahn and Arthur Rosenthal, *Set Functions* (Albuquerque, N.M., 1948), pp. 11, 61.

Combinations (a) and (b) could be included together in the set S without any restriction. The set thus defined will be referred to as a *complete* set of the individual's experience and denoted by $\bar{S}$.[11] In equation (1) above, S can be replaced by $\bar{S}$.

Evidently, the individual's position is always inside $\bar{S}$, for as soon as he reaches a new relevant position the set $\bar{S}$ will change so as to include this combination also.

There seems to be no reason to assume discontinuities of φ in respect to $\bar{S}$, except when a *new* commodity is introduced in $\bar{S}$.[12] The set function φ will be indeterminate for $S = 0$, for the individual must have some economic experience in order to have any indifference map at all.

When drawing the indifference map for a given $\bar{S}$, account should be taken of whether the two combinations C^1 and C^2 do or do not belong to $\bar{S}$. If neither C^1 nor C^2 belongs to $\bar{S}$, their ophelimity ordering corresponds to a *mentally* projected comparison based however on the experience $\bar{S}$. If only (say) C^1 belongs to $\bar{S}$, then the comparison is mentally projected at one end only. So long as at least one of the two combinations does not belong to $\bar{S}$, the ophelimity ordering of C^1 and C^2 will not necessarily be the same *before* and *after* experimenting with the one not included in $\bar{S}$.[13]

Consequently, it is no longer possible, as in the Pareto scheme, to derive from the indifference map an *absolute* ophelimity ordering of all combinations of commodities. From this point of view, although moving to a mentally preferred (according to his present $\bar{S}$) position, the consumer might not feel better off after he has actually experimented with this new combination.

Solely with the view of illustrating some characteristics of the model presented in this paper, let us assume that there are only two commodities, X and Y, and that

$$(2) \qquad \varphi(C; S) = X^{\xi(\bar{S})} Y^{\eta(\bar{S})}.$$

11. It may perhaps seem legitimate to assume that if two combinations C^1, C^2 belong to $\bar{S}$, any combination C on the straight line C^1C^2, between C^1 and C^2, should also belong to $\bar{S}$. This would make all complete sets convex. Though this might be so if C^1 and C^2 are not too far apart, it is doubtful that this will be the case in general.

12. In the latter case, the indifference map may change either because the individual has for the first time the possibility to experiment with a commodity already existing on the market or because an entirely new commodity has been introduced as the result of technological progress in trade or industry.

13. This aspect should not be confused with the famous nonintegrability case which affects only the transitivity condition. Even if nonintegrability, but nothing else different from the classical scheme, is assumed, the individual returning to any previous situation (i.e., the same price constellation and the same income of any previous stage) will choose the same equilibrium position as held before, independently of his new experience during the interval. Neither should the same aspect be interpreted as a discontinuity of φ.

Let us assume further that the past economic experiences of the individual are represented by the combinations α, β, γ, δ (Fig. 2–1). For determining the values of the set functions $\xi(\overline{S})$, $\eta(\overline{S})$, the following scheme may be used (again as mere illustration):

The set $\overline{S}$ is represented by the smallest convex contour which contains all points of the set S as well as two other fixed points I and J. Let a be the ratio between the area of this contour (the shaded area) and that of the triangle OIJ. Then write, for example,[14]

$$(3) \qquad\qquad \xi(\overline{S}) = \frac{1+a}{2+3a}, \qquad \eta(\overline{S}) = \frac{2+5a}{1+2a}.$$

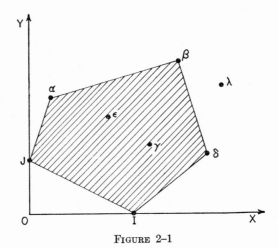

FIGURE 2–1

Supposing now that the individual experiments with a new relevant combination, say λ. The value of a will change and consequently the indifference map will change also. Even if the individual comes back to the same situation which led him to (say) α, he will no longer find his economic equilibrium in α. There are, however, positions (such as ϵ in the present schematic example) the experimenting with which will bring about no change in the indifference map.

14. Or take the following alternative example: $\overline{S}$ is the smallest convex contour containing S; x_0, X_0, y_0, Y_0 are the smallest and greatest values of X and Y in $\overline{S}$; $m(Y_0)$, $M(Y_0)$, $m(X_0)$, $M(X_0)$, are the smallest and greatest values of X for $Y = Y_0$ and of Y for $X = X_0$ in $\overline{S}$. Then,

$$\xi(\overline{S}) = \frac{\Sigma A X_0 + \Sigma A_1 m(Y_0)}{\Sigma a X_0 + \Sigma a_1 m(Y_0)}$$

$$\eta(\overline{S}) = \frac{\Sigma A' X_0 + \Sigma A_1' m(Y_0)}{\Sigma a' X_0 + \Sigma a_1' m(Y_0)}$$

where A, A_1, a, a_1, ... are physical constants.

In reality, the structure of the set function φ is exceedingly complex, especially since the time element has to be introduced as a measure of the intensity of each economic experience both with respect to the duration of the latter and to the interval of time elapsed since this experience took place. There is little hope that the complex structure of φ will ever make possible any actual investigations aimed at discovering some of its main features. The model does not however exclude the theoretical justification of a possible similarity between the indifference maps of different individuals to such a point that their grouping together would be legitimate for practical purposes. But only actual investigations of a deeper nature than those carried out so far in connection with consumers' budgets could decide whether such a hypothesis would also be realistic.

IV. THE CONSTANCY OF ECONOMIC LAWS

The immediate consequences of the model discussed here refer to the question of the constancy of economic laws and to that of the determinateness of consumer equilibrium.

According to this model, at a given position (M_1), the consumer has a definite demand (D_1) for X. If a change in price brings him to a new position (M_2) capable of altering his indifference map, his demand for X will implicitly be changed into a new curve (D_2). Save for an irrelevant coincidence, no new shift in prices *alone* could bring the consumer back to the former position (M_1). This irreversibility, mentioned in passing by Trygve Haavelmo in an admirable paper,[15] appears with this model to be the normal case, whereas reversibility is the exceptional one.

The classical theory of choice depicted *homo oeconomicus* as having an invariant behavior in quantitative terms, independent of his past experiences. The intermediate positions adopted before the ultimate equilibrium was reached had, in the classical theory, no influence upon the process by which the equilibrium was obtained. Such a point of view led to a theory that was not only a poor approximation to the economic reality but also essentially different from it. Demand curves were considered as depending solely upon income and prices of the other commodities, and when attempts to obtain statistical demand curves valid for the next period failed, economists turned to challenge the constancy of economic laws and implicitly the foundations of any economic theory. The above analysis shows that those attempts were vitiated not by a false stand regarding the constancy of economic laws, but by the ignorance

15. Trygve Haavelmo, "The Probability Approach in Econometrics," *Econometrica*, Supplement, July 1944.

of other possible shifts of the demand curves, distinct from those caused *directly* by a variation of prices or of income.

The derivation of statistical demand curves by the already known methods is therefore bound to be meaningless, for such an attempt is caught between two equally unfavorable alternatives: either (1) the variation of prices and income over the observed period are very small so that they may be assumed to bring about no change in the indifference map; but then the statistical errors of the best estimates will be so great that the latter will be meaningless; or (2) prices and income have sufficiently large oscillations to ensure a small error of the best estimates; but then these estimates will no longer have a meaningful correspondence in the model.

V. DETERMINATENESS OF CONSUMER EQUILIBRIUM

Does the preceding analysis justify the negation of the constancy of economic laws? The right answer seems to be that, on the contrary, it eliminates the variability of consumer's behavior as an eventual argument against such a constancy. However, the micro-approach is deprived to a large extent of any quantitative predictability over finite (i.e., important) changes.

It may yet be argued that the hereditary influence included in the model is not compatible with the usual static assumption of absence of learning. If adopted, such a point of view would deprive static theory of even a "home work" usefulness. It is more plausible to think that classical static theory did not take into consideration the hereditary factors, as distinct from the evolutionary ones, because the rigid theory of consumer's choice was looked upon as a satisfactory and complete explanation.

A second argument which could be advanced here is that the static theory assumes perfect knowledge. In the case of consumer theory, perfect knowledge could mean that the individual has experienced all possible combinations. But this would be a highly unrealistic assumption. There is, however, another possible interpretation of perfect knowledge. This may mean that $\varphi(C; \overline{S})$ has such a structure as will enable us to derive, in a finite number of steps, any of its values knowing only some of them.

The function φ may be given, but this does not necessarily imply that we can operate with it in the described manner. The function being given might simply mean that there is a way of determining its values. Thus, the number π (= 3.14159 . . .) is in this sense given, but its structure does not allow us to write down all its decimal digits. The numerical illustration, (3), towards the end of Section III, above, does correspond to this interpretation of perfect knowledge. But the real structure of consumer's behavior could hardly be as simple as this.

Perfect knowledge in the case of (3) is equivalent to making $a = \infty$, which leads to

$$\varphi = X^{\frac{1}{3}}Y^{\frac{5}{2}},$$

which represents the indifference map of perfect knowledge.

This was possible here because the definition adopted for the particular function (3) included, in a concealed way, the assumption that the individual is able to extrapolate up to infinity his present and limited experience, without actually experimenting with all possible combinations. The example in note 14, above, proves that this is not always so.[16]

If perfect knowledge, in one sense or the other, is not assumed, the uniqueness of equilibrium is no longer a normal feature of the model. Indeed, the definition of equilibrium is in this case susceptible of different formulations:

A. The equilibrium is that position which is mentally foreseen by the individual before experimenting with it.

B. The equilibrium position is that which satisfies the equilibrium conditions if the individual hits upon it at once.

C. The equilibrium position is that which satisfies the equilibrium conditions, but which is reached after a number of normal trials and errors have taken place.

As a mere support for the argument, let us refer to the example of Section II, and suppose that the only experience of the consumer is represented (Fig. 2–2) by the point $M(\mu, 0)$, where μ is his income ($\mu > 1$), and that the prices $p_x = p_y = 1$. Let us also take $I(1, 0)$, $J(0, 1)$. After the individual moves to any position $P(x, y)$ on the budget line MN, his indifference curves will be given by (2) and (3) where

$$a = (\mu - 1)(y + 1).$$

16. In economic literature, one finds very often the implicit assumption that indifference curves are expressed by analytic functions, i.e., by such functions that from the knowledge of their values over a small region it is possible to compute *without further observations* any other value of the same functions. Cf. Arthur Smithies, "The Boundaries of the Production Function and the Utility Function," *Explorations in Economics*, Notes and Essays in Honor of F. W. Taussig (New York, 1936), p. 326.

Properties not supported by evidence, however weak, cannot be attributed to indifference curves. The numerical results of the experiments sporadically carried out on individual preferences display neither the regularity nor the stability needed for supporting the hypothesis that the functions involved are analytic. And, from all we know, the stringent requirements for evidence that the functions are analytic are inconsistent with the groping manner in which man decides what he prefers. Of course, for a rough analysis of a concrete and restricted problem one has no choice other than to work with quite simple analytical formulae, such as those used by Modigliani and Duesenberry in their macro-economic approach.

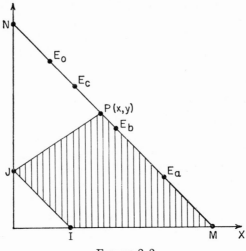

FIGURE 2-2

The position, $E_p(X_p, Y_p)$, of the equilibrium mentally foreseen from P is then given by

$$(4) \qquad X_p = \frac{\mu\xi(a)}{\xi(a) + \eta(a)}, \qquad Y_p = \frac{\mu\eta(a)}{\xi(a) + \eta(a)}.$$

If in these relations we make $a = \mu - 1$ (corresponding to $y = 0$) we obtain the equilibrium position, $E_a(X_a, Y_a)$, as in definition A for the initial position M. And if we put $a = \mu^2 - 1$ (corresponding to $y = \mu$) we obtain $E_\infty(X_\infty, Y_\infty)$, which represents the equilibrium under the assumption that the individual's past experience covers, directly or indirectly, *all* budget combinations.

Clearly, $0 < Y_a$, $Y_\infty < \mu$. On the other hand, a little algebra shows that $dY_p/dy > 0$, and also that the equation $Y_p = y$ has only one root in the interval $0 \le y \le \mu$. Hence, if Y_b is this root, we have $Y_a < Y_b < Y_\infty$ and

$$(5) \qquad \begin{array}{lll} y \le Y_p \le Y_b & \text{if} & y \le Y_b, \\ y > Y_p > Y_b & \text{if} & y > Y_b. \end{array}$$

The point $E_b(X_b, Y_b)$ corresponds to the equilibrium as stated in definition B, but it also fits definition C. Indeed, a simple cobweb diagram shows that E_b is approached in the limit if the individual, starting from M, always moves to the next equilibrium mentally foreseen by him.

Finally, let us consider the case where, due to the stochastic factor (Postulate A), from M the individual moves first to $E_0(X_0, Y_0)$, $Y_0 > Y_b$. According to (5) the equilibrium mentally visualized from X_0 corresponds to $E(X_c, Y_c)$, $Y_c < Y_0$. This means that E_c belongs to $\overline{S}$ now. And if the individual moves to E_c — as he very likely will — the move will not alter the body of his experiences: E_c does not represent a *new* experience. Once the individual reaches E_c, there is no reason for him to move away from it. E_c represents an equilibrium according to definition C. However, since its position depends upon that of E_0, this sort of equilibrium is indeterminate. It is rather unfortunate that C seems to be the most realistic assumption to be made.

One might think of avoiding such an uncomfortable conclusion by requiring that the consumer will not decide upon his ultimate equilibrium before trying out *all* combinations compatible with the budget restrictions. But such a tryout would be an unrealistic assumption to make if the great number of commodities with which a consumer is usually confronted in real life is taken into consideration. Yet even such a concession would be made in vain, for the time element of the experiment will constitute sufficient grounds for the indeterminateness of equilibrium, unless we take refuge in a still more singular assumption, namely that of infinite time.

VI. CONCLUSIONS

Despite the insuperable complexity of the complete picture of consumer's behavior, some "truths" of static theory remain unaffected by the model introduced in this paper.

Such is, for instance, the statement that an increase in price will bring about a decrease in the quantity demanded. However, let us not be misled. This statement is not a quantitative law in the usual sense of the word, for it does not imply a reversible relationship between prices and quantities, but is simply a common feature of all demand laws, whether reversible or not. From this to the representation of the demand law by a curve relating prices and quantities is a very long way. Yet, the step is commonly taken in connection with all similar truths in which economic theory abounds.

Could we feel justified in concluding that reversible laws, if they are to be discovered at all in economics, are possible only in macro-economics, especially since a similar structure based on a set function could also be introduced in a theory of investment decisions?

3

Choice, Expectations,
and Measurability

For arguments about matters concerned with feelings and actions are less reliable than facts: and so when they clash with facts of perception they are despised, and discredit the truth as well.

Aristotle, *Ethica Nichomachea*, 1172a–b

I. INTRODUCTION

Economists learned only recently, and with no little surprise, of the revolutionary consequences that could be derived for our science if the basic ideas used by Daniel Bernoulli and Gabriel Cramer in their solution of the Saint Petersburg paradox were further developed by an extensive mathematical analysis. The event was marked by the publication in 1944 of the epoch-making work of Professors von Neumann and Morgenstern, *Theory of Games and Economic Behavior*,[1] which among other things informed economists — at least — that utility *is* measurable. However, this discovery soon spread general uneasiness. The reason for this has its origins in the following conflict: It is only "we" who can say what "measure" should mean, and it is only "we" who *immediately* should — according to the theory — arrive at the measure of utility. And when "we" end our search without finding in "us" what "we" have defined in advance, certainly something must be wrong.[2] Two symptoms found

NOTE: This paper is reprinted from *Quarterly Journal of Economics*, LXVIII (1954), 503–534.

1. A complete list of references appears at the end of this essay, p. 213.

2. This is, perhaps, a more precise way of expressing, for this particular purpose, the idea of Professor Hayek, who in his recent book insists with such timeliness that what distinguishes the essence of moral from that of natural phenomena is the fact that the moral phenomenon implies man's awareness of his participation in it (Hayek, 2, part I, chap. iii). It is with regard to moral sciences, more than to anything else, that it is appropriate to follow one of William Blake's *Proverbs of Hell:* "Truth can never be told so as to be understood, and not be believed." (Quoted by Ramsey, p. 156.)

in recent economic literature show that this problem is far from being satisfactorily answered. In the first place, almost every economic periodical takes up with remarkable frequency the subject of explaining once more the rationale of the measurability of utility. The recent publication in successive issues of the *American Economic Review* of two very meritorious articles by Professors Alchian and Strotz,[3] both aiming to explain in an "elementary" way the meaning of utility measurements, shows clearly how badly some type of explanation is needed. Professor Alchian (at his p. 26n) even confesses his "early impulses to abandon the attempt to understand recent utility theory." Last, but certainly not least, there is the disagreement regarding cardinalism, not among the rank and file, but among those economists who, because of their important contributions to the theory of the consumer, are veterans in the field.

As Professor Wold (p. 661) has "always felt some doubt as regards the relevance of the idea of measurable utility, the new [Samuelson's] argument adds to [his] hesitation." Dr. Manne (p. 667) follows Blake's mentioned proverb and states categorically that he does not "feel convinced by this line of reasoning — *reasonable though it appears.*" [4] But the most eloquent case is offered by Professor Samuelson, whose important contributions to a noncardinalist theory of the consumer earned him a great part of his world-wide esteem. While writing in support of the cardinalist point of view, he still finds reasons for blowing both cold and hot on the theory (6, p. 677) and reserves for his concluding remarks the admission that he also is not sure about being able to find what he has defined in advance (p. 678). Professors Friedman and Savage, to whom we owe a version of the cardinalist doctrine slightly different from the Neumann-Morgenstern one, lost a great deal of their aplomb between their first and second papers dealing with the same subject (1 and 2), and end their recent defense of the cardinalist doctrine with cautious reservations (2, pp. 473–474). On the other hand, Professor Strotz believes that just a little simple coaching will suffice to convert to the cardinalist faith "any person not deemed insane" (p. 393). In this, Strotz follows Professor Marschak (1, p. 111–112; 2, p. 493), who has adopted the attitude that the theory can be saved by simply treating as irrational behavior that which does not offer a support for its conclusions, on the ground that "the rational man does not make logical and arithmetical errors" — i.e., is always right. The only difficulty with such a stand is that it decides authoritatively that, between the two conflicting positions, that of the consumer who cannot believe the theory and that of the cardinalist, the latter's is the rational one.[5] Is it not, though, the greatest

3. See list of references.
4. Italics added.
5. Yet the cardinalist may admit, as Samuelson frankly did, that his other side, the consumer, is not "rational"! (6, p. 678).

of all irrationalities to assume that any given individual, be he a cardinalist, is *ex definitione* rational in the above sense?

It is true that no error can be found in the mathematical proofs, either in that of Morgenstern-Neumann or in that of Marschak. But this alone does not justify the conclusion that the cardinalist is rational: *his error may reside in his choice of assumptions.* It was mainly this thought which prompted the author to take a second look, not at the proofs, but at the assumptions on which the cardinalist argument rests.

Most of these assumptions were borrowed from the ordinalist theory of the consumer, as developed mainly by the contributions of the thirties which followed the path-breaking "Reconsideration" of Professors Allen and Hicks. However, the basic ideas involved in these assumptions were much older, very old indeed, as we shall presently see. And since the ordinalist assumptions and conclusions developed no difficulties of their own, the foundations of the ordinalist theory acquired ultimately the prestige of unchangeable truth. This explains why when the ordinalist-cardinalist controversy developed later, both camps thought it vital to attack and defend only that (unique) assumption which they considered as specific to the cardinalist edifice:[6] the *Strong Independence Axiom*, as Samuelson called it. If this trodden path is abandoned and if the pure ordinalist assumptions are resubmitted to critical examination, one reaches the conclusion that it is these rather than the Strong Independence Axiom which are the seat of the main trouble.

For historical, and also accidental, reasons, in the folds of the ordinalist description of consumer's choice there was couched much more than the ordinalists intended to accept. It is doubtful that they ever have been aware of this; thus, their very own arguments became the chrysalis of the cardinalist doctrine.

II. BASIC POSTULATES OF CHOICE

How much of the complexity of real phenomena is reflected by a scientific model is a matter which depends upon various and varying factors. And although, more especially in moral sciences, there is no general agreement as to which factors should be included in the model, no inquiry can proceed in a useful way unless the model is clearly circumscribed. This is necessary in order to avoid synthetic truths being introduced *ad libitum* during the successive stages of the logical argument. The fact that many economic controversies had really no object and were caused only by the confusion between synthetic and analytic truths has

6. Cf. Friedman and Savage (2, pp. 468–469); Alchian (p. 37); Samuelson (6, p. 677).

paved the way for an increasing use of axiomatization.[7] Even in the case, such as the present analysis, where the main interest is focused on the realistic aspects of the assumptions, rather than on their purely logical implications, axiomatizing is highly indicated.[8]

When it came to deciding which factors are *essential* to the economic actions of *man*, it was found that two aspects are likely to create insuperable analytical difficulties if accounted for in the model *homo oeconomicus*. The first is that *man* is a continuously changing structure; the second, that his reactions to the environing universe are affected by a psychological threshold.[9] Although these aspects are very frequently mentioned in the literature — if not explicitly at least implicitly — their inclusion in any theory of the consumer has been systematically avoided.[10] Since both ordinalist and cardinalist arguments have been thus propounded in terms of a *homo oeconomicus* identified with a choice mechanism which is both *invariant and perfectly exact*, it will be necessary to preserve the same background for the first part of the present analysis. Following the usual line of the ordinalist argument, all situations involving risk or uncertainty will not be considered, at least for a while. This will aid in presenting a clearer picture of the new points.

For convenience of expression, let us refer to this model of the consumer as *homo oeconomicus A*, or *HOA*. It is described by the following postulates:[11]

(II.1) *HOA is confronted only with alternatives represented by combinations of various commodities that involve no risks and no uncertainties. The commodities are quantity-measurable[12] and every point $C(x_1, x_2, \ldots, x_n)$ in the commodity space is an alternative.*

(II.2) *When confronted with two alternatives C^1, C^2, HOA will either (a) prefer one (C^k) to the other (C^j), or (b) regard the two alternatives as indifferent. Indifference is a symmetric relation; preference is not. (For preference we write $C^k P C^j$, for indifference, $C^1 I C^2$.)*

7. However, it would be a fatal error to encourage this use for the sake of logical elegance alone, lest the significance of economic science be reduced to that of some "magic squares."

8. It is almost certain that Sir Arthur Bowley (pp. 1–2) was the first to preface a demand theory by a formal axiomatization of the consumer. He was soon followed by Professor Frisch (1, pp. 3–5).

9. Pareto (p. 174).

10. The only start in the opposite direction seems to be that made by two earlier papers of the author, Georgescu-Roegen (1 and 2) [in the present volume, the two essays preceding this one].

11. There are various approaches to consumer theory. Most of them have their roots in the works of Pareto, even when they do not seem so prima facie. The following scheme follows, however, the line inaugurated by Frisch (1).

12. This means that once a certain quantity is chosen as the *unit*, all other quantities have a unique measure.

(II.3) *The preference of HOA is the same every time he is confronted with the same C^1 and C^2.*

(II.4) *Around the origin of coordinates, there is a region where C^1 is preferred to C^2 if C^1 is obtained by adding to C^2 more of at least one commodity.*

(II.5) *The relation $\overline{P}$ of nonpreference is transitive, i.e., if $C^1\overline{P}C^2$, $C^2\overline{P}C^3$, then $C^1\overline{P}C^3$.*

(II.6) *If $C^1\overline{P}C^2$, $C^1\overline{P}C^3$, then $C^1\overline{P}[\alpha C^2 + (1 - \alpha)C^3]$, where $0 \leq \alpha \leq 1$.*

It should be remarked, in the first instance, that the above postulates include also some important properties of the "commodities." Thus (II.1) states not only that they are continuous and quantity-measurable but also that they can always be measured exactly. The first part is found in the literature under "divisibility"; the latter is seldom expressed explicitly.[13]

Changes of tastes are excluded by (II.3).[14] This latter postulate in conjunction with (II.2) excludes also the psychological threshold, thus making *HOA* a *perfect* choosing instrument. Postulates (II.2) and (II.5) are specific to the theory of choice, as distinct from the utility theory of the consumer, and have been used for the first time by Frisch (1).[15] Postulates (II.4)[16] and (II.5)[17] may be regarded as representing the rational traits of *HOA*.

Postulate (II.6) describes an internal structure of the choosing individual: it may be called the *Principle of Complementarity*. It simply says that 100 per cent mix of some preferred alternatives is still a preferred situation, no matter what the composition of the mix is. This may seem at first entirely trivial; however, it is easily seen that this is not so by trying the principle on nonpreferred alternatives. If $C^2\overline{P}C^1$, $C^3\overline{P}C^1$,[18] the

13. See, however, Georgescu-Roegen (1, p. 572n) [i.e., "The Pure Theory of Consumer's Behavior," note 24, in the present volume].

14. The importance of this postulate was realized much earlier than that of the others. Cf. Marshall (p. 94), Pareto (pp. 260 ff).

15. Pareto, despite his analysis of the open and closed cycles (p. 556), never mentioned transitivity. The fact that postulate (II.2) is too strong for the theory of consumer's equilibrium is proved by the author in Georgescu-Roegen (3).

16. Frisch (1 and 2) does not mention (II.4). This leads him into difficulties which he overcomes by making additional metric assumptions (1, p. 5). The idea expressed by (II.4) is found in Jevons (p. 24) and Pareto (p. 152). Its incorporation into a consumer's theory appears in Georgescu-Roegen (1, pp. 550, 557n) [in the present book, pp. 137, 142n16].

17. The "integrability condition" is no longer troublesome once the transitivity postulate is accepted, Georgescu-Roegen (1, pp. 568, 584–585) [in the present book, pp. 150, 163–164]. This relationship was apparently misunderstood for some time since the problem of integrability was considered as either not being "really an important problem," Samuelson (1, p. 68), or just a "chasing a will-o'-the-wisp," Hicks (2, 19n). Recently, however, the trend seems to have been reversed. See Samuelson (5); also the short but significant remarks in Corlett and Newman.

18. $\overline{P}$ means the negation of P. Thus $C^2\overline{P}C^1$ means "either C^1PC^2 or C^1IC^2."

alternative formed by half of C^2 and C^3 may very well be preferred to C^1, because some additional *complementarity* can reveal itself in the new combination.

It is highly interesting to note that the Complementarity Postulate would have no meaning if the commodities were only ordinally measurable, *because "half" of C would not be then uniquely defined.* In fact, any law describing the structure of consumer's behavior depends upon the *type* of measure used for commodities. The postulate of decreasing marginal rate of substitution, whose place is taken here by (II.6), is in this category, too. The two postulates are, however, not necessarily equivalent. Their logical equivalence depends upon the actual existence of the indifferent alternatives, and, as surprising as this may seem, the above postulates do not necessarily imply it.[19]

That the basic ideas of postulates (II.2) and (II.5), with some of (II.1), (II.4) implicitly present in the background, are sufficient to build an ophelimity index in Pareto's sense (hence, a system of indifference varieties), was, and still is, an article of common faith for almost every ordinalist.[20] However, in order to be able to construct an ophelimity index, i.e., an ordinal measure of the preferences of *HOA*, we need an additional postulate.

For the sake of clarity, we need first to introduce the concept of a preferential set. An example of such a set is that formed by all combinations where the quantities of all but one commodity remain constant; another one is offered by $\omega\omega^1$ in Georgescu-Roegen (1, p. 550, Fig. 1) [Fig. 1–1 in the present volume]. A preferential set consists of a continuous sequence of alternatives arranged in their order of preference; its more precise definition is

(II.a) (C^α) *is a preferential set if α takes all the values of an interval of rea numbers and if $C^\beta P C^\gamma$ whenever $\beta > \gamma$.*

If a preferential set (C^α) contains at least one alternative $C^\beta P C$, and at least one such that $C P C^\gamma$, then it is easy to prove on the basis of Postulates (II,1–5)[21] that (C^α) is divided into three other preferential sets: (C^a) containing all combinations preferred to C; (C^b) containing all combinations with respect to which C is preferred; (C^c), containing all combinations indifferent to C. The set (C^c) contains *at most* one combination, i.e., either one or none.[22]

The question of whether or not (C^c) is empty naturally arises at this time. One alternative is represented by the following postulate:

19. Furthermore, (II.6) may still have a meaning even if in some cases the consumer cannot make a choice between two alternatives, Georgescu-Roegen (3).

20. E.g., Frisch (1); Samuelson (3, p. 94; 2, p. 67); Weldon (p. 228).

21. It should be observed that from (II.5) it follows that there, $C^1\overline{P}C^3$ can be interpreted as C^1IC^3 if and only if C^1IC^2, C^2IC^3.

22. A similar idea is found in Samuelson (3, p. 151 *et passim*).

(II.7) *If the preferential set* (C^α) *contains* C^β, C^γ *and if* $C^\beta PC$, CPC^γ, *then it contains* $C'IC$.

Postulate (II.7) not only introduces the existence of indifferent alternatives but also offers a definite procedure for obtaining an alternative indifferent to any given one (provided the latter is not completely empty). This is in fact Postulate A used by this author sometime back (1, p. 549) [p. 136 of the present volume],[23] but it has been hardly noticed, despite the fact that no ordinal measure of utility can be obtained without its assumption. The ordinalist's error consists precisely of ignoring this fact.

Obviously one may propose to ignore completely the question that led to (II.7) on the ground that the theory of demand is not affected in the least by the way we would decide to answer that question. Market behavior, exactly like any other behavior, is determined by preference only. Preferences, therefore, are all we need for a rationale of the theory of demand. Whether or not the Dedekind cut determined by (C^a), (C^b) belongs to one of these sets is immaterial for our purposes, although it is more convenient to set it in a class by itself.[24] Mr. Little is more emphatic on this point than all the others and decides "to get rid of [the indifference concept] altogether" (2, p. 22).[25]

However, for the special problem of arriving at an adequate theory of consumer's behavior Occam's razor is particularly counterindicated at this stage of the argument.[26] The whole story is not told yet. And it happens that the entire support of the cardinalist doctrine depends, as we shall show, upon whether or not (C^c) is empty, i.e., upon whether or not we accept Postulate (II.7), and not upon the Strong Independence Axiom.

III. WANTS VS. UTILITY

One particular aspect distinguishes Postulate (II.7) from all the others introduced previously: it does not seem to express any particular "self

23. Marschak's Postulate II (1, p. 117; 2, p. 500) represents essentially the same idea as Postulate A, though he seems not to be aware of this.

24. Cf. Armstrong (1, p. 457), Samuelson (4, p. 248).

25. I should like to take this opportunity to point out that Little's claim (stated repeatedly, p. 25 *et passim*) to have arrived at the behavior lines "solely by the use of the relation 'taken rather than' " defined on "index number formula" (Definition 2, p. 33) is entirely unfounded. In the argument of the appendix to his chap. ii, "chooses A rather than B" of Axiom II is indiscriminately used not with the meaning of Definition 2 but with that of *the classical extra-market choice between A and B, defined independently of whether or not B could have been bought when the price-income situation puts the consumer in A*.

26. What is involved in the application of this principle may be exemplified by the existence of pedate animals with more than two feet, though only two are required for walking.

evident" trait of *HOA*. At the same time, any type of experiment designed to test it *directly* seems out of the question. It is commonplace that there are no means for testing assertions involving the *continuum*. The existence of the psychological threshold in actual choice adds to this problem difficulties not present in similar problems of the natural sciences.

It appears, therefore, that if the problem of whether or not we should adopt Postulate (II.7) has to be answered, one must attack it with means entirely different from those already mentioned. Thus, we are led to give up the guidance of purely formal arguments and to re-examine the facts, particularly those that have been cited in favor of the modern theory of value.

The modern theory of the consumer has become such an integral part of our thinking that we are predisposed to overlook the significance of many facts even when these are mentioned in the current literature. In order to free ourselves from this institutional influence upon our own perspective,[27] we have to go back, 'way back, before the Jevons-Walras team, under the influence of the rapid developments in the natural sciences, had oversimplified with great enthusiasm the behavior of the individual by placing all his economic actions under a *unique* motor: utility.[28] This was a second way, different from and opposed to that of Marx, of ending the bicephalic status which value had in classical economics.

The roots of the idea that economic value must finally be reduced to a unique element are usually traced back to Aristotle who, after arguing that "all things that are exchanged must be somehow comparable," and "must therefore be measured by one thing," concludes that there can be no "exchange if there were not equality, nor equality if there were not commensurability."[29] It is in fact the last passage which is invoked by Marx (p. 68) in support of his theory of the *Equivalent form of value* (pp. 64 ff). Marx and his followers went to great pains to explain in detail the reasons why all commodities must have a common factor (Marx, pp. 43–45), because they realized that this point must be firmly established before a one-cause explanation of value could be offered. In this respect they adopted a more scientific procedure than Jevons, Walras, Wieser, *et alii* who attributed to all commodities an essential but identical property, without investigating whether or not this was supported by

27. "The marginal utility analysis is firmly established in the economic thought of all important countries." Knight (2, p. 357).
28. The influence of the models used in the natural sciences upon their thinking is evident. "The theory here given may be described as *the mechanics of utility and self-interest*," Jevons (p. 21). "And this pure economics is a physico-mathematical science," Walras (p. 51). (My translation.)
29. *Ethica Nicomachea*, 1133a–b.

the facts.[30] In fact, they simply replaced by a sharper concept, *utility*, that of Aristotle's evasive *demand*: "Now this unit is in truth demand, which holds all things together." [31] However, Aristotle remained evasive here because he very probably realized the extent of the difficulty as shown by his later cautioning that "in truth it is impossible that things differing so much become commensurate, but with reference to demand they become so sufficiently." [32] No wonder that it is on this statement that Marx (p. 68) parts with Aristotle. So do, in fact, the founders of the doctrine of marginal utility.

Interestingly enough, the point that "our wants vary indescribably [so that] commensurability is never really attained" and that "money prices are a compromise, and conceal rather than remove the real unfixity, the individual relativity of value" (Joseph, p. 111), has been frequently used in criticisms of Marx, but almost never in the attacks against modern utility theory. Yet this would have been a more normal thing to do, since a simple re-examination of the pre-Jevonsian writings would have disclosed that they contained the essence of the incommensurability of the Aristotelian χρεία.

There are sufficient reasons why the mid-nineteenth century could be considered as that period when economic thought was concerned almost exclusively with finding an acceptable theory of value. And thus, this problem was in the center of the preoccupations of all, including the best minds among economists. It is because of this that the pre-Jevonsian writings have a definite significance for our problem.

There, we find that before anyone speaks of utility, of value, or of how the individual behaves, one mentions *needs, wants, uses*, etc. These latter concepts are, it is true, far from being precisely defined, *but so is utility or satisfaction*, if we care to look into the matter. Lack of precise definition should not, however, disturb us in moral sciences, but improper concepts constructed by attributing to man faculties which he actually does not possess, should. And utility is such an improper concept, supported by other undefined concepts such as wants, uses, etc.[33] If the latter cannot

30. Walras in particular (p. 97) is remarkably frank in admitting his liberties: "The absolute intensity of utility escapes us . . . Very well, then! This difficulty is not insurmountable. Let us assume that this relationship exists . . ." (my translation).

31. *Ethica Nicomachea*, 1133a.

32. *Ibid.*, 1133b. The translation of χρεία by *demand*, both here and in the passage quoted above, is not a fortunate one, since in an economic context demand may suggest something entirely different from *need, want, use* which are the basic meanings of the Greek word. The only commentator on Aristotle who relates χρεία to "demand and need," but without making of this an important issue, is H. H. Joachim, *The Nicomachean Ethics: A Commentary* (Oxford, 1951), p. 150*n*.

33. E.g., Jevons (p. 45), Marshall (p. 92). Also Plato, *Republic*, IV. 438A: "Each desire in itself is of that thing only of which it is its nature to be."

be properly defined, *a fortiori*, "utility" or "satisfaction" cannot be defined.[34]

And, thus, we can indeed direct against utility the same objection as Joseph directed against money: that it conceals the real problem. As will become clear in the sequel, the great simplification which the concept of "utility" introduced in economic theory was not entirely gratuitous since its general adoption led to many unfounded controversies and also influenced us in refusing to accept some elemental facts because they did not fit in the theoretical scheme.

The reality that determines the individual's behavior is not formed by utility, or ophelimity, or any other single element, but by his wants, or his needs. Even that economic literature which considers utility theory so undisputable and so established that it dispenses with any discussion of wants or needs as entirely superfluous, cannot offer an independent presentation of utility.[35] This becomes particularly clear in the case of the principle of decreasing marginal utility, without which utility would not have become so famous. It is, however, impossible to find one single economist whose argument in favor of this principle would not in fact head towards entirely different principles. Yet, all arguments claim to prove that "marginal utility decreases."

IV. STRUCTURE OF WANTS

Any discussion involving man's wants may be met *ab initio* with the objection that they do not lend themselves to an easy classification, to say nothing about their definition. Other concepts among the objects studied by economics are in the same position; yet this does not prevent us from theorizing fruitfully about them and classifying them in broad

34. It is, perhaps, because of this impasse that some economists consider the approach offered by the theory of choice as a great progress which cuts the Gordian knot. This is simply an illusion because even though the postulates of the theory of choice do not use the terms "utility" or "satisfaction," their discussion and acceptance require that they should be translated into the other vocabulary. Otherwise, one is forced to admit that the postulates have neither a rational explanation nor an experimental justification. E.g., Hicks (2, p. 25). A good illustration of the above point is offered by the ingenious theory of the consumer constructed by Samuelson. His Postulate III is not a bit "perfectly clear," in the sense that if deprived of any connection with the classical choice, it is not a *transparent* property. In his system, the definition of "ψ is chosen over ψ'" is given in terms of *equilibrium* at ψ. This, clearly, is not *choice* in the classical sense, i.e., independent of market equilibrium, so that between ψ and ψ' there may not necessarily be a choice. Cf. Samuelson (3, p. 152).

35. The *Encyclopaedia of the Social Sciences* is the most eloquent sample in this respect.

categories for purposes of actual observation. We need mention as belonging to this category such basic concepts as those of "goods," "capital," "labor," etc. Wants can be handled in the same way,[36] although at times they are replaced by "needs," from which they generally evolved, or by "uses," to which they lead.

It has long since been observed that human needs and wants are hierarchized.[37] In fact, as the reader may convince himself by looking at random in the literature, this hierarchy is the essence of any argument explaining the principle of decreasing marginal utility. Whether the argument refers to the isolated farmer whose uses for corn are arranged in the well-known order of importance: food, seeds for next season, alcoholic beverages, fodder, growing parrots;[38] or to the needs for water: drinking, cooking, washing, laundering, watering the grass;[39] or simply states in general terms "all these successive units have for their possessor an intensity of utility decreasing from the first unit which responds to the most urgent need to the last, after which satiety sets in";[40] or in a still more laconic way observes that there is a "unit which is employed for *the most important use*" and one "in *the least important employment*";[41] all point out that separate "satisfactions of concrete needs have different degrees of importance to us." [42] Even Pareto, whose name is generally identified with the simplification of the theory of the consumer to the point of denying any meaning even to utility,[43] could not ignore the plain facts which reveal the hierarchy of wants and introduced instead the hierarchy of commodities (pp. 281–282), a less acceptable position for a satisfactory theory of value. For the argument developed here, it is very important to remark that this is the only case when Pareto did not use a Cartesian graph to describe his ideas, but offered instead an entirely different type of graph (p. 282), much more like a Hasse diagram.[44]

Closer investigations will show that what can be generally described as the hierarchy of human wants involves several distinct principles.

Banfield, one of the main sources of Jevons' inspiration, remarks as

36. Compare, for instance, Marshall's classification of goods (pp. 54–59) with Hermann's classification of wants (Marshall, p. 91*n*). Furthermore, there seems to be no definition of economic goods that does not refer to wants (Menger, pp. 51 ff).

37. "Now the first and chief of our needs is the provision of food for existence and life. The second is housing and the third is raiment and that sort of thing," Plato, *Republic*, II. 369D. Also Marshall (pp. 86–91).

38. Menger (p. 129).

39. Menger (pp. 133 ff); Jevons (p. 44); Wieser (p. 88).

40. Walras (p. 98). My translation.

41. Knight (2, p. 359). Italics added.

42. Menger (p. 122).

43. The discovery that utility is a superfluous concept for the logic of demand law was, however, made by Irving Fisher (pp. 14, 88), a fact clearly recognized by Pareto (p. 159).

44. Birkhoff (p. 6).

early as 1844 (pp. 11–21): "The first proposition of the theory of consumption is that *the satisfaction of every lower want in scale creates a desire of a higher character* . . . The removal of a primary want commonly awakens the sense of more than one secondary privation: thus a full supply of ordinary food not only excites to the delicacy in eating, but awakens attention to clothing. The highest grade in the scale of wants, that of pleasure derived from the beauties of nature or art, is usually confined to men who are exempted from all the lower privations . . . It is the constancy of a relative value in objects of desire, and the fixed order of succession in which this value arises, that makes the satisfaction of our wants a matter of scientific calculation . . ."

To all this, Jevons (p. 54) raised only one objection; he preferred to say that "the satisfaction of a lower want . . . merely permits the higher want to manifest itself" which is a more fortunate way of describing the *Principle of the Subordination of Wants*.[45]

This principle clearly implies that all our wants are finally satiable. It is true that one may speak of satiety being reached after a continuous decrease in the "intensity" of the corresponding want, but the only thing of which one can be sure in general is that satiety exists. It is that state of mind where any addition of the object of previous desire is no longer wanted. Irrespective of whether or not we accept the idea that wants have an "intensity" or an ordinal sequence of states, we have reached the second principle, namely Gossen's *Principle of Satiable Wants*.[46] Now, as a rule, not only does one want have to reach satiety before the next one can manifest itself, but it appears that there is always a *next* want. The number of wants seems to know no end. This is the *Principle of the Growth of Wants*,[47] which is tantamount to the absence of absolute saturation. It should, however, be observed that the nature of this postulate is rather dynamic and, therefore, its significance for a static scheme may be questioned.

The marginal utility theory has ignored the first and the last of the above principles and based its entire edifice on the Principle of Satiable Wants which was rechristened for the occasion as the *Principle of Diminishing Marginal Utility*. Apparently, only Marshall (p. 93) had no hesitation in recognizing openly this passage from one point of view to the other. Other neoclassical economists, including the founders of marginal utility theory, presented their arguments in a manner which veiled the passage from a theory of many wants to that of a unique want to which all others could be reduced. For this is in fact what *utility* represents; the common essence of all wants, the unique want into which all wants can

45. See also Menger (p. 125).
46. Banfield's *Lectures* appeared, however, before Gossen's *Entwickelung*.
47. Menger (pp. 82–83), Marshall (p. 86).

be merged. It is the implicit assumption that such a common want exists which made it possible to construct a description of the individual's behavior by using only one single postulate out of the total of three mentioned above. As a consequence of this procedure, very important problems which cannot be answered except in terms of the ignored principles, were gradually moved into the category of meaningless questions.

It is true that, with time, the theory of utility has undergone many changes which transformed it gradually into the theory of choice, which avoids any reference to utility and, hence, to wants, needs, or uses. However, the changes introduced were concerned mainly with the relations existing among goods themselves and with axiomatic aspects of the formulation. The existence of a common denominator for all wants has never been brought under discussion. New terms have been introduced, new diagrams too, but there is nothing in market behavior which can be explained by one theory and not by the other. Such theories of choice are only axiomatic molds of utility theories and retain all the consequences of the belief in the reducibility of all wants.

But not *all* human wants can be reduced to a common basis. In contrast with the principles already mentioned, the *Principle of the Irreducibility of Wants* seems to have escaped the attention of neoclassical economists. The standard literature on the subject mentions it only indirectly. Here and there, remarks are found which may be interpreted as indicating that their author had something like it in mind.[48] A highly instructive example is offered by Jevons (p. 24), who after quoting Paley as stating that "pleasures differ in nothing but in continuance and intensity," ends by admitting that "motives and feelings are certainly of the same kind to the extent that we are able to weigh them against each other; but they are, nevertheless, almost incomparable in power and authority" (p. 26). It is difficult to explain how Jevons, after reaching such a conclusion, could still follow it with a scheme where all wants are reduced to a single common base, utility.

In support of the Irreducibility of Wants, one may refer to many everyday facts: that bread cannot save someone from dying of thirst, that living in a luxurious palace does not constitute a substitute for food, etc. But there is another very important argument in favor of the irreducibility. If all wants were reducible we could not explain why in any American household water is consumed to the satiety of thirst — and therefore should have a zero "intensity" of utility at that point — while,

---

48. E.g., Pareto (pp. 252, 256–258), and, especially, Marshall (pp. 14–28). The only clear statement in this direction seems to be that of the British philosopher Joseph (p. 111). The question is essentially identical with that of the difference between *sweet* and *white* discussed by Aristotle, *De Anima*, 426b–427a.

since water is not used to satiety in sprinkling the lawn, it must have a positive "final degree of utility." Yet, no household would go thirsty — no matter how little — in order to water a flower pot. In other words, if a commodity satisfies several wants, it may very well happen that its "marginal utility" with respect to some wants may be zero (because these wants are completely satisfied) and yet the "utility" of the last unit be not null. As the quantity of that commodity decreases, only the units previously allocated to the least important want will be eliminated and not those that are contributing to the saturation of the other wants.[49] This is true of *numéraire-money*. Anyone with an average income can vouch for the fact that expenditure reaches the satiety level in many directions, i.e., it has a zero marginal utility from such points of view, and yet in many other uses it is still far from satiety. I believe that nobody would doubt that a normal *homo oeconomicus*, earning about $5,000 a year, would have no use for any dollars earmarked as valid for potatoes only (say),[50] even though the same *homo oeconomicus* would certainly welcome some other type of dollars which he could spend for the satisfaction of other wants. It appears, therefore, that the analysis of the distribution of a commodity (water, money, etc.) between different uses as presented by Jevons (pp. 58–60) and Marshall (pp. 117–118) is applicable only to those wants which have a common basis.[51] This analysis does not tell the whole story, which is no doubt much more complicated.

The pattern of our wants as reflected by the economic goods which satisfy them appears inextricably complex because there is no one-to-one correspondence between *wants* and *goods*. "Usually not a single good but a *quantity* of goods stands opposite not a single concrete need but a *complex* of such needs" (Menger, p. 129). Fuel is connected with shelter and nourishment, while nourishment requires also water, wheat, meat. Most foods which are used for nourishment are also used for the need of fellowship or social distinction. The classification of commodities, a necessary procedure for any price theory, is therefore likely to increase our difficulty of seeing through the mosaic of wants.

Despite all this, and despite the fact that the want pattern differs from one individual to another, most of these patterns have a great deal in common. (1) *The hierarchy of wants seem to be for all men identical up to a*

49. This is how Menger proceeds with his Robinson Crusoe (pp. 133 ff). It is only Jevons-Walras who have replaced ordered importance with decreasing utility. Thus, although not for identical reasons, I side with Professor Stigler on the fact that "Menger's Theory was greatly superior to that of Jevons." See Menger (p. 11n3); also Hayek (1, pp. 400–401).

50. The static assumptions and the interdiction of trading either potatoes or potatoes-dollars is a necessary part of the above analysis — as of any similar theory of the consumer.

51. The existence of such wants is in no way denied, especially among the highest ones.

certain rank. One may be almost sure that this refers at least to: thirst, hunger, leisure, shelter. (2) *Individuals belonging to the same culture are likely to have in common still a greater number of wants at the top of the hierarchy than those common to all men.* It is because of these facts that we can *understand* the wants of other fellow men, still more the wants of those who belong to the same cultural pattern, while being aware at the same time of the fact that tastes differ among individuals. It is because of this that the proposition "the marginal utility of money is greater for the poor than for the rich" makes sense in some cases and does not do so in others. Cases differing in this way were used by the opposing camps in the controversy over the interpersonal comparison of utility, which explains why both stands appear to be right.

Our evaluation of other people's pleasures by "extended sympathy" (Edgeworth, p. 475) has a definite limit: it has no valid support beyond those wants which are not identically ordered for the group of people and for the observer. *Any* observer understands why there is welfare sense (not necessarily implying obligatory policies) in taxing people who spend their summers on the Riviera, in order to help undernourished people; but most observers will question that there is sense in taxing the same people to help others go each Sunday to a football game.

No doubt, cultural patterns are reflected in the type of wants which are identically ordered for the people belonging to the same "community." For a Spaniard pities the fellow who cannot go to a bullfight on Sunday but not the one who walks to his working place, while an American will do the contrary. It is obviously such cases which are at the basis of Professor Robbins' contention (1, p. vii) that "the comparison of the different satisfactions of different individuals involves judgments of value rather than judgments of fact." Just because Diminishing Marginal Utility does not provide a meaning for the intercomparison of wants, an operation whose importance for economic life is as important as buying and selling, it does not follow that "it has no place in pure science" (*ibid.*, p. 139), unless one sides peremptorily with the theory, rather than with the facts (*ibid.*, p. 141).

V. WANTS AND CHOICE

Choice aims at satisfying the greatest number of wants starting with the most important and going down their hierarchy. Therefore, choice is determined by the least important want that could be reached. This is why when we ask for the reason of choice we get answers which seem, *prima facie*, silly. An individual may give as the reason for which he bought a particular car "the nice emerald green color of the panel"; another would say that he bought his house because "it offered a nice location for a bird house." But what both individuals mean is that after comparing all available cars and all available houses from the viewpoint of more important wants, they finally came down to the color of the panel and to the bird house. No matter what we choose, houses, cars, or combinations of commodities C, the procedure is the same.[52] *The choice between two combinations is always decided by the lowest relevant want that can be reflected in any of the two combinations.*

To illustrate this mechanism of choice by a very simple, hence crude, scheme, let us *imagine* a *homo oeconomicus B* who would have at least the following wants — for food, for palatable taste, and for the companionship of his friends — and who would know only two commodities — X_1, margarine and X_2, butter. Let us also assume, for simplicity, that the "food" as seen by our *HOB* in any combination (x_1, x_2) is the corresponding amount of calories $k = x_1 + x_2$; taste may be also assumed to be offered only by butter, $t = x_2$. If the want for food is more important than for palatable taste, up to a certain limit $k \leq K$ our *HOB* will feel that food is more urgent than taste and will make his choice on the basis of the size of k. Only if two combinations have the same value of k would he take into consideration his next want and choose that combination containing more butter. If, however, $k > K$, the importance of the first want will become secondary and the choice will be made on the basis of the greatest value of t; but if the two values of t are equal, the third want will come up to decide the choice. We may assume that the third want is shaped according to how many friends *HOB* has who need simply "food" or who can enjoy "taste," and reflected by $e = ax_1 + bx_2$, where $a, b > 0$.

In his choice, therefore, our *HOB* will prefer $C'(x_1', x_2')$ to $C''(x_1'', x_2'')$ according to the following scheme

$$\text{(A)} \quad \begin{array}{l} 1)\ k'' < k' < K, \\ 2)\ k'' = k' \leq K;\ t'' < t', \\ 3)\ k'' < K < k', \\ 4)\ K < k', k'';\ t'' < t', \\ 5)\ K < k', k'';\ t'' = t';\ e'' < e'.^{53} \end{array}$$

It is clear now that the behavior of our *HOB* fulfills Postulates (II, 1–6); therefore he is an *HOA*, but he does not fulfill Postulate (II.7). We have

---

52. Why otherwise would the advertisers go to the trouble of pointing out to us that brand A of cheese comes in an attractive package, that brand B saves space in the refrigerator, etc., etc.?

53. These criteria are easily recognized to be based on the logical product of some ordered sets; see Birkhoff (p. 9).

thus arrived at the conclusion that there are certainly cases which do not fulfill the latter postulate.[54] *This conclusion was reached not because it would constitute a more convenient approach or lead to a simpler scheme, but because it offers a more adequate interpretation of the structure of our wants.*

One may now replace the functions k, t, e used in the above example by other more general functions that would generate three families of convex curves; the criteria (A) will still be valid. A graph representing the choice mechanism of such an *HOB* is shown in Fig. 3–1. There,

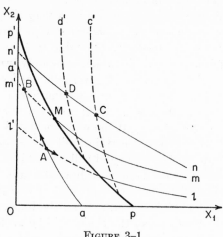

FIGURE 3-1

lines aa', pp' represent curves of (k) family (pp' corresponds to $k = K$), lines ll', mm', nn' represent (t) family curves, and the others, Dd', Cc', are curves from family (e).[55] Any point above the curve aa' is preferred to any on it, which in turn is preferred to all points below aa'. Yet B is preferred to A. The curves aa' have much in common with the indifference lines; we may adopt for them a convenient term introduced by Little (1), that of *behavior curves*. For the region above pp', they are formed by the e family; C is preferred to D, and this is preferred to M. However, $p'Mm$ is also a behavior curve.[56]

---

54. That there may be cases when the postulate is nevertheless fulfilled is not an excluded possibility. This depends upon the number of wants and of commodities.

55. The fact that x_2 is the tastier food would make the slope of the tangent to the t curve less steep than that to the k curve passing through the same point. This bears a striking analogy to Johnson's definition of "x_2 more urgent" (p. 498), which makes one wonder whether he had not in mind a hierarchy of wants rather than complementarity when introducing this concept.

56. This shows that there is no logical necessity for assuming that behavior curves do not meet, an important assumption in Little's argument. Cf. Georgescu-Roegen (3).

The ordering of the alternatives by *HOB* is exactly that of a chain, i.e., they are simply ordered.[57] For convenience we may refer to this property of the alternatives as that of *comparability*. Further, we may use the term *ordinal measurability* to denote that particular comparability which can be represented unambiguously by a chain of real numbers, i.e., *where to each element one can assign a real number which will completely identify its relative ranking*.

The important fact is that because of the plurality of wants of *HOB* the alternatives although *comparable* are not *ordinally measurable*.[58] The ordinalist's error consists precisely in ignoring this possibility.[59] But once this point is established, one cannot even think of introducing a *cardinal measure* of choice unless by a distorted stretching of the meaning of words.[60]

VI. INDIFFERENCE AND SUBSTITUTABILITY

There are a few other important aspects common to most modern theories of *homo oeconomicus* which are profoundly affected by the Banfield-Menger theory of wants.

In the first place, we are no longer sure that indifference exists. It is this conclusion which is most likely to arouse the greatest objections against the theory. Why, it will be asked, as we pass continuously from the "rejection" class to the "preference" class[61] do we not pass through an intermediary class of "indifference"? This way of looking at the matter has such a long tradition with us that the absence of the intermediary class may be considered *ab initio* as absurd. For, it will be argued, to deny the existence of this intermediary class would mean simply to go counter to the principle of continuity. But let us consider for a minute the question: to what does continuity apply here? Clearly, continuity does not apply either to "rejection" or to "preference" which are *states* described in no other way. And although it is true that we cannot pass from a negative number to a positive one without passing through zero, because *numbers* have the property of varying continuously, it is not

57. Birkhoff (p. 10). In economic literature, the same concept is very frequently called a *completely* ordered set.

58. This follows from Theorem 2 or Ex. 6*, Birkhoff (p. 32).

59. The belief so clearly expressed by Mr. Armstrong (3, p. 119), that "when we have one dimension . . . complete ordering is impossible," may account for this error. Cf. also Armstrong (1, p. 456n).

60. I do not deny that sometimes one does so for the purpose of saving "a point," but this is only a vacuous success.

61. As in the preferential set considered in Section II.

at all necessary that between two states there should always be an intermediary one. For if we accept the principle that there is one, we are getting into the most paradoxical situation. What is then the state between preference and indifference? It is thus clear that the above principle — the existence of the intermediary state *by necessity* — must be rejected.

States are many times related to a continuous variable, but this has no bearing upon the existence of intermediary states.[62] Thus, life and death are two states related to a continuous variable, time, and yet there is no intermediary state. The same is true, for instance, for decisions which depend upon a continuous variable. The decision to act may appear without first going through an intermediary state of pure undecidedness.[63]

If indifference exists at all, its existence has to be proven directly by other means. A closer analysis of wants leads us, however, to the conclusion that indifference may exist but not as a rule.[64]

We should also be aware of the pitfalls of interpreting as "indifferent states" those which *man* cannot order without a great deal of hesitation or without some inconsistency. Such cases are the symptoms of imperfections in the mechanism of choice caused by a psychological threshold, which is absent in HOA. It is true that, in real life, even a hungry individual will give up a very small amount of margarine in exchange for some butter, but this may be due to his not being able to feel in any case the loss of a *small* amount of food.[65]

And if some of us would still maintain a grudge against a theory that undermines such esteemed concepts as Indifference and Substitutability, one may point out that this very theory offers, in exchange, a clarification of a theoretical impasse as old as the ordinalist doctrine itself. Introspection seems to show beyond doubt that if C^1PC^2, C^2PC^3, one can always tell whether C^2 is "nearer" to C^1 or to C^3. Therefore, by trial and error, one can find the "midpoint" between C^1 and C^3. Generally, this has been regarded as a sufficiently solid foundation on which to construct

---

62. Unless the intermediary states have *ex hypothesi* a continuous ordinal measure.

63. The unjustified transmission of the property of continuity from the variable to the states determined by the variable is a very common error. "It is obvious that utility passes through inutility [*sic*] before changing into disutility, these notions being related as +, 0, −." Jevons (p. 58). Another example is the argument offered by Friedman-Savage (2, p. 468) in support of their postulate P_2. Indeed, if no small variation in the traffic could affect our decision with regard to crossing the street, this decision should be the same for all intensities of traffic!

64. Clearly, this affects also the famous Principle of Substitution, which is found as a posit in Neoclassical literature. Cf. Marshall (pp. 341, 356), Pareto (pp. 256, 270–271), Frisch (1, pp. 3–4), Hicks (2, p. 20). In fact only Marshall mentions it by name, adding merely that its "applications [are] extended over almost every field of economic inquiry."

65. With pure types of HOB, this is no longer possible. "The soul of the thirsty then, in so far as it thirsts, wishes nothing else than to drink, and yearns for this and its impulse is towards this," Plato, *Republic*, IV. 439A.

a cardinal measure of utility.[66] Confronted with this argument, the ordinalists have adopted two entirely different tactics: (1) to deny that "distances" are comparable[67] or (2) to adopt a metaphysical point of view which, if generally accepted, would deny the possibility of any cardinal measure.[68] However, one can introduce as an additional assumption the faculty of HOB to compare adjacent differences $\Delta(C^2, C^1)$, $\Delta(C^3, C^2)$ without modifying at all the mechanism of his choice as described in Section V. Since *comparability* of differences does not necessarily imply their *ordinal measurability* this additional faculty does not raise any new problems for the rationale of HOB, who could compare differences without his preference scale being ordinally measurable.

VII. CHOICE AND PSYCHOLOGICAL THRESHOLD

The description of the behavior of HOB, although based on an introspection argument, was finally formulated in such a way that a behavioristic type of experiment could conceivably be used for testing all the assumptions, with one very important exception. As already pointed out, the questions involved in Postulate (II.7) cannot be settled in this way. This raises the question whether indifference could not be defined in such terms as to avoid all references to introspection and so as to base the definition only on observable factors. In a previous paper,[69] the author has shown that because of the psychological threshold which must be present in choice as everywhere else, the choice between C^1 and C^2 may not always show the same preference, so that a new *homo oeconomicus*, HOC, has to be considered. New concepts have to be introduced.[70]

1. Observed preference: $C^1 \mathcal{P} C^2$ if $C^1 P C^2$ in all cases.
2. Observed indifference: $C^1 \mathcal{I} C^2$ if $C^1 P C^2$ happens with probability
   $$0 < \pi < 1.$$
3. Defined indifference: $C^1 = C^2$ if $\pi = \tfrac{1}{2}$.

66. Cf. Pareto (p. 264), Edgeworth (p. 473), Bowley (p. 2), Lange (1 and 2), Armstrong (1), Weldon. Samuelson, in a very interesting paper (3), has pointed out that comparability of adjacent differences is not enough to construct a cardinal measure. At the same time he gives, I believe, for the first time a precise definition of this type of cardinality. But there he twice commits the ordinalist's error; the second time, when he assumes that his Postulate 2 entails ordinal measurability of differences.

67. E.g., Pareto (p. 264).

68. "The fact that I can find a point B . . . such that the move from A to B is rated *just* as highly as the move from B to C does *not* seem to me to be the same thing as saying that the interval A-C is *twice* the interval A-B," Robbins (2, p. 104). Cf. Russell (pp. 281–282).

69. Georgescu-Roegen (1, Section VI).

70. *Ibid.*, pp. 570, 575 [in the present volume, pp. 152, 156].

Then between the "rejection" class ($\pi = 0$) and the "preference" class ($\pi = 1$) there is always, in a preferential set, an "indifference" class which is not empty. This leads to two types of "limiting curves," the boundaries of the "rejection" and of the "preference" classes which contain in between them the "indifference" class.[71] In such a scheme, however, indifference is no longer transitive.[72]

From here one can continue along two different lines. One may ask what would happen to the indifference class if the psychological threshold went down to nil and the "limiting curves" became identical. Since the indifference class is represented at all times by an open set (so that the Borel-Heine Theorem cannot be invoked),[73] we cannot arrive at an answer solely on the basis of this argument: the limit of the indifference class may be empty.

Another further way is to analyze the concept of defined indifference in order to see whether it could not replace the indifference concept of the ordinalists. It is soon discovered that to test the transitivity of this "defined concept" — which has no transparent meaning — involves us in as many insuperable difficulties as testing the existence of indifference by Postulate (II.7).[74] It seems, therefore, that one has to fall back on introspection.

VIII. CHOICE AND RISK

Before we proceed to examine the results of introducing *risk* and *uncertainty* in the alternatives facing the individual, some preliminary remarks — not so extensive as one might wish — are necessary in view of the fact that these concepts continue to be confused, in spite of the arguments of Professor Knight (1) and Lord Keynes (1 and 2).

Many times we are faced with the impossibility of predicting the outcome of a particular phenomenon. But the cases where this is so belong

71. *Ibid.*, pp. 572 ff [in the present volume, pp. 153 ff]. These "limiting curves" are the generalization of Mr. Little's upper and lower behavior lines. Our region of indifference is his ignorance region, Little (1, pp. 94–95). Furthermore, it is seen that the concept of indifference *was* after all defined in terms of choice, Little (2, p. 25).

72. Georgescu-Roegen (1, p. 579) [in the present volume, p. 158 f]. The fact that the psychological threshold destroys transitivity of indifference is the main point of Armstrong's paper (1); very likely, he must have ignored the author's paper, to argue further that the threshold implies cardinal utility.

73. R. Courant, *Differential and Integral Calculus* (New York, 1938), II, 99.

74. Transitivity means in this case that if Prob $(C^1PC^2) = \frac{1}{2}$ and Prob $(C^2PC^3) = \frac{1}{2}$, then Prob $(C^1PC^3) = \frac{1}{2}$. Without some additional and very powerful postulates (Georgescu-Roegen 2, p. 130) [in the present volume, p. 175 f], this is far from being so. On the basis of the postulates introduced so far, Prob (C^1PC^3) can have absolutely any value from 0 to 1.

to two essentially different categories. The first contains those processes which consist of using a known mechanism under varying initial conditions. Tossing a die, spinning a roulette, picking up a ball in an urn, or any other isomorphic mechanisms belong to this category. The result of any individual outcome cannot be predicted, although the mechanism is known. This is a *risk*. The second category includes the cases where the individual outcome cannot be predicted *and* the mechanism is not completely known. In general, what we know here are the results of a few observations of the past. A procedure known to have succeeded ten times and failed twice is a simple example of *uncertainty*. In most cases we do not know even that, because the procedure we are interested in is quite new; so we have to rely on the past observations of another, though somewhat similar, procedure. In the case of risk, but not in the case of uncertainty, we can define the probability of the outcome.[75] A risk proposition is then defined by

$$\Gamma = \begin{pmatrix} C^1, C^2, \ldots, C^n \\ p_1, \ p_2, \ \ldots, \ p_n \end{pmatrix}, \qquad (\Sigma p_i = 1),$$

where C^i is a *sure* alternative and p_i is the probability of its outcome.[76]

It has always been implicitly assumed that the risk propositions can be transformed according to the basic postulates of probability. With this there has been little quarrel.[77] The choice among risk propositions has been explained in terms of substitution between the mathematical expectation and other higher moments, considered to reflect different dimensions of risk.[78] The real surprise has been caused, however, not so much by the Morgenstern-Neumann scheme, but by some consequences of the Strong Independence Axiom:[79]

(VIII.1) If $C^1 P C^2$ and if $p_1' > p_1$, then

$$\begin{pmatrix} C^1, C^2, C^3, \ldots, C^n \\ p_1', \ p_2', \ p_3, \ \ldots, \ p_n \end{pmatrix} \ P \ \begin{pmatrix} C^1, C^2, C^3, \ldots, C^n \\ p_1, \ p_2, \ p_3, \ \ldots, \ p_n \end{pmatrix}.$$

75. This is a physical characteristic of the mechanism. It is not appropriate to discuss here whether this can be estimated in practice by the Laplace-Poincaré procedure or by that proposed by Professor Mises. The important factor is that the individual have a complete understanding of what probability actually means to *him*, because it is *his* reactions vis-à-vis risk that we want to find out.

76. Considering a more general type of risk proposition, involving an infinite number of alternatives, will not materially affect the argument.

77. Von Neumann and Morgenstern, Axioms (3:C), introduced the compounding of probabilities as a distinct assumption of their scheme. This is perfectly commendable for elegance of precision (although they did not mention the principle of total probabilities which is implied in the construction of U), but it has induced some economists to speculate that the weak spot of the cardinalist doctrine may be located in it. Alchian (p. 37) seems to have considered this possibility.

78. E.g., Hicks (1), Makower-Marschak, Tintner.

79. As shown by Marschak (1, p. 136), this axiom is implied by Neumann-Morgenstern. Its explicit formulation is due to Friedman-Savage (1).

Its meaning is so transparent that only the belief that cardinality has its roots in it could possibly explain the extensive analysis to which it has been submitted in comparison with the other postulates.[80]

In the hands of the cardinalists, this axiom not only produced cardinality but at the same time did away with the classical multidimensionality of risk propositions. All aspects of risk are melted into a single criterion, that of "expected utility"; nothing else concerns the individual. It is this result which ought to have pointed from the beginning towards the inadmissibility of the doctrine. For even *if* we admitted that utility of sure combinations were measurable, everyday facts show beyond doubt that other aspects of a risk proposition besides the first moment of the utility function are taken into consideration. Since, on the other hand, the Strong Independence Axiom cannot possibly be rejected it simply means that another postulate is the cause of all this discrepancy. The only other postulate which seems to invite challenge is (II.7). Thus, even if (II.7) were true for riskless alternatives, it should not be included in a theory of risk-choices. No matter how many distinct criteria intervene in a riskless choice, risk adds an essentially distinct one: a sure alternative and a risk proposition, being heterogeneous, can in no case be indifferent.[81]

The preceding remarks deprive Marschak's Theorem 1 (1, p. 122) of neither its beauty nor its significance. (I am inclined to think that they would even enhance its relevance for our theory.) Interpreting any risk proposition based on C^1, C^2, C^3 as a point in a triangular system of coordinates A^1, A^2, A^3 (Fig. 3–2), it is easy to prove that the "behavior lines" are parallel straight lines, (aa'), given by the equation

$$u = p_1u_1 + p_2u_2 + p_3u_3 = \text{const.}[82]$$

Thus if $u < u'$, then $\Gamma'P\Gamma$, but if $u = u''$, Γ'' is not necessarily indifferent to Γ. In the latter case, the choice between Γ and Γ'' will be determined by the next criterion, very likely the "variance" of the risk proposition. Exactly as in the mechanism described in Section V, a second family of parallel lines (bb'), represented by the equation $p_1u_1^2 + p_2u_2^2 + p_3u_3^2 = $ *const.*, determines the choice between two risk propositions whenever the first criterion fails. All risk propositions form a chain but are not even ordinally measurable. Other moments have to come in if the number of C^i is greater than three. Thus choice among risk propositions is not independent of other elements considered by the early Marschakians. If A^2B is the behavior line passing through A^2, and if A^2PB, then the individual

80. Cf. Marschak (1, pp. 120–121), Friedman-Savage (2, pp. 468 ff). Also several articles in *Econometrica*, October 1952.
81. Cf. Postulate II of Marschak (1) and Postulate 3 of Friedman-Savage (1).
82. Although this proposition is based on a weaker system of assumptions, proofs much more elementary than Marschak's are readily available.

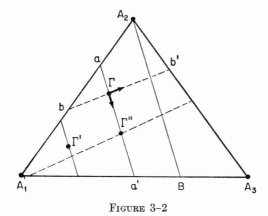

FIGURE 3-2

dislikes danger; the direction from B to A^2 is a preference direction. The opposite is true if the individual *loves* danger.[83]

Thus even if sure alternatives were measurable, so that the utility of C^i is measured by u_i, u itself could not be a measure for the entire field of alternatives which includes risk propositions also. Utility could not be regarded as measurable — without stretching the meaning of the words to the point of creating dangerous confusions — even if we would adopt the definition of cardinality given by Friedman-Savage in their recent paper (2, Theorem, p. 468).[84]

It is very interesting to observe that these authors have adopted here a much more reserved attitude than in their previous paper, where cardinal measurability was presented as an economically meaningful concept that could be used for such important problems as that of constructing an econometric definition of socio-economic classes (1, pp. 87 ff). Their analysis of this latter problem is an extremely interesting *trouvaille,* and one should feel rather happy that it can be preserved even if cardinality goes overboard. The attitude of many individuals who find themselves just between two socio-economic classes can be perfectly described in terms of behavior lines alone. At the intermediary level between two socio-economic classes, leisure, for instance, or the quiet relaxation of a family atmosphere, or the esteem of other fellow men, becomes very often *inferior goods.* How much one would walk (or take) for a "Camel" is a more realistic description of what is involved in the so-called increasing marginal utility of income than one based on measurable utility.

83. If he loves danger, then, as in Fig. 3–2, $\Gamma''P\Gamma$.
84. In fact, even there, the use of the term "measure" for labeling the property described by Theorem 1, without further relevant additions, seems not very fortunate: "triangulability" instead of "cardinal measurability" would have been a better choice, perhaps.

IX. CHOICE AND UNCERTAINTY

The real acid test of any behavior theory is provided by the choice among uncertainties. It is on this basis, for instance, that Armstrong attempted to prove that "the preference theory, as at present formulated, fails, therefore, completely to provide a theory of choice" (2, p. 10). He does not tell us how he reached the conviction that preference theory deals only with sure alternatives (p. 4). However, this *ad hoc* limitation of preference theory does not suffice to prove that utility theory can succeed where the former failed. Armstrong has to assume further that both utility and uncertainty are measurable, more exactly that uncertainty is treated as probability (p. 8).[85]

True, Armstrong offers an argument for the measurability of uncertainty. After assuming that uncertainty is simply ordered, he assigns 0 and 1 respectively to complete uncertainty and complete certainty, then $\frac{1}{2}$ to the "neutral" uncertainty between the previous two positions, and so on with every new interval. We know by now (*supra*, Section VI) that this argument is based on an illusion — common to the ordinalist's error — namely that *there are enough numbers between 0 and 1 to tag a different one on each different uncertainty*. To assume, as Armstrong implicitly does, that uncertainty forms a one-parameter chain is an unwarranted position.[86]

Others have tried to prove that uncertainty is cardinally measurable by identifying it with the betting quotient.[87] This procedure is in no way more satisfactory since it ignores the fact that to the same betting quotient may correspond several *distinct* uncertainties; the latter would thus lose their individuality because of the way in which the question forced upon the individual was formulated. In the same fashion one may prove that we live in a two-dimensional world by asking questions which would demand that the answer include only the latitude and the longitude of a location.

Apparently, the cause of persisting in the idea that uncertainty and probability are identical resides in the fact that both concepts are related

85. If this point is granted there is no need for other argument than the one already available in Note IX of Marshall's *Principles*, p. 843.

86. That this is a common fallacy of most theories of subjective "probability" was pointed out a number of years ago by an "American Reviewer" (still unidentified by the author's search) who detected the ordinalist error in Professor Jeffreys' *Scientific Inference*. See Harold Jeffreys, *Theory of Probability* (Oxford, 1939), p. 19.

87. The most scientific approach along these lines seems to be that of Ramsey (1, p. 176 *et passim*). It is interesting to note that Ramsey's setup contains the first indication that measurability of utility and measurability of subjective probability have to share the same fate.

to the impossibility of an exact prediction. Nevertheless, they are entirely different essences.[88] Probability is a physical characteristic of a certain type of mechanism and its value depends upon the latter's physical properties: symmetry, number of sectors, of faces, etc. Probability belongs to the same type of concepts as mass, speed, angle, *et al.* Uncertainty is quite something else; it is a *state of mind*. States of mind are almost without exception correlated with more than one physical measure, and even when they form a chain, the latter is not necessarily a one-parameter ordering.

The simplest scheme for illustrating the preceding arguments, as well as those that follow, is to consider all uncertainty situations related to a Bernoullian urn.[89] Any of these uncertainties may be represented either by $T(m, n)$, where $N = m + n$ is the number of past observations and $m = p'N$, the number of white balls observed, or by $t(p)$, where p is the true proportion of white balls in the urn. In this case, p' is the observed measure of a true measure, p. The number N is clearly an index of our *credibility* in the observed measure. It reaches its highest state when p is known, *when uncertainty is identical with probability*, and its lowest state, when $N = 0$ and p' is unknown, when uncertainty is in its purest form.[90] Uncertainty is composed of two distinct elements: the observed measure and the credibility.

Assuming that the occurrence of a white ball is what the individual desires, no objection could possibly be raised against the following ordering for $N \neq 0$:

(A) $t(p_1)Pt(p_2)$ for $p_1 > p_2$ and $T(m_1, n_1)PT(m_2, n_2)$ for
$$N_1 = N_2, p_1' > p_2'.$$

(B) $t(p)PT(m, n)$ or $T(m, n)Pt(p)$ for $p = p' \geq$ or $\leq \frac{1}{2}$.

(C) $T(m_1, n_1)PT(m_2, n_2)$ or $T(m_2, n_2)PT(m_1, n_1)$ for $N_1 > N_2$ and $p_1' = p_2' \geq$ or $\leq \frac{1}{2}$.

This alone does not make uncertainty a chain. $T(0, 0)$ raises difficult problems; it seems not to be actually comparable with any other uncertainty.[91] Even if we leave $T(0, 0)$ definitely out as a singular situation, uncertainty still does not form a chain.[92]

88. Knight (1) preceded Keynes (1) in noticing and analyzing this difference.

89. The extreme simplicity of the scheme used here strengthens the position that uncertainty connected with a real situation is far from being measurable.

90. The identification of $t(p)$ with $T(m, n)$ where $N = \infty$, $p' = p$, constitutes a problem whose discussion lies beyond the scope of this paper.

91. Does it represent the midpoint between $t(0)$ and $t(1)$? If so, what is the relative position of $t(\frac{1}{2})$? Could an individual whose life depended upon the drawing of a white ball be completely indifferent between $T(0, 0)$ and $t(\frac{1}{2})$? Are these states of mind identical in any relevant way? Is the complete absence of knowledge equivalent to any other complete or incomplete knowledge? Such questions seem to be ignored by most oversimplified treatments of uncertainty.

92. It seems that Keynes was first to argue that uncertainty not only has no cardinal measure, but it is not even comparable (1, p. 34 *et passim*).

Several "Rules of Succession" have been offered in answer to this impasse. In the last analysis, they all amount to the introduction of substitutability between *observed probability* and *credibility* with the inherent family of indifference curves. Thus, for Laplace's First Rule[93] the indifferent uncertainties are determined by the family of straight lines

$$\frac{m+1}{m+n+2} = \text{const.}$$

But both Laplace and his followers have arrived at their results by assuming that the individual has a kind of knowledge that no man possesses:[94] namely that he knows the cosmogonic distribution of the urns.[95] Furthermore, their universe is essentially different from the one we live in since they assumed that all attributes are represented by a finite number of independent predicates.[96] Without such artificial assumptions, it is difficult to think of reasons which would justify the existence of any substitutability between p' and N. This argument appears clearer if the equivalent problem is considered for the case where the variable observed is, say, the I.Q. Is there any reason why there should be a compensation between the value of the I.Q. and the number of tests on which it was established?

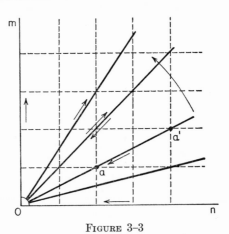

FIGURE 3–3

The most acceptable solution seems to be suggested by Laplace's Second Rule.[97] Its indifference *loci* are the straight lines (aa'), given

93. Generalizations have been offered by Professor Carnap (1, p. 568; 2, p. 30).
94. Keynes (2, p. 222).
95. Carnap (1, pp. 262 ff).
96. *Ibid.*, pp. 58, 73.
97. Laplace (pp. 280 ff). Known also as the Most Probable Value, as Fisher's Maximum Likelihood, or as the Straight Rule.

by m/n = const. (Fig. 3–3). In other words, uncertainty is ordered according to one criterion only, the observed probability. This gives

(D) $t(p_1)Pt(p_2)$ for $p_1 > p_2$, $T(m_1, n_1)PT(m_2, n_2)$ for $p_1' > p_2'$.

Since credibility must also play a role in the ordering of uncertainty we may add to the preceding preference (B) and (C). As indicated by the arrows in Fig. 3–3, a *two-parameter* ordering is thus established for the Bernoullian uncertainty, similar in structure to the ordering of sure alternatives and of risk propositions. All alternatives belonging to m/n = const have the same "betting quotient," $p' = m/N$, and yet they are not equivalent.[98]

If we consider now the alternative of obtaining C^1 with the uncertainty T and if C^1PC^2, then the choice between (C^1, T) and C^2, for instance, could not be explained in terms of maximizing moral expectation even if the utility of sure alternatives were measurable.

X. CONCLUDING REMARKS

As pointed out in the preceding analysis, the cardinalist argument rests upon two unwarranted assumptions. In the first place, the oversimplified pattern of human behavior ignores the irreducibility of wants. It is this assumption which leaked also into the ordinalist argument and which made it possible to use the real number system for ordering all non-equivalent alternatives.[99] In the second place, credibility is also ignored with the consequence that the background for unpredictable results is reduced to probability. This implies that man has an almost demiurgic knowledge. These, I believe, are sufficient reasons to explain why the cardinalist doctrine even though analytically true is synthetically false, i.e., why, even if it is understood, it cannot be believed.

It is, however, possible to avoid this conflict and still have a consistent theory of consumer's choice, such as that presented here. Incidentally, this theory is equally valid for explaining not only consumer's choice, but choice *in general*. Thus, it could provide an adequate background for a theory of entrepreneurial decisions under multiple criteria, some of

98. This seems to be the furthest we can go without ignoring the necessity of relevant explanations.

99. The problem involved here is similar to that of the famous Pythagorean conflict caused by the discovery of incommensurability. It was this discovery which revealed that the edifice of natural numbers does not suffice for a complete treatment of measure (i.e., of quantity measure), and that the hypothesis that there is only a countable number of points on a line had to be abandoned. Cf. John Burnet, *Early Greek Philosophy* (4th edn., London, 1930), §50 *et passim*.

which, not having a purely economic implication, ought to be classified as social factors.

I do not want to end this paper, however, without adding some final remarks concerning another possible course of conflict between theory and choice in the real world. Most theories, perhaps all, consider only alternatives of the kind used in the preceding sections, derived by super-imposing risk and uncertainty upon sure alternatives $C(x_1, x_2, \ldots, x_n)$. This is clearly an oversimplification. In the real world man is confronted with alternatives where every commodity α is represented not by a quantity x_α but by a function $f_\alpha(t)$ representing the expected flow of that commodity at time t, or by $F_\alpha(t)$, representing the total income of α up to t. Such alternatives cannot be adequately represented by Cartesian diagrams in a Euclidean space.[100]

It is perfectly true that the illusion that the consumer chooses between an apple and an orange is not entirely unfounded. The real choice is, however, one between $[F_1(t) + 1, F_2(t), \ldots, F_n(t)]$ and $[F_1(t), F_2(t) + 1, \ldots, F_n(t)]$, where $[F_1(t), F_2(t), \ldots, F_n(t)]$ is the situation of the individual before he is asked to choose between an additional apple (commodity 1) and an orange (commodity 2). This observation raises an additional objection against the relevance of many experiments performed now and then with the view of arriving either at the indifference curves or at the utility function. It is the value of the latter experiments that is particularly doubtful. If in an experiment aimed at determining the utility function from the betting quotient, the first 50-50 bet was, say, 8 against 10, i.e., if it was a choice between $F(t)$ and the 50-50 chance for $F(t) - 8$ or $F(t) + 10$, the second betting coefficient would be determined not from a situation represented by $F(t)$ but by either $F(t) - 8$ or $F(t) + 10$ following the result of the first bet. The betting quotient could not be the same, unless the stakes were very small in comparison with $F(t)$, but then the experiment would be irrelevant for other reasons.

Finally, we also assume that the number of commodities is finite. However, taking into consideration the great variations of brands, of the methods of selling and financing them, this assumption is another oversimplification. It would be better to assume that α has the power of continuum. But even if we leave α finite, the cardinal power of the alternatives $C[F_\alpha(t)]$ is greater than that of continuum. This power has baffled scientists in other fields.[101] As observers of "man" we find that our intuition fails completely to guide us through so *many* alternatives, but as "man" *per se* we can manage somehow to find our way through

100. E.g., Wold's argument in *Econometrica* would have gained in clarity if it had not been supported by such a diagram.

101. E.g., in probability. See Paul Lévy, *Théorie de l'addition des variables aléatoires* (Paris, 1937), p. 25.

many trials and many more errors. To explain the essence of this mechanism it is perfectly legitimate to use some schematic support for our ideas, exactly as some arguments concerned with the fourth dimension are presented in two-dimensional graphs, provided we do not overlook the fact that such graphs not only cannot represent all the features of which they are an image but, at times, also display some alien properties.

REFERENCES

Alchian, Armen A., "The Meaning of Utility Measurement," *American Economic Review*, XLIII (1953), 26–50.
Allen, R. G. D., and J. R. Hicks, "A Reconsideration of the Theory of Value," *Economica*, I (1934), 52–76, 196–219.
Armstrong, W. E. (1) "The Determinateness of the Utility Function," *Economic Journal*, XLIX (1939), 453–467.
———— (2) "Uncertainty and the Utility Function," *Economic Journal*, LVIII (1948), 1–10.
———— (3) "A Note on the Theory of Consumer's Behavior," *Oxford Economic Papers*, II (1950), 119–122.
Banfield, T. C., *Four Lectures on the Organization of Industry*, London, 1845.
Birkhoff, Garrett, *Lattice Theory* (rev. edn.), New York, 1948.
Bowley, A. L., *The Mathematical Groundwork of Economics*, Oxford, 1924.
Carnap, Rudolf (1) *Logical Foundations of Probability*, Chicago, 1950.
———— (2) *The Continuum of Inductive Methods*, Chicago, 1952.
Corlett, W. J., and P. K. Newman, "A Note on Revealed Preference and the Transitivity Condition," *Review of Economic Studies*, XX (1952–53), 156–158.
Edgeworth, F. Y., *Papers Relating to Political Economy*, vol. II, London, 1925.
Fisher, Irving, *Mathematical Investigations in the Theory of Value and Prices*, New Haven, 1925 reprint.
Friedman, Milton, and L. J. Savage (1) "The Utility Analysis of Choices Involving Risk," *Readings in Price Theory* (American Economic Association Series, vol. VI), pp. 57–96, Homewood, Ill., 1952 (a revised version of a paper published in 1948).
———— (2) "The Expected-Utility Hypothesis and the Measurability of Utility," *Journal of Political Economy*, LX (1952), 463–474.
Frisch, Ragnar (1) "Sur un problème d'économie pure," *Norsk Matematisk Forening Skrifter*, series 1, no. 16, Oslo, 1926, pp. 1–40.
———— (2) "General Choice-Field Theory," *Report of the Third Annual Research Conference on Economics and Statistics*, Cowles Commission for Research in Economics, Colorado Springs, 1937, pp. 64–69.
Georgescu-Roegen, Nicholas (1) "The Pure Theory of Consumer's Behavior," *Quarterly Journal of Economics*, L (1936), 545–593 [Essay 1 in the present book].
———— (2) "The Theory of Choice and the Constancy of Economic Laws," *Quarterly Journal of Economics*, LXIV (1950), 125–138 [Essay 2 in the present book].
———— (3) "Choice and Revealed Preference," *Southern Economic Journal*, XXI (1954), 119–130 [Essay 4 in the present book].

Hayek, F. A. (1) "Carl Menger," *Economica*, I (1934), 393–420.

——— (2) *The Counter-Revolution of Science: Studies on the Abuse of Reason*, Glencoe, Ill., 1952.

Hicks, J. R. (1) "The Theory of Uncertainty and Profit," *Economica*, XI (1931), 170–189.

——— (2) *Value and Capital* (2nd edn.), Oxford, 1948.

Jevons, W. Stanley, *The Theory of Political Economy* (4th edn.), London, 1924.

Johnson, W. E., "The Pure Theory of Utility Curves," *Economic Journal*, XXIII (1913), 483–513.

Joseph, H. W. B., *The Labour Theory of Value in Karl Marx*, London, 1923.

Keynes, John M. (1) *A Treatise on Probability*, London, 1921.

——— (2) "The General Theory of Employment," *Quarterly Journal of Economics*, LI (1937), 209–223.

Knight, Frank (1) *Risk, Uncertainty, and Profit* (reissue), London, 1948.

——— (2) "Marginal Utility Economics," *Encyclopaedia of the Social Sciences*, V (1931), 357–363.

Laplace, P. S. de, *Oeuvres complètes*, vol. VII, Paris, 1886.

Lange, Oskar (1) "The Determinateness of the Utility Function," *Review of Economic Studies*, I (1933–34), 218–225.

——— (2) "Notes on the Determinateness of the Utility Function," *Review of Economic Studies*, II (1934–35), 75–77.

Little, I. M. D. (1) "A Reformulation of the Theory of Consumer's Behavior," *Oxford Economic Papers*, I (1949), 90–99.

——— (2) *A Critique of Welfare Economics* (corrected edn.), London, 1950.

Makower, H., and J. Marschak, "Assets, Prices and Monetary Theory," *Economica*, V (1938), 261–288.

Manne, Alan S., "The Strong Independence Assumption — Gasoline Blends and Probability Mixture," *Econometrica*, XX (1952), 665–668.

Marschak, Jacob (1) "Rational Behavior, Uncertain Prospects and Measurable Utility," *Econometrica*, XVIII (1950), 111–141.

——— (2) "Why 'Should' Statisticians and Businessmen Maximize Moral Expectation?" *Proceedings of the Second Berkeley Symposium on Mathematical Statistics and Probability*, University of California Press, 1951, pp. 493–506.

Marshall, Alfred, *Principles of Economics* (8th edn.), New York, 1949.

Marx, Karl, *Capital*, vol. I, Chicago: Kerr & Co., 1932.

Menger, Karl, *Principles of Economics*, transl. and ed. by J. Dingwall and B. F. Hoselitz, Glencoe, Ill., 1950.

Pareto, Vilfredo, *Manuel d'économie politique*, Paris, 1927.

Ramsey, Frank P., *The Foundations of Mathematics and Other Logical Essays*, New York, 1931.

Robbins, Lionel (1) *An Essay on the Nature and Significance of Economic Science* (2nd edn.), London, 1935.

——— (2) "Robertson on Utility and Scope," *Economica*, XX (1953), 99–111.

Russell, Bertrand, *Human Knowledge: Its Scope and Limits*, New York, 1948.

Samuelson, Paul A. (1) "A Note on the Pure Theory of Consumer's Behavior," *Economica*, V (1938), 61–71 and 353–354.

——— (2) "The Numerical Representations of Ordered Classifications and the Concept of Utility," *Review of Economic Studies*, VI (1938–39), 65–70.

——— (3) *Foundations of Economic Analysis*, Cambridge, Mass., 1947.

——— (4) "Consumption Theory in Terms of Revealed Preference," *Economica*, XV (1948), 243–253.

—— (5) "The Problem of Integrability in Utility Theory," *Economica*, XVII (1950), 355–385.

—— (6) "Probability, Utility and the Independence Axiom," *Econometrica*, XX (1952), 670–678.

Strotz, Robert H., "Cardinal Utility," *American Economic Review*, XLIII (1953), 384–405.

Tintner, G., "The Theory of Choice Under Subjective Risk and Uncertainty," *Econometrica*, IX (1941), 298–304.

Von Neumann, John, and Oskar Morgenstern, *Theory of Games and Economic Behavior* (2nd edn.), Princeton, 1947.

Walras, Léon, *Eléments d'économie politique pure* (3rd edn.), Paris, 1896.

Weldon, J. C., "A Note on Measures of Utility," *Canadian Journal of Economics and Political Science*, XVI (1950), 227–233.

Wieser, F. von, *Social Economics*, New York, 1927.

Wold, H., "Ordinal Preferences or Cardinal Utility," *Econometrica*, XX (1952), 661–664.

Choice and Revealed Preference

1. It is impossible to go through the literature of modern contributions to the ordinalist theory of consumer's behavior without noticing that two approaches which seem not only different, but almost opposite, have evolved. The first is the Theory of Choice, a refinement of the Fisher-Pareto indifference-preference construction. Its modern foundations, however, were laid down by Professor Frisch in his 1926 paper "Sur un problème d'économie pure." [1] In the Fisher-Pareto theory, choice merely reflects ophelimity; therefore, ophelimity is the primary and choice is the secondary concept. [2] With Frisch, choice becomes the basic element of the theory of the consumer; the indifference curves are a derived concept introduced to facilitate the rationalization of choice. [3] The second approach is that of Professor Samuelson, for which he has chosen the term of Revealed Preference Theory. Its pivot is one of the most interesting contributions to the general theory of the consumer, even though it was Samuelson's first *oeuvre de jeunesse*. [4] Recently, he told us that his

NOTE: This paper is reprinted by courtesy of the publisher from *Southern Economic Journal*, XXI (1954), 119–130. With the original appearance I explained in a footnote that the paper was a part of a larger research project sponsored by the Institute of Research and Training in the Social Sciences of Vanderbilt University. I also wrote: "Although the author acknowledges the valuable criticism of Mr. H. S. Houthakker and Professor Paul A. Samuelson on the paper which follows, he alone is responsible for any faults it might contain."

1. Ragnar Frisch, "Sur un problème d'économie pure," *Norsk Matematisk Forening Skrifter*, series I, no. 16 (1926), 1–40. I am glad to point out here this important contribution of Frisch which, because it came to my attention only after my paper on "The Pure Theory of Consumer's Behavior" appeared in *Quarterly Journal of Economics*, L (1936), pp. 545–593, was not mentioned there.

2. Irving Fisher, *Mathematical Investigations in the Theory of Value and Prices* (New Haven, 1925 reprint), p. 12, §2; Vilfredo Pareto, *Manuel d'économie politique* (Paris, 1927), chap. iii, para. 55; Appendix, para. 4, 8.

3. Frisch, "Sur un problème," p. 5 *et passim*.

4. Paul A. Samuelson, "A Note on the Pure Theory of Consumer's Behavior," *Economica*, V (1938), 61–71, 353–354.

inspiration had its roots in a discussion with Professor Haberler and in Professor Leontief's analysis of indifference curves.[5]

Nevertheless, the basic idea goes back to Pareto who in the *Manuel*[6] and, especially, in his article "Economie Mathématique," [7] not only tells us that the family of indifference curves and the family of offer curves provide two equally good foundations for a theory of the consumer, but also shows us how we could pass from one to the other. The essence of his ideas could not be more explicitly expressed than by the following statement: "Au point de vue exclusivement mathématique, il est indifférent, pour la détermination d'équilibre, de connaître les actions de l'individu au moyen des fonctions d'offre et de demande ou au moyen des fonction indices. Ce choix résulte de considérations d'opportunité expérimentale." [8]

In further elaborating on these ideas, Pareto envisaged two schemes by which one could pass from the market behavior of an individual to his preferential pattern. In the first case, the observed equilibrium quantities at varying prices are used to determine the marginal rate of substitution at every point of the commodity space.[9]

This yields the total differential equation

$$(1) \qquad \varphi_1 dx_1 + \varphi_2 dx_2 + \cdots + \varphi_n dx_n = 0$$

from which Pareto once thought that the ophelimity potential

$$(2) \qquad \psi(\varphi) = \text{const}$$

could be derived.[10]

The second aspect of market behavior on the basis of which Pareto thought a preference field could be constructed is the more general proposition that if the initial position of the consumer, $X^0(x_1^0, x_2^0, \ldots, x_n^0)$, and the price constellation, $P(p_1 = 1, p_2, \ldots, p_n)$ are given, the equilibrium, $\overline{X}(\overline{x}_1, \overline{x}_2, \ldots, \overline{x}_n)$, is completely determined. In other words, one may write

5. Paul A. Samuelson, "The Problem of Integrability in Utility," *Economica*, XVII (1950), 355–385. Very likely, he had in mind Leontief's paper, "Composite Commodities and the Problem of Index Numbers," *Econometrica*, IV (1936), 39–59, which, I believe, provides the first discussion of how revealing index numbers are for the preference field of the consumer.

6. P. 184*n*; Appendix, para. 42.

7. *Encyclopédie des Sciences Mathématiques*, tome 1, vol. 4, fascicule 4, pp. 591–640.

8. *Ibid.*, p. 597. See also p. 595.

9. *Ibid.*, pp. 593, 596.

10. Vito Volterra, in reviewing the Italian edition of *Manuel*, reminded Pareto that this equation alone does not ensure the existence of an ophelimity index. Cf. *Manuel*, p. 546. In this way, the nonintegrability case was born to economics, but it caused no trouble until much later. For a very instructive history of the case, see Samuelson, "Problem of Integrability," quoted above.

(3) $\overline{X} = F(X^0; P)$,

which is equivalent to the n-equation system of Pareto.[11] Here again, Pareto committed a second mathematical slip which enabled him to conclude that the system (3) and the equation (1) are equivalent. For this to be true some additional conditions are necessary; their economic interpretation is that the equilibrium $\overline{X}$ does not depend upon the initial position X^0, but depends instead upon "income," i.e., some function of prices and of X^0 which includes as a particular case the usual definition of income. Consequently, if (3) is to ensure (1), it must be of the form[12]

(4) $\overline{X} = F[I(X^0, P); P]$.

Even though Pareto's attempts at clarifying the relationships existing between the four molds, (1)–(4), into which consumer's theory could be cast, were not completed,[13] some of the most interesting problems have thus been raised to provoke the sagacity of economists.

None of the problems involved are simple. It is not surprising, therefore, that it took so long to make some progress. Professor Allen, in a contribution which usually receives less credit than it deserves, opened a new era by showing that for the "stability" of consumer's equilibrium, equation (1) does not need to be integrable, but neither can it be absolutely arbitrary.[14] Thus a new restriction, dictated by economic considerations, was imposed upon equation (1). *But this could still be nonintegrable.* It was only then that the nonintegrability of (1) became a real problem of the theory of the consumer. For if the principle of the persistence of nonpreference is sufficient to account for consumer's equilibrium, why (in economic terms) does not the same principle allow the construction of a preference field? The fact that (1) was usually interpreted as the "local indifference element" — i.e., X and $X + dX$ were indifferent com-

11. *Encyclopédie*, p. 595.

12. This was pointed out by the author in "A Note on a Proposition by Pareto," *Quarterly Journal of Economics*, XLIX (1935), 706–714.

13. Pareto was not aware of the difference between (3) and (4) and he denied a fifth alternative — recently revived by the cardinalists.

14. R. G. D. Allen, "The Foundations of a Mathematical Theory of Exchange," *Economica*, XII (1932), 197–226. His formulation of the principle involved was rather "difficult to translate" (p. 219). In their path-breaking "Reconsideration of the Theory of Value," *Economica*, I (1934), 52–76, 196–219, Professors Allen and J. R. Hicks presented the same principle as the decreasing marginal rate of substitution. Finally, a presentation where the number of dimensions did not matter was offered by the author in what may be called the principle of persisting nonpreference direction, Georgescu-Roegen, "The Pure Theory of Consumer's Behavior," *Quarterly Journal of Economics*, L (1936), 553, 556 [in the present volume, pp. 139, 141]. See also J. R. Hicks, *Value and Capital* (2nd edn., Oxford, 1946), p. 25. In this latter work, however, Hicks assumed integrability throughout. Later, Hicks presented it as the "ultimate generalization of demand" law. *Ibid.*, pp. 51–52, 329. Also Paul A. Samuelson, *Foundations of Economic Analysis* (Cambridge, Mass., 1947), p. 109.

binations provided dX satisfied (1) — made the answer more difficult.[15] The problem was to explain in economic terms why a succession of infinitesimal indifferent displacements could not lead to an indifference variety of some sort. The answer was shown to rest in the absence of transitivity.[16]

Spiral "integral curves" around the *absolute* or *relative* saturation point were used to prove two points:

(a) The "integral curves" do not necessarily represent indifference varieties — unless transitivity is assumed. Hence, integrability is a necessary but not a sufficient condition for a theory of choice in the Fisher-Pareto-Frisch sense, even though it may allow the construction of an *analytical* ophelimity index.[17]

(b) The equilibrium position is a point of relative saturation.[18]

At that time I overlooked a very important point which has since remained unnoticed. If integrability is denied to (1), then we have to reject not only transitivity but *choice in the large* as well. Choice in the Fisher-Pareto-Frisch sense, i.e., between *any* two combinations X^1 and X^2, no longer exists. In their sense, man's choice is similar to that of a bird which can select between any two distant alternatives. The Fisher-Allen[19] rationale, where choice is possible only between X^1 and $X^1 + dX^1$, depicts man's choice as similar to that of a worm who at each moment selects the direction of its journey by comparing only infinitesimally close situations. In the latter case, however, *the principle of persisting non-preference direction can no longer suffice to explain equilibrium.* Indeed, without additional assumptions, it is possible that the individual may always move in a *preferred*, yet *wrong*, direction, i.e., away from the equilibrium position. This is illustrated by the path $ABC \cdots L$ in Fig. 4–1(a).[20]

2. Before proceeding further, some familiar concepts have to be made more precise.

A preference, an antipreference, or an indifference direction, denoted by AB^+, AB^-, AB^0 respectively, means that, if the individual were at A,

15. There is no doubt that Pareto adopted this interpretation in *Encyclopédie*, p. 609, although in the *Manuel* he carefully avoids this error (e.g., pp. 542, 547–548).

16. Georgescu-Roegen, "Pure Theory," pp. 567–568 [in the present volume, pp. 149–150] and Section IV.

17. *Ibid.*, pp. 562, 565, 566 [in the present volume, pp. 146, 148, 149]. That "transitivity" in the revealed preference sense also requires integrability was pointed out recently by H. S. Houthakker, "Revealed Preference and Utility Function," *Economica*, 1950 (XVII), 173–174.

18. Georgescu-Roegen, "Pure Theory," p. 563 [in the present volume, p. 147].

19. Fisher too considered this possibility, *Mathematical Investigations*, p. 88.

20. This represents the budget plane; S is the equilibrium position.

he would favor, would reject, or would regard as indifferent a move towards B. We shall also use AB^n to denote a nonpreference direction, i.e., either AB^- or AB^0. The basic assumptions of a Fisher-Allen theory of the consumer may now be stated as follows:

(A) *The direction dX is one of preference, of nonpreference, or of anti-preference depending upon whether*

(5) $L \equiv \Sigma\varphi dx >$, $=$, or < 0.[21]

(B) *The functions φ in (5) are continuous.*

(C) *An antipreference direction extended to an infinitesimally close point remains one of antipreference whereas a limiting direction becomes a non-preference one.*

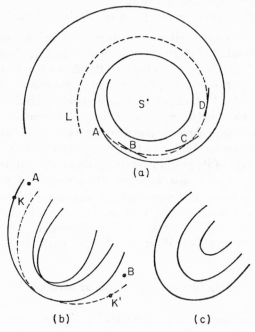

(a)

(b)   (c)

FIGURE 4–1

In other words,

(6)   If $\Sigma\varphi\Delta x =$ or < 0,   then $\Sigma(\varphi + \Delta\varphi)\Delta x \leq$ or < 0.[22]

From this we also obtain

(7)   If $\Sigma\varphi\Delta x > 0$,   then $\Sigma(\varphi + \Delta\varphi)\Delta x > 0$,

provided that $|\Delta X|$ is sufficiently small.

21. The directions for which $L = 0$ may be referred to as limiting directions and denoted by AB^1. By definition, if AB^0, then there exists M on AB such that MA^0.

22. The above postulates differ slightly from the conditions used in my "Pure Theory," pp. 551–553, 556 [in the present volume, pp. 137–139, 141].

Postulate C is the principle of persisting nonpreference; from this we can easily prove that

(i) *If AB^-, then BA^+ and if AB^0, then either BA^0 or BA^+.*[23]

It can hardly be overemphasized that to assert the converse of (i), i.e., that BA^+ implies either AB^- or AB^0, would be an error. However,

(ii) *If AB^+, then BA^- provided that B is sufficiently close to A.*

Therefore, on any straight line there is at most one single closed interval AB such that both directions away from any point on AB are of non-preference.

We already intimated that if all directions away from S are nonpreferred, then S is an absolute saturation point. If all directions away from S to any other alternative of a set (X) — such as that determined by fixed income and constant price — are nonpreferred, then S is a point of saturation relative to (X), and will satisfy

$$(8) \qquad \Sigma\varphi^s(x^s - x) \geq 0^{24}$$

or, in other words, SX^n, for all X of (X). According to (i), we also have that either XS^0 or XS^+, hence

$$(9) \qquad \Sigma\varphi(x - x^s) \leq 0.^{25}$$

It can be further proved that

(iii) *The set formed by the alternatives available with a fixed income and constant prices contains a saturation point.*

(iv) *Any convex set of alternatives contains at least one saturation point.*[26]

It follows that we always have at least one absolute saturation point. It is natural to assume that the origin of the commodity-coordinates is not a saturation point. Then (9) expresses the usual condition that X^1 is preferred to X^2 if all $x^1 \geq x^2$ (but not all $x^1 = x^2$).[27]

In order to offer an explanation for the equilibrium of the consumer one may introduce at this point the additional postulate:

(D) *Out of a set of alternatives, the consumer always selects a relative saturation point, if this exists.*

But why should he, if his choice is assimilated to that of a worm? To assume that he is "attracted" in some way by the saturation point would simply mean that he is endowed with some homing-pigeon instinct also. It is this point which renders the Fisher-Allen approach unsatisfactory as a consumer theory. This can be better illustrated by the fact that even

23. The possibility of simultaneously having AB^- and BA^n is thus excluded. *Infra* (10).

24. Georgescu-Roegen, "Pure Theory," p. 558 [in this volume, p. 143]. As shown there the saturation points form a convex set.

25. Fisher, *Mathematical Investigations*, pp. 74, 88, and Allen, "Foundations," pp. 218–219, related the saturation point to a similar property.

26. See Mathematical Note at the end of this paper.

27. Cf. my "Pure Theory," p. 557n [this volume, 142n16]. Also I. M. D. Little, *A Critique of Welfare Economics* (Oxford, 1950), p. 33, Axiom I.

if the attraction of *homo oeconomicus* to the saturation point is accepted as an additional hypothesis, he still may not always be able to choose between any two combinations.

The result of the comparison between two combinations, A and B, can lead only to three distinct relationships:

(10)
- (a) AB^+, BA^+ denoted by $AsnB$;
- (b) AB^0, BA^0 denoted by $AsiB$;
- (c) AB^+, BA^n denoted by $BspA$;
- (d) BA^+, AB^n denoted by $AspB$.

$AspB$ means that A is "*s*-preferred" to B because it is a saturation point of A and B; if $AsiB$, the choice between A and B is "*s*-indifferent"; if $AsnB$, there is no choice.

The preceding analysis puts "the nonintegrability case" into an entirely new perspective. In the past, this case has been viewed rather from personal affinities.[28] Now, it appears clear that there are strong reasons for regarding nonintegrability as unsatisfactory, if not as unacceptable.

3. The theory which places the equilibrium at the saturation point has a perfect structural similarity to Samuelson's revealed preference. Indeed, if the absolute saturation points are beyond the means of the consumer, (9) is obviously fulfilled, while (8) corresponds to Samuelson's Postulate III,[29] alias his Weak Axiom.[30] The fact that X "is revealed to be preferred to Y," or that XrY, implies $XspY$. The addition of (D) to (10) makes the two schemes equivalent. Therefore, whatever pitfalls the nonintegrability case might involve, they will be present also in revealed preference. Recently, Samuelson admits that he has always been aware of the imperfections of his theory[31] and he rightly salutes Houthakker's important contribution in the form of the Strong Axiom.[32]

28. Cf. Samuelson, "A Note" (my note 4, above), p. 68; Hicks, *Value and Capital* (note 14, above), p. 19*n*.

29. Samuelson, "A Note," p. 65. The transparence of this postulate presented as the irreversibility of the revealed preference does not seem to be as immediate as Samuelson claims. In *Foundations of Economic Analysis*, p. 110, he justifies the postulate on the basis of the impossibility of reversing the numerical inequality of ophelimity. This would, however, deprive the revealed preference of its usefulness as an *independent* analytical tool. Incidentally, the position in the Addendum to the "Note," where he discards as redundant his Postulate I — equivalent to (4) above — is incorrect. (See also *Foundations*, p. 111.) If (4) is not postulated, one cannot use (1) without repeating Pareto's second error. Cf. *supra*, note 12.

30. Samuelson, "Problem of Integrability," p. 370.

31. *Ibid.* There are other indications of this. In the *Foundations*, p. 109 *et passim*, for instance, Samuelson leans simultaneously on ophelimity. There are also the remarks on pp. 151–152, which touch the very core of the matter and which, if further explored, might have led him to the solution of the problem. However, the impossibility of explaining *why* the saturation point is actually reached in the nonintegrability case has not been pointed out.

32. H. S. Houthakker, "Revealed Preference," p. 163.

The Samuelson-Houthakker Axiom is that

(11)   If $ArB, BrC, \ldots, MrN$, then $N\bar{r}A$ [33]

or, in terms of the saturation principle:

(12)   If $ApB, BpC, \ldots, MpN$, then $N\overline{sp}A$.

Here p stands for "either sp or si," but not every time for si.

As pointed out by Samuelson[34] and recognized by Houthakker[35] these relations do not represent true transitivity. But in the ultimate analysis, this construction reduces to a very simple principle which brings the essence of the "consistency assumption" to the surface. This principle is worded exactly as Samuelson's initial Postulate III:

(E) *If A is preferred to B, B can never be preferred to A, i.e., if APB, then* $B\bar{P}A$.

In this postulate, P = "preferred" is defined as a sequence of p's as in (12):

(13) APB if $ApL, LpM, \ldots, NpB$

or, in other words, that one can go from B to A following always a preferred or an indifferent direction.

It is obvious that

(14)   If APB, BPC, then APC,

which brings us back to *full* transitivity.[36]

A relationship between the above construction and the integrability condition has been established rather hurriedly. In fact, as already pointed

33. Samuelson, "Problem of Integrability," p. 370. Here $\bar{r}$ stands for the negation of r. This extends the chain of combinations from three, in Samuelson's initial postulate, to any finite number.

34. Samuelson, *Foundations*, p. 152.

35. Houthakker, "Revealed Preference," p. 163.

36. Superficially, this condition reads exactly as that introduced by I. M. D. Little in his "A Reformulation of the Theory of Consumer's Behavior," *Oxford Economic Papers*, I (1949), 90–99, reproduced with corrections in *A Critique*, chap. ii and Appendix. Although very inspiring, Little's analysis contains many inaccuracies. His Axiom II (*A Critique*, p. 33) is incorrect, as Samuelson had pointed out in advance, *Foundations*, p. 151 (74). Had Little arrived at the correct formulation, such as that behind our (14), he would have arrived at the solution of the puzzle of revealed preference before Houthakker. The fact that Little's analysis was confined to two commodities would not have been an obstacle. As is argued throughout this paper, the puzzle exists in the two-commodities case also. The importance of Houthakker's contribution does not lie in his use of a three-dimensional illustration, but in his Strong Axiom.

One should also remark that Little's Axiom III cannot be true for combinations which do not include all commodities. Since this type of combination is the most frequent in reality, the objection cannot be ignored. Furthermore, Little seems to regard the weak axiom as a mere tautology and, accordingly, does not state it explicitly; nevertheless, he makes use of it in note 2 on p. 36 of *A Critique*. Clearly, in that note he does not prove what he intends.

out,[37] integrability is not as important for consumer theory as the concept of a potential.[38] Indeed, the preceding construction leads to something which is more than mere integrability but less than the existence of a potential.

4. Let L and M be any two points such that LPK, MPK. On the segment LM there is a relative saturation point S. For any point N of the segment LS, NL^n; for any point Q of MS, QM^n. Therefore, any point on LM is preferred to K. This simple remark leads to the important result that

(v) *The set* (K) *of all combinations* R, *such that* RPK, *is convex*[39] (Fig. 4–2).

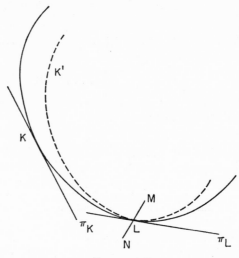

FIGURE 4–2

Because of (E), the boundary of (K), denoted by $b(K)$, contains K. This illustrates how (E) works: *without it, the entire space may be in the relation* P *to* K. Moreover,

(vi) *The hyperplane* π_k, $\Sigma\varphi^k(x^k - x) = 0$, *is a supporting plane of* (K)[40] (Fig. 4–2).

37. Georgescu-Roegen, "Pure Theory," Section III.

38. It is because of this that I cannot agree with Samuelson and Houthakker who — along with Pareto, *Encyclopédie*, p. 559 — think that the case of two commodities cannot raise any difficulty. Samuelson, "Problem of Integrability," pp. 360, 380; Houthakker, "Revealed Preference," p. 172.

39. If (K) consists only of K, K is a saturation point.

40. A supporting plane of a convex set is a plane having at least one point in common with the set and containing no interior point of the set. It is a generalization of a tangent plane.

Other important properties of (K) can easily be proved.

(vii) *If L belongs to $b(K)$, then π_L, $\Sigma\varphi^1(x^1 - x) = 0$, is a supporting plane of (K).*

If (vii) were not true, then there would be a point $M = L + dL$, interior to (K), not belonging to π_L, and such that LM^-. But then, LPK (Fig. 4–2). Now, if N on ML is sufficiently close to L, we would have LN^+ and, therefore, NPK, which is impossible.[41]

From this it follows that

(viii) *The $b(K)$'s are integral varieties of (1).*

And further,

(ix) *If K' is an interior point of (K), then $(K') \subset (K)$.*

Indeed (K') cannot contain a point K'' exterior to (K), because from $K''PK'$, $K'PK$, it would follow $K''PK$.[42]

The relationship between (K) and (K') is no longer as simple as this, if K' does not belong to (K). Clearly, both (K') and (K) must contain the set of saturation points. As this set is not empty, (iv), (K) and (K') are not disjoint. If neither set contains the other, $b(K)$ and $b(K')$ must have some common points at which these two varieties must have the same "tangent" plane. See Fig. 4–1(b).

Thus, it appears that the integral varieties $b(K)$, although they have many properties in common with the indifference varieties, cannot possibly be related to the latter in any truly significant way. There are many aspects in which the structure of Fig. 4–1(b) differs from that of classical indifference varieties. In the first place, *not all combinations can be compared one with another.* This is the case of A and B, in Fig. 4–1(b). There is no path connecting them and going always in a nonpreferred direction. The scale of preferences does not necessarily form a *chain*; it may simply be a *tree*.[43] Nevertheless, and this is the most important fact, a scheme such as that in Fig. 4–1(b) *offers a complete explanation of why the consumer reaches the equilibrium point, even though his choice is concerned with infinitesimally close situations only.* For this, we require more than integrability but less than the existence of a potential.

The order structure of consumer's choice cannot be made simpler than this without introducing some additional assumptions. In order to avoid

41. L itself is not necessarily preferred or indifferent to K. This analytical aspect of the problem was pointed out by Samuelson, *Foundations*, p. 152, but Little provided the actual model for it; cf. "A Reformulation."

42. However, $b(K)$ and $b(K')$ may have points in common, such as L in Fig. 4–2; but there, they would have the same "tangent" plane.

43. Cf. Garret Birkhoff, *Lattice Theory* (rev. edn., New York, 1948), pp. 31, 47. In fact, the only restriction to be imposed on preferences in addition to their forming a partly ordered set is that they possess a greatest element (*ibid.*, p. 7).

intersecting integral varieties, for instance, one has to assume that (1) has no singular points other than those of absolute saturation.[44]
Then,

(x) *If K' is an exterior point of (K), then $(K) \subset (K')$,*

with the consequence that

(xi) *If K' belongs to $b(K)$, then $(K) = (K')$.*

The (K)'s form a sequence of nested sets, as in Fig. 4–1(c). An index of market behavior, $\varphi(K)$, can be constructed *analytically* so that

(15) $\qquad\qquad$ If $\varphi(K) > \varphi(K')$, then KPK'.

However, this is not an index of ophelimity because $\varphi(K) = \varphi(K')$ does not necessarily imply that K is indifferent to K', *not even that they are comparable.*

This not only completes the proof of the mechanism by which Samuelson thought a preference field could be constructed from market data, but it also sets off the salient implications of the argument. The most important finding — which deserves emphasis — is that actual market behavior, contrary to what one might have thought, does not require even that all alternatives be comparable.[45]

5. Finally, I wish to point out a different way of looking at the relation between the theory of choice and that of revealed preference. Instead of letting the income vary, as Samuelson does, one may consider the income equal to unity, by writing any budget equation under the form $xp_x + yp_y = 1$. The individual choices can then be described as concerning price constellations (p_x, p_y), instead of combinations (x, y). The problem would be thus transferred into the price-space, where we could introduce the usual axioms, similar to those used in the theory of choice. Passing from one scheme to another is achieved by a simplified form of (4), $\overline{X} = F[I = 1; P]$, or by $p_i = \varphi_i / \Sigma x \varphi$.[46]

The two pictures — the theory of choice in the commodity-space and

44. Little alone pointed out the necessity of imposing some analytical conditions on the φ's in order to secure nonintersecting integral varieties. Cf. *A Critique*, p. 35. However, to assume that the integral varieties do not intersect without offering some supporting evidence is a purely analytical feat without relevance to reality. In a paper read at the meeting of the Econometric Society in September 1953, I argued that there are good reasons for introducing the contrary assumption: that integral varieties do meet. This assumption is not (as it may seem *prima facie*) incompatible with the chain structure of preferences.

45. Cf. Nicholas Georgescu-Roegen, "Choice, Expectations and Measurability," *Quarterly Journal of Economics*, LXVIII (1954), 503–534 [in the present volume, Essay 3].

46. H. S. Houthakker, in "Compensated Changes in Quantities and Qualities Consumed," *Review of Economic Studies*, XIX (1952), 3, introduces the terms "direct" and "indirect" utility to denote φ and $F(I; P)$ respectively.

the theory of revealed preference in the price-space — are the dual formulation of the same structure. The first looks at the indifference curve as the *loci* of a moving point, the second, as the *envelope* of its tangents. This perfect symmetry is shown plainly in the Hicks-Samuelson general law of demand: $\Sigma \Delta p \Delta x < 0$.[47]

The decision as to which of these two alternatives to use as the cornerstone of the theory of value seems to be almost a matter of personal taste. Though the theory of revealed preference has not yet fulfilled the claims of its initiator, namely, to derive the preference structure from *actual* market data, it inspired fruitful research which finally showed how *little* of the rationale of classical theory of choice is reflected by the market behavior. The theory of choice seems, however, more transparent and easier to manipulate: it is easier to talk of a circle in terms of its points than of its tangents.

MATHEMATICAL NOTE

From the principle of persisting nonpreference, it follows that any one-dimensional simplex, i.e., any segment AB, possesses at least one saturation point. Let us consider now a two-dimensional simplex, ABC. Let C_1 be on AB, B_1 on AC, such that B_1C_1 be parallel to BC. Let S be a saturation point of B_1C_1 and let $[S]$ be the set of all S's when B_1C_1 varies remaining parallel to BC. Let $[A]$ be the set of all points D belonging to ABC such that DA^n. If $[S]$ and $b[A]$ have only point A in common, it means that for the points belonging to $[S]$ either all SA^+ or, all SA^-. In the first case A, in the second, a saturation point of BC is a saturation point of ABC. If $[S]$ and $b[A]$ have other points in common, besides A, they are saturation points.

The passage to an n-dimensional simplex can now be easily obtained by complete induction.

Furthermore, the principle of persisting nonpreference being preserved by a conical projection, we can project the positive closed orthant onto a simplex and thus arrive at the result that there is always at least one absolute saturation point.

The proof of (iv) is easily obtained by using the above result in a triangulation of any convex set.

47. Samuelson, "A Note," p. 68; Hicks, *Value and Capital*, p. 329.

5

Threshold in Choice
and the Theory of Demand[1]

1. The question why the starting point in the modern analysis of market phenomena is not Demand but Choice may only seem idle. For in favor of taking Demand as the fundamental concept of this analysis one may invoke the fact that Demand, in contrast with Choice, relates the two basic coordinates which constitute the primary concern of the economist. Even the theory of the consumer can be entirely cast in terms of Demand alone: P. A. Samuelson's theory of Revealed Preference needs only a little amending for the completion of this project. Clearly, it would not help to argue that the ensuing theory stops short of explaining the consumer's acts. If economics was supposed to offer such an explanation, the theory of choice would not suffice either: economics should then cover the physiology of our biological and cultural wants. Actually, the cognitional algorithm would never end, and economics would have to become *the* science.

The *raison d'être* of the theory of choice as a chapter of economics is above all the simplification it brings to the theory of demand. True, both the map describing the consumer's behavior in terms of Choice and that describing the same thing in terms of Demand consist of a one-parameter family of curves: for two commodities, for instance, the theory of choice leads to the curves $u(x_1, x_2) = k$, the theory of demand, to $x_1 = D(r, I)$, where r is the exchange-ratio and I the income measured in *numéraire*. Yet the indifference map is not only the simpler but of the simplest type possible: the curves do not intersect and, what is more, all display the same shape-uniformity. In contrast with this, the shape of demand curves

NOTE: This paper is reprinted by courtesy of the publisher from *Econometrica*, XXVI (1958), 157–168. Note 1, below, along with the other footnotes, appeared in the article as there published.

1. I wish to acknowledge the privilege of having J. Marschak's constructive criticism of an early version of this paper. However, I do not want to imply that Marschak shares the responsibility of whatever faults the paper may have.

may vary broadly, and the curves may even intersect each other in the region where X_1 is an inferior good. Simplicity, however, is the concern of all scientific disciplines, not only because a simpler framework helps theorizing, but also because such a frame reduces the amount of observations necessary to determine any individual structure. Thus, because of the simple structure of the indifference map, considerably fewer observations are needed to obtain an approximate drawing of this map by experiment than would be necessary for the map of all demand curves. It is important to note also that the types of questions asked of the consumer in the two experiments differ fundamentally. For the indifference map, it is sufficient to find out the result of *binary choices*, i.e., choices between any two points of the commodity-space. In the experiment for determining the demand curves, the question "How much would you buy at the price p?" requires the consumer to name a quantity, unless we would be prepared to increase further the number of observations by asking instead "Would you buy more or less than x_1 at the price p?"

The argument in favor of the theory of choice collapses, however, if the indifference directions are not integrable, or if — a closely related circumstance — choice is not transitive. It collapses simply because the indifference map vanishes. No wonder then that the explosion of the nonintegrability case caused such a stir among economists. True, some thought of defending the old line by authoritatively dismissing nonintegrability as a will-o'-the-wisp, but the question continued to tax the sagacity of other economists. Only recently attempts have been made to decide the issue by the only valid procedure: actual tests. One such test led to the conclusion that "at most, it may be argued that no evidence has been revealed which [stochastically] contradicts the hypothesis that individual preference systems satisfy the transitivity axioms." [2]

2. Actual testing of any assumption regarding Choice raises some points common to all experimental verifications of laws formulated in ideal concepts. Before anything else, we must have a clear picture of the error laws affecting the observations. In the case of Choice, we must begin with a theory of the individual not as a perfect choosing-instrument, but as a stochastic one. With these points in mind, some time ago I offered a model of the consumer's behavior which introduced threshold in *binary choice*.[3] Among other results, such as the nontransitivity of "indifference," [4] the paper included the following:

2. Andreas G. Papandreou, "An Experimental Test of an Axiom in the Theory of Choice," *Econometrica*, XXI (1953), 477.
3. Nicholas Georgescu-Roegen, "The Pure Theory of Consumer's Behavior," *Quarterly Journal of Economics*, L (1936), 568–584 [in the present volume, Essay 1, above].
4. *Ibid.*, pp. 578–580 [this volume, pp. 158–159].

R. *A demand penumbra (i.e., a stochastic distribution of the quantity demanded at every price) can be derived from the map describing the binary choices, exactly as the map of binary choices in the classical model leads to the one-valued demand schedule.*[5]

According to this result, it appeared that, despite the introduction of threshold in choice, no essential change in the relationship between Choice and Demand would be necessary, and that, at least for theoretical problems, the classical model would retain its validity in representing the "typical" as opposed to the "particular" decision of the consumer. The similarity between some of my results concerning the effect of threshold upon order and transitivity, and those recently presented by R. Duncan Luce,[6] induced me to reread my own paper. On this occasion, I discovered that the result R regarding the relationship between the binary choice map with the demand penumbra contains a mathematical blunder, which for twenty years had escaped notice. In all probability, the result must have been regarded as trivial on the ground that one naturally would expect most one-value elements of a rigid model to be replaced by stochastic distributions if a stochastic structure is grafted on such a model. The error in my argument would not be worth mentioning by itself. It happens, however, that the nature of that error unveils some unsuspected difficulties concerning the relationship between the binary choice map and demand. And since these difficulties concern the core of the present behavioristic foundations of demand theory, they will be discussed here.

3. Let us first re-examine the main properties of the threshold model.[7]

AXIOM I: *The result of a binary choice between two points $A(a_1, a_2, \ldots, a_n)$ and $B(b_1, b_2, \ldots, b_n)$ in the commodity-space E_n is either A or B.*

This axiom excludes the case where the consumer is unable to make a choice. Therefore, if $\omega(A, B)$ is the probability that A be chosen, Axiom I simply states that

(1) $$\omega(A, B) + \omega(B, A) = 1.$$

At this point, we may introduce the following convenient terminology and notations:

5. *Ibid.*, pp. 580 ff [this volume, pp. 159 ff].
6. R. Duncan Luce, "Semi-Orders and a Theory of Utility Discrimination," *Econometrica*, XXIV (1956), 178–191.
7. In Georgescu-Roegen, "The Pure Theory," the basic properties of the model were not stated as explicit axioms, but introduced in the course of the argument. See, however, Georgescu-Roegen, "The Theory of Choice and the Constance of Economic Laws," *Quarterly Journal of Economics*, LXIV (1950), 129–130 [the present volume, pp. 175–176].

Strong Preference:	$A \Pi B$	if	$\omega(A, B) = 1$;
Preference:	APB	if	$\omega(A, B) > \frac{1}{2}$;
Weak Preference:	$A\pi B$	if	$\omega(A, B) = p \geq \frac{1}{2}$;
Complete Indifference:	AIB	if	$\omega(A, B) = \frac{1}{2}$;
Indifference:	AiB	if	$\omega(A, B) \neq 0, 1$.

AXIOM II: *If $A \geq B$, then $A\Pi B$.*[8]

This axiom implies that the quantities of those commodities absolutely necessary to life are measured from the necessary minimum. It also excludes "saturation," a condition which may be fulfilled by considering only a specified domain of E_n. But its most important implication is that the quantities of all commodities involved are not estimated by the consumer's senses alone, but determined by outside scales. On the basis of the quantities thus determined, a rational individual will always choose A if $A \geq B$, however small the difference $A - B$.[9] This means that in the model under consideration the *threshold in choice* is completely isolated from the *ordinary sensorial threshold*, that is, from a phenomenon irrelevant to the economic behavior of the consumer in a world where quantities exchanged are determined with the aid of physical instruments. Indeed, if the process of choice between two *given* vectors would involve no threshold at all, no particular problem other than that of the errors of observation in measuring the physical amounts of commodities would arise in the theory of the consumer.[10]

AXIOM III: *The function $\omega(X, A)$ is continuous in X, except for $X = A$, where it can have any value in the closed interval $(0, 1)$.*[11]

The last part of this axiom reflects the transparent fact that if A and B are *identical* combinations a rational consumer may choose A between A and B with *any frequency* whatever. Analytically, it is justified by the fact that in the neighborhood of A, $\omega(X, A)$ takes any value of the closed interval $(0, 1)$.

8. The relation $A \geq B$ means that at least one component of the vector A is greater than that of B, and none is smaller.

9. Cf. "The Pure Theory," p. 572n [this volume, p. 154n24].

10. In connection with the above remarks, it may well be pointed out that Luce's model (Luce, "Semi-Orders") differs essentially from that of "The Pure Theory," for it views the threshold in choice as a sensorial threshold in discriminating a *cardinal* variable: utility. Another important difference between the two models is that Luce's assumes that at a certain point the choice changes from one gradeless state, "indifference," to another gradeless state, "preference." This discontinuity seems hardly compatible with the very idea of threshold, which is basically a stochastic concept.

11. Because of the indeterminateness of $\omega(A, A)$, some subsequent statements become false if the points involved are allowed to coincide. Since in this case the statements become also irrelevant, to shorten the diction it will be assumed that points denoted by distinct letters remain distinct whenever the statement would otherwise become false.

AXIOM IV: *If $A \leq B$, then $\omega(A, C) \leq \omega(B, C)$, the equality sign holding only if $\omega(A, C) = 1$ or $\omega(B, C) = 0$.*[12]

This axiom extends to the threshold in choice a characteristic property of the sensorial threshold: the individual responds to the smallest increase in the difference between two stimuli by an increase in the frequency of right guesses, excepting the case where the difference is such that he always guesses right.[13] Accordingly, $\omega(X, C)$ is monotonically increasing except in the domains for which its value is either zero or unity.

AXIOM V: *If $\omega(A, B) = \omega(B, C) = p \geq \frac{1}{2}$, then $\omega(A, C) \geq p$.*[14]

This axiom expresses what may be called a pseudo transitivity; as will be seen, it constitutes, however, the basis for other transitivity properties of the model (Corollaries ii–v).[15]

AXIOM VI: *If $C = \lambda A + (1 - \lambda)B$, with $0 > \lambda > 1$, then $\omega(A, B) \leq \omega(C, B)$.*

This axiom states that if X moves on a straight line away from B, the value of $\omega(X, B)$ cannot increase. It is easily seen that this property generalizes the Principle of Persisting Nonpreference Direction.[16] In the case of a rigid model $\omega(A, B)$ can assume only three values, 0, $\frac{1}{2}$, 1. The Principle of Persisting Nonpreference Direction states that if $\omega(C, B) = \frac{1}{2}$, then $\omega(A, B) \leq \frac{1}{2}$ for any A such that C is between A and B, and if $\omega(C, B) = 0$, then for the same situation $\omega(A, B) = 0$.

4. The derivation of the main results of the preceding system of axioms presents little difficulty.

THEOREM 1: *The set $S(Y; p)$ of all X such that $\omega(X, Y) \geq p$, is closed and its boundary forms an $(n - 1)$-dimensional variety in E_n.*

Because of Axiom III, $S(Y; p)$ is closed. By Axiom IV, if $X \in S(Y; p)$, then $X + dX \in S(Y; p)$ for any $dX \geq 0$. Therefore, $S(Y; p)$ forms an n-dimensional domain in E_n, and its boundary is an $(n - 1)$-dimensional variety.

12. "The Pure Theory," p. 571 [this volume, p. 153].
13. Cf. "The Pure Theory," pp. 569–570 [this volume, pp. 151–152].
14. *Ibid.*, pp. 572–574 [this volume, pp. 154–156].
15. A stronger axiom seems difficult to find. The following may suggest itself:

AXIOM Va: *If $A\pi B$ and $B\pi C$, then $\omega(A, C) \geq \max [\omega(A, B), \omega(B, C)]$.*

It is easy to see that this would not do. Indeed, if $A = B + \Delta B$, with $\Delta B \geq 0$, then $\omega(A, B) = 1$, and the proposed axiom leads to $\omega(A, C) = 1$, even if $\omega(B, C) < 1$. In other words, only a small increment ΔB suffices always to pass from weak preference to strong preference. Clearly, this in in violent contradiction with the very essence of threshold. Actually, as will be seen later (Corollary vi), it is *minimum*, not *maximum* that fits in the right-hand side of the relation in Axiom Va.

16. "The Pure Theory," pp. 556, 575–576 [this volume, pp. 141, 156].

THEOREM 2: *If $\omega(A, B) > \omega(B, C) \geq \frac{1}{2}$, then $\omega(A, C) > \omega(B, C)$.*

PROOF: (1) $B \leq A$. The theorem is a direct application of Axiom IV. (2) $B \not\leq A$. Let us consider the function $f(k) = \omega(kA, B)$, of the scalar $k \geq 0$. Since also $A \not\leq B$ (for otherwise $\omega(A, B)$ would be zero), there is no value of k for which $kA = B$. Therefore, according to Axiom III, $f(k)$ is everywhere continuous. In addition, $f(0) = 0$ and $f(1) = \omega(A, B) = p'$. According to Axiom IV, $f(k)$ is monotonically increasing in an interval $(k', 1)$, $k' < 1$. Hence, there is a unique value $K < 1$, such that $f(K) = p = \omega(B, C) < p'$. Let $A^* = KA$. Clearly, $A^* \leq A$, and $\omega(A^*, B) = p$. According to Axiom V, $\omega(A^*, C) \geq p$, and according to Axiom IV, $\omega(A, C) > \omega(A^*, C)$. Hence $\omega(A, C) > p = \omega(B, C)$.

THEOREM 3: *If $\omega(B, C) > \omega(A, B) \geq \frac{1}{2}$, then $\omega(A, C) > \omega(A, B)$.*

PROOF: As in the preceding proof, if $C \not\leq B$, then there is a $B^* \leq B$, such that $\omega(B^*, C) = p = \omega(A, B) < p' = \omega(B, C)$. According to Axiom IV, $\omega(B, A) > \omega(B^*, A)$. From Axiom I, it further follows that $\omega(A, B^*) > \omega(A, B) = \omega(B^*, C)$. Hence, according to Theorem 2, $\omega(A, C) > \omega(B^*, C)$, or $\omega(A, C) > \omega(A, B)$.

COROLLARY i: *If AIB and BPC (or CPB), then APC (or CPA).* This is an immediate application of Theorems 2 and 3.

COROLLARY ii: *If AIB and BIC, then AIC.* Corollary i excludes both APC and CPA.

COROLLARY iii: *If APB and BPC, then APC.* This is a direct application of Axiom V, and Theorems 2 and 3.

COROLLARY iv: *If $A\pi B$ and $B\pi C$, then $A\pi C$.* This follows from Corollaries i–iii and the symmetry of relation I.

COROLLARY v: *If $A\Pi B$ and $B\Pi C$, then $A\Pi C$.* By Axiom V.

COROLLARY vi: *If $A\pi B$ and $B\pi C$, then $\omega(A, C) \geq \min[\omega(A, B), \omega(B, C)]$, with the equality sign implying $\omega(A, B) = \omega(B, C)$.* This is a condensed form of Theorems 2 and 3, and Axiom V.

THEOREM 4: *$S(A; \frac{1}{2})$ is convex toward the origin.*

PROOF: Let $B, C \in S(A; \frac{1}{2})$, and assume the notation such that $B\pi C$. By Axiom VI, $M\pi C$ for any $M = \lambda B + (1 - \lambda)C$, $0 < \lambda < 1$. By Corollary iv, $M\pi A$; hence $M \in S(A; \frac{1}{2})$.[17]

5. The preceding results enable us to establish equivalence classes on the basis of relation I (Complete Indifference). We may refer to these

17. From Axiom VI, there follows immediately that $S(A; p)$ is semiconvex with A as a pole, which means that if $B \in S(A; p)$ and $C = \lambda B + (1 - \lambda)A$, $0 < \lambda < 1$, then $C \in S(A; p)$.

equivalence classes as "complete indifference" varieties.[18] They are convex, nonintersecting, $(n-1)$-dimensional varieties in E_n,[19] and possess other similar properties with the classical indifference varieties. These properties are:

THEOREM 5: *A one-to-one correspondence exists between the real number system $[\nu]$ and the complete indifference varieties, $[S]$, such that if $A \in S(\nu)$ and $B \in S(\nu')$, $\nu >$, $=$, $< \nu'$ implies APB, AIB, BPA, respectively.*

PROOF: Let us consider the base kA of a vector $A > 0$, where $k \geq 0$. Since $k_1 A$ and $k_2 A$ cannot belong to the same equivalence class if $k_1 \neq k_2$, we can exhaust all positive real numbers by writing $S(k)$ for the equivalence class containing kA. We can prove that every equivalence class contains one point of the form kA. Indeed, let S be any equivalence class, and let $A' \in S$, such that $A' \neq kA$ for all values of k. We can find k' such that $k'A > A'$. Let $f(k) = \omega(kA, A')$. We have $f(0) = 0$, $f(k') = 1$; therefore, there is a $k = K$ such that $f(K) = \frac{1}{2}$. The point KA belongs to S. The set $[S(k)]$, for $k \leq 0$, includes therefore all equivalence classes. It is easy to see that the correspondence $S(k)$ fulfills the other properties of Theorem 5.

THEOREM 6: *Let $[Q]$ be the intersection of a complete indifference variety with one of its supporting planes, $[L]$. If $A \in [Q]$, $B \in [L]$ and $B \notin [Q]$, then APB. Conversely, if $A \in [L]$ and $A\pi B$ for any $B \in [L]$, then $A \in [Q]$.*

PROOF: For the first part let us remark that because of the convexity of the complete indifference varieties, $B \notin S(A; \frac{1}{2})$. Hence $\omega(B, A) < \frac{1}{2}$, or by (1), APB. For the second part, if A were not in $[Q]$, then A would not be preferred to all the points in $[L]$, namely to those belonging to $[Q]$.

THEOREM 7: *Any triplet A, B, C satisfies the inequalities*

$$(2) \qquad 1 \leq \omega(A, B) + \omega(B, C) + \omega(C, A) \leq 2.$$

PROOF: Because of (1), reversing the cyclic ordering of the points A, B, C, does not alter the relation to be proved. We may choose, therefore, either of the two cyclic orders. After establishing the manner in which the various pairs of points stand with respect to π, let us choose the notation such that $A\pi B$, $B\pi C$, $A\pi C$. Let also $\omega(A, B) = p$ and $\omega(B, C) = q$. By Corollary vi and (1), $\omega(C, A) \leq 1 - \min(p, q)$. Hence

$$1 \leq p + q \leq \omega(A, B) + \omega(B, C) + \omega(C, A) \leq 1 + p + q - \min(p, q) \leq 2.$$

COROLLARY vii: *The n-tuple $A_1, A_2, \ldots, A_n$ satisfies the inequalities*

$$1 \leq \omega(A_1, A_2) + \omega(A_2, A_3) + \cdots + \omega(A_{n-1}, A_n) + \omega(A_n, A_1) \leq n - 1.$$

6. The consistency of the system of Axioms I–VI is proved by a

---

18. "The Pure Theory," p. 580 [this volume, p. 159 f].
19. These properties follow immediately from Theorem 4, Corollary ii, and Theorem 1 respectively.

particular two-commodity model used in "The Pure Theory." This model proceeds from the one-parameter family of differential equations in the plane of $X(x_1, x_2)$,

(3) $$\varphi_1(X; p)dx_1 + \varphi_2(X; p)dx_2 = 0.$$

It assumes that the coefficients φ_1, φ_2 satisfy the classical conditions of indifference directions (convexity), and that in addition

(4) $$0 < \frac{\varphi_1(X; p)}{\varphi_2(X; p)} < \frac{\varphi_1(X; p')}{\varphi_2(X; p')} < \infty$$

for $0 < p < p' < 1$. Let $H(X; p) = C$ be the general integral of (3). A binary choice map can be completely described in terms of this two-parameter family of curves. Let us choose

(5) $$H(X; \tfrac{1}{2}) = C$$

as the complete indifference curves; this implicitly determines the pairs for which $\omega(X, Y) = \tfrac{1}{2}$. Assuming the signs are chosen so that $H(X'; p) < H(X; p)$ for $X < X'$, the other values of $\omega(X, A)$ are defined according to the following rules:

for $p \neq 0, 1$: $\omega(X; A) = p$, if either $H(X; p) = H(A; p)$, or

(6) $$H(X; 1 - p) = H(A; 1 - p), \text{ and}$$

$$H(X; \tfrac{1}{2}) \gtrless H(A; \tfrac{1}{2}) \text{ according to whether } p \gtrless \tfrac{1}{2};$$

(7) $$\omega(X, A) = 1, \quad \text{if } H(X; 1) \geq H(A; 1), H(X; 0) \geq H(A; 0) \text{ and}$$
$$H(X; \tfrac{1}{2}) > H(A; \tfrac{1}{2});$$

(8) $$\omega(X; A) = 0, \quad \text{if } H(X; 1) \leq H(A; 1), H(X; 0) \leq H(A; 0) \text{ and}$$
$$H(X; \tfrac{1}{2}) < H(A; \tfrac{1}{2}).$$

It is easy to see that the model just described satisfies Axioms I–VI.

In Fig. 5–1, the curve A_pAA_p' represents $\omega(X, A) = p$. There is such a curve passing through A for every value of p, $0 \leq p \leq 1$. This is true of any point in the commodity plane. The region of the threshold in choice relative to A is represented by the set of all X for which

(9) $$[H(X; 1) - H(A; 1)][H(X; 0) - H(A; 0)] < 0.$$

In Fig. 5–1, this is represented by the shaded area between A_1AA_1' and A_0AA_0'. If B is in this region, then BiA, and since i is a symmetrical relation, it follows that AiB, i.e., A is in the threshold region of B.

However, "indifference" in this sense, i.e., as the negation of clear-cut choice, *is not transitive*. As shown in Fig. 5–1, CiB and BiA but $C\Pi A$.[20] Another feature of the model is that the threshold relative to any point widens as we move away from that point. This reflects the transparent

20. "The Pure Theory," pp. 578–579 [this volume, pp. 158–159].

fact that we find two "distant" alternatives (i.e., involving radical changes in consumption) more difficult to compare than two "near" ones (such as the changes in consumption due only to small variations in prices or income).

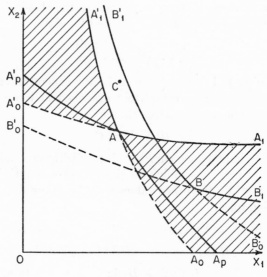

FIGURE 5-1

7. Let us now turn to the relation between the binary choice map and demand, and in order not to obscure the main line of the argument by irrelevant technicalities, let us refer to the case of only two commodities. Under all circumstances, the budget equilibrium involves a choice not between two alternatives, but theoretically at least between infinitely many alternatives. And, as emphasized by K. O. May in this journal,[21] there is no *a priori* relationship between the first type of choice, *the binary choice*, and the second one, *the multiple choice*. In the case of the rigid model, however, the two choices are related: the point N on the budget line which is strongly preferred to any other budget allocation, M, in the binary choice (N, M), is also the chosen allocation if all alternatives are considered together.[22] Since from N we derive a point on one of the demand curves, the binary choice in the rigid model automatically leads to the map of demand schedules.

When threshold is present, the budget equilibrium involves a probability distribution: the probability $\omega(A \mid \mathcal{B})$ of choosing any allocation A

21. Kenneth O. May, "Transitivity, Utility, and Aggregation in Preference Patterns," *Econometrica*, XXII (1954), 2–3.
22. For simplicity I assume here that N is unique, which means that the indifference lines are strictly convex.

out of the set $\mathfrak{B}$ of all alternatives, or if perfect divisibility is assumed, the elementary probability $\mathcal{P}(A)ds$ that the chosen alternative should lie between A and $A + dA$. From this distribution, we can derive the *demand penumbra*, i.e., the probability $\mathcal{P}_1(x)dx$ that the quantity demanded at the assumed price should lie between x and $x + dx$. The fate of the argument that the binary choice provides a simpler basis for consumer's theory than the theory of demand depends, therefore, upon the possibility of deriving $\omega(A \mid \mathfrak{B})$ or $\mathcal{P}(M)$ from the knowledge of $\omega(X, Y)$.

My proof that this can be done took for granted that

$$(10) \qquad \omega(A, C, B) = \omega(A, C) - \omega(B, C)$$

whenever APB,[23] where $\omega(A_1, A_2, \ldots, A_n)$ is the probability that the consumer's preference ordering is $A_1, A_2, \ldots, A_n$. Actually, (10) cannot be justified unless $A\text{II}B$.[24] Additional assumptions are necessary to clarify this point and also to place in better perspective the difficulties of relating multiple to binary choice.

8. Two axioms suggest themselves on elementary intuitive grounds:

AXIOM A: *For any* $A_1, A_2, \ldots, A_n$

$$(11) \qquad \begin{aligned} \omega(A_2, A_3, \ldots, A_n) &\equiv \omega(A_1, A_2, \ldots, A_n) + \omega(A_2, A_1, A_3, \ldots, A_n) \\ &+ \cdots + \omega(A_2, A_3, \ldots, A_1, A_n) + \omega(A_2, A_3, \ldots, A_n, A_1).\text{[25]} \end{aligned}$$

AXIOM B: *If* $\mathfrak{A}$ *is the set* $[A_1, A_2, \ldots, A_n]$, *then*

$$(12) \qquad \omega(A_1 \mid \mathfrak{A}) \equiv \Sigma\omega(A_1, A_{i_2}, A_{i_3}, \ldots, A_{i_n}),$$

where the sum extends over all permutations $(i_2, i_3, \ldots, i_n)$ *of* $(2, 3, \ldots, n)$.

An iterative application of (11) leads to

$$(13) \qquad \omega(A_1, A_2) \equiv \Sigma\omega(A_{i_1}, A_{i_2}, \ldots, A_{i_n}),$$

where the sum extends over all permutations $(i_1, i_2, \ldots, i_n)$ of $(1, 2, \ldots, n)$ such that if $1 = i_k$, $2 = i_l$, then $k < l$. We need only to invoke Axiom I to obtain from (13),

23. "The Pure Theory," p. 573 [this volume, p. 154].

24. *Infra*, note 27.

25. I have heard, at times, the view expressed that preferential ordering, in contrast with binary choice, is not an operational concept, either because the consumer may be unable to rank given alternatives or because he may not understand what is meant by preferential ordering. The view that the consumer may not always be able to rank, say, three points in the commodity-space is undoubtedly legitimate, but not more so than the assertion of the existence of incomparable pairs of points. In fact, it is because of this latter alternative that we need Axiom I. As to the actual tests, several alternatives are available as specific instructions to the subject. The individual may be asked, for instance, to place A, B, C in relation to the numbers 1, 2, 3, and be told that he will have the combination related to 1 if a true die shows more than "three," the combination related to 2 if the die shows "three" or "two," and the combination related to 3 if the die shows "one."

(14) $$\Sigma\omega(A_{i_1}, A_{i_2}, \ldots, A_{i_n}) \equiv 1,$$

where the sum extends over all permutations $(i_1, i_2, \ldots, i_n)$ of $(1, 2, \ldots, n)$. And from (12) and (14), it follows that

(15) $$\Sigma_i\omega(A_i \mid \alpha) \equiv 1.$$

The following relations are immediate for three alternatives:

(16) $$\omega(A, C, B) - \omega(B, C, A) \equiv \omega(A, C) - \omega(B, C).$$

(17) $$1 \leq \omega(A, B) + \omega(B, C) + \omega(C, A) \leq 2.^{26}$$

The important fact, however, is that from the knowledge of all $\omega(A_i, A_j)$, not much can be said in general about $\omega(A_1 \mid \alpha)$. In the first place, we can prove

THEOREM 8: *A general relation between $\omega(A_1 \mid \alpha)$ and $\omega(A_i, A_j)$ is incompatible with Axioms* A *and* B.

PROOF: Because of the linearity of the *identities* (12) and (13), the relation cannot but be of the form

(18) $$\omega(A_1 \mid \alpha) \equiv \Sigma l(i, j)\omega(A_i, A_j),$$

where the sum extends over $1 \leq i, j \leq n$, $i \neq j$, and the l's are numerical constants. From (13), we obtain

(19) $$\omega(A_1 \mid \alpha) \equiv \Sigma L(i_1, i_2, \ldots, i_n)\omega(A_{i_1}, A_{i_2}, \ldots, A_{i_n}),$$

where

(20) $$L(i_1, i_2, \ldots, i_n) = \Sigma l(i_a, i_b) \quad \text{for all} \quad 1 \leq a < b \leq n.$$

From (19) and (12), we obtain

(21) $$L(i_1, i_2, \ldots, i_n) = \begin{cases} 1 & \text{if} \quad i_1 = 1, \\ 0 & \text{if} \quad i_1 \neq 1. \end{cases}$$

Hence,

(22) $$L(1, i_2, \ldots, i_{n-2}, i_{n-1}, i_n) = L(1, i_2, \ldots, i_{n-2}, i_n, i_{n-1}) = 1,$$
$$L(i_1, i_2, \ldots, i_{n-2}, i_{n-1}, 1) = L(i_1, i_2, \ldots, i_{n-2}, 1, i_{n-1}) = 0.$$

If we replace L by (20), relations (22) yield

(23) $$l(i, j) = l(j, i) \quad \text{for all} \quad i \neq j.$$

Therefore, (18) becomes

(24) $$\omega(A_1 \mid \alpha) \equiv \Sigma l(i, j),$$

the sum being extended over all $i < j \leq n$. But (24) shows that $\omega(A_1 \mid \alpha)$

26. This is the same as Theorem 7. But it should be observed that (17) is derived from Axiom I and Axiom A alone, while (2) is derived from Axioms I–VI. This may be taken as an indication — not sufficient though — that Axiom A does not contradict the system of Axioms I–VI.

is a numerical constant, or in other words, that all A_i are chosen with the same probability independent of $\mathcal{C}$. This absurdity proves the theorem.

9. The remark that if $\omega(A_1, A_2) = 0$ all the terms in the right-hand side of (13) are zero, leads because of (12) to

THEOREM 9: *If* $\omega(A_1, A_2) = 0$, *then* $\omega(A_1 \mid \mathcal{C}) = 0$.[27]

If $\mathcal{C}$ contains some alternatives that will definitely not be chosen in a multiple choice, Theorem 9 provides a convenient rule for eliminating those alternatives only on the basis of the results in binary choice.[28] In contrast with this rule, nothing in general can be inferred about $\omega(A_1 \mid \mathcal{C})$ from the knowledge that $\omega(A_1, A_2) = 1$. Actually, $\omega(A_1 \mid \mathcal{C})$ may even be zero, as is easily illustrated by the situation where $A_1 \Pi A_2$ and $A_3 \Pi A_1$. Further investigation of possible inequality relations between the $\omega(A_i \mid \mathcal{C})$'s and the $\omega(A_i, A_j)$'s runs at the outset against a discouraging sign from the following example:

(25)
$$
\begin{aligned}
\omega(A, B, C) &= \tfrac{1}{6} - a, \\
\omega(A, C, B) &= \tfrac{1}{6} - b - c, \\
\omega(B, C, A) &= \tfrac{1}{6} - b - c - d - e, \\
\omega(B, A, C) &= \tfrac{1}{6} + a + 2b + c + d + e, \\
\omega(C, A, B) &= \tfrac{1}{6} + a + 2b + c + d, \\
\omega(C, B, A) &= \tfrac{1}{6} - a - 2b - d,
\end{aligned}
$$

where $a, b, c, d, e > 0$, $b + c + d + e < \tfrac{1}{6}$, $a + 2b + d < \tfrac{1}{6}$. The ω's thus determined fulfill condition (14); they also yield:

(26a)
$$
\omega(A, B) = \tfrac{1}{2} + b + d, \qquad \omega(B, C) = \tfrac{1}{2} + b,
$$
$$
\omega(A, C) = \tfrac{1}{2} + b + d + e
$$

and, denoting by $\mathfrak{I}$ the set (A, B, C),

(26b)
$$
\omega(A \mid \mathfrak{I}) = \tfrac{1}{3} - a - b - c, \qquad \omega(B \mid \mathfrak{I}) = \tfrac{1}{3} + a + b,
$$
$$
\omega(C \mid \mathfrak{I}) = \tfrac{1}{3} + c.
$$

In this example, APB, APC and moreover $\omega(A, B)$, $\omega(A, C) > \omega(B, C) > \tfrac{1}{2}$, yet $\omega(A \mid \mathfrak{I}) < \omega(B \mid \mathfrak{I})$, $\omega(C \mid \mathfrak{I})$. In other words, though the binary choice indicates A as the most frequently chosen alternative, A is the least frequently chosen in the multiple choice. Because (26a) satisfies Corollary vi, which is the basic nonmetric feature of the system of

27. The same remark shows that if $A \Pi B$, then $\omega(B, C, A) = 0$. This shows that if $A \Pi B$, relation (10) follows from (16).

28. Thus, the segment TS in Fig. 9, "The Pure Theory," p. 581 [in the present volume, Fig. 1–9], represents the only alternatives likely to be chosen out of all possible budget allocations.

Axioms I–VI, the illustration (25) retains its significance for the system obtained by the addition of Axioms A–B.[29]

In particular, the above illustration shows that *although N, the point of tangency between the budget plane and the complete indifference varieties, is the only alternative strongly preferred to any other allocation, N may be the "least preferred" alternative in the multiple choice between all budget allocations.* Consequently, the argument of my earlier paper is incorrect;[30] the map of complete indifference varieties offers no basis for deriving the "typical demand." Strictly speaking, according to the present state of the problem no relation can be established between the binary choice map and the demand penumbra. In other words, the binary choice cannot be regarded as a foundation of the demand theory if we accept the view that the individual is not a perfect choosing-instrument. Other assumptions than those analyzed in this paper may change our conclusion, but until such assumptions are formulated, the only way by which we can arrive at a behavioristic determination of the "average" demand is to observe the quantities demanded, not the results of binary choices.

Two primordial problems stand, therefore, before the econometricians and the behavioral scientists:

I. *What axioms are logically necessary and experimentally justified, to relate the multiple choice probability to that of the binary choice?*

II. *Granted these axioms, what is the reflection of the "average" demand into the binary choice map?*

Theorem 8 shows, however, that if the first problem is to have a solution, one must abandon Axiom A or B, transparent though they are.

29. This is an important point, for it is known that the nonmetric properties of the Axioms A and B do not entail the nonmetric property expressed by Corollary vi. An example which satisfies Axioms I and A, but which does not fullfill the transitivity of π, is given in J. Marschak, "Norms and Habits of Decision Making under Certainty," in *Mathematical Models of Human Behavior*, Dunlap and Associates (Stamford, Conn., 1955), p. 50.

30. Cf. *supra*, Section 2.

6

The Nature of Expectation
and Uncertainty

The kinetics of social phenomena cannot be fully grasped without an understanding of the nature of *that* uncertainty involved in the most fateful decisions in the life of an individual or a community. Most social scientists and decision theorists reason on the assumption that such an uncertainty can be represented by some kind of numerical probability. Warnings issued from time to time against the futility, nay the danger, of treating all decisions as if they referred to "a gamble on a known mathematical chance" [1] have had little effect. Probability theorists, too, have generally bent their efforts to justify this view. Yet, long ago, von Kreis argued that the relation between uncertain events being only that of "more or less probable," it cannot always be expressed numerically. [2] A few decades later, Keynes took an even stronger position, that expectations may not always be compared: "On some occasions none of these alternatives ['equal,' 'more,' 'less'] hold, and . . . it will be an arbitrary matter to decide for or against an umbrella." [3] Even Fréchet, the outstanding spokesman of the Laplacian school, has recently admitted that not all expectations can be "measured" and has called upon the econometrician not to proceed on the assumption that all cats are grey. [4] How-

---

NOTE: This paper is reprinted by courtesy of the publisher from *Expectations, Uncertainty and Business Behavior*, ed. Mary Jean Bowman, a publication of the Social Science Research Council (New York, 1958), pp. 11–29.

1. Frank H. Knight, *Risk, Uncertainty, and Profit* (reissue, London, 1948), p. xiv and chap. vii.
2. J. von Kreis, *Die Prinzipien der Wahrscheinlichkeitsrechnung* (Freiburg, 1886), pp. 26 ff.
3. John Maynard Keynes, *A Treatise on Probability* (London, 1921), p. 30. More recently, B. O. Koopman, in a series of remarkable articles visibly inspired by Keynes' views, showed that the most generous set of transparent postulates does not make all probabilities numerically measurable: B. O. Koopman, "The Axioms and Algebra of Intuitive Probability," *Annals of Mathematics*, XLI (1940), 269–292; "Intuitive Probability and Sequences," *ibid.*, XLII (1941), 169–187; "The Bases of Probability," *Bulletin of the American Mathematical Society*, XLVI (1940), 763–774.
4. Maurice Fréchet, "Sur l'importance en économétrie de la distinction entre les probabilités rationnelles et irrationnelles," *Econometrica*, XXIII (1955), 303–306.

ever, most students have preferred to advance on the less thorny passage of measurable uncertainty — a passage more fertile in analytical results — rather than to face the delicate complexity of expectation and be content with making smaller, yet more relevant, strides. Witness Jeffreys' reaction to Keynes' views: "Keynes is merely creating difficulties [and his] postulate is . . . one of those attempts at generality that in practice lead only to vagueness." [5] Savage also conjectured that the idea of noncomparability of expectations "would prove a blind alley losing much in power and advancing little, if at all, in realism." [6]

The controversy over the structure of expectation is in many ways similar to that of Pythagoras' time over the structure of length. As we recall, the Pythagoreans believed that the length of a line is proportionate to the "number of points" on it and, consequently, any length can be expressed by a rational number. No abstract argument could rock this doctrine, which could not survive, however, the first concrete example of a line of irrational length (the diagonal of the square). I hope that the concrete examples given in Section VI on the Map of Expectations below will show clearly that the real number system does not suffice to represent all expectations and that those expectations which can be meaningfully connected with a real number form only a special class.[7] A larger but still restricted class can be represented by complex numbers. As we get nearer pure uncertainty, however, no number system seems to fit into the picture. And a brief analysis of the main schools of thought will reveal that two expectations represented by the same number according to the various rules proposed are not necessarily equivalent.

I. THE BASIC ASPECTS OF EXPECTATION

1. Many idle controversies involving the nature of expectation could be avoided by recognizing at the outset that man's conscious actions are

5. H. Jeffreys, *Scientific Inference* (New York, 1931), pp. 223–224. Jeffreys' later work, *Theory of Probability* (Oxford, 1939), contains only a scant remark about Keynes' views, in a footnote (p. 25).

6. L. J. Savage, *The Foundations of Statistics* (New York, 1954), p. 21.

7. The first basic example was presented in my "Utility, Expectations, Measurability, and Prediction," Institute of Research and Training in the Social Sciences, Vanderbilt University, 1953 (mimeo.), a paper read at the 1953 Kingston meeting of the Econometric Society. The concrete examples offered by Ernest Nagel, "Principles of the Theory of Probability," *International Encyclopedia of Unified Science*, vol. I, no. 6 (Chicago, 1939), pp. 68–69, only illustrate the difficulty of assigning a significant number to the degree of confirmation of a scientific hypothesis in some typical situations already mentioned by Keynes, *A Treatise*, pp. 29–30. They do not prove that such a number could not be found.

the reflection of his beliefs and of nothing else. This is true independently of whether he aims at choosing the shortest route between two points on a map or at guessing the color of the ball to be drawn from an urn. The fact that in one case geometry offers an undisputed answer, while in the second case such an answer is not available, does not affect the statement. In both cases the decision of the individual depends entirely on his state of mind at the time of the decision. This is why I propose to define expectation by paraphrasing De Morgan's definition of probability: Expectation is "the state of the mind [of a given individual] with respect to an assertion, a coming event, or any other matter on which absolute knowledge does not [necessarily] exist." [8] With this definition the spurious properties that might easily be hidden in such terms as "degree of belief," "degree of doubt," or "degree of potential surprise," cannot become *ab ovo* a part of the concept of expectation.

Expectation as defined above does not reflect all aspects of the state of the mind produced by an assertion, such as would be the pleasure, the gloom, or the ethical reaction caused by the anticipation of a future event. These aspects do bear upon decisions, for without the anticipating of the future most, if not all, decisions could not even be conceived. But by keeping expectation and motive as two separate elements, many seeds of possible confusion later in the argument are eliminated. The concept of expectation can now be formally expressed in terms of its objective elements by the notation $\mathcal{E}(I, E, P)$, where I identifies the individual, E stands for the "evidence" available to him, and P denotes the "prediction." [9] In its deep waters, however, lies a difficulty that ought to be mentioned at this stage.

Though the variables I and P can hardly give rise to any complication, the same is not true of E. Since there is no way of telling *a priori* what part of the individual's knowledge does not bear upon a given expectation, E must stand for *all knowledge of the individual* at the time. This Principle of Total Knowledge differs in many respects from the Requirement of Total Evidence, which has been continuously emphasized in the literature since the time of Bernoulli and which says that no factual evidence should be omitted from E if available. [10] In the first place, the Principle of Total Knowledge does not presuppose that E must contain all the knowledge obtained by pooling together the knowledge of all individuals, as the Requirement of Total Evidence has usually been interpreted. [11] On the

8. Augustus De Morgan, *Formal Logic* (London, 1847), p. 173.
9. "Prediction," as used here, does not necessarily mean a statement only about future events, but includes also such statements as "Rome was founded by Romulus."
10. Cf. Rudolf Carnap, *Logical Foundations of Probability* (Chicago, 1950), pp. 211–213.
11. *Ibid.*, p. 212.

contrary, it allows E to comprise even opinions that are known or may eventually prove to be wrong.[12] Needless to say, E may contain more opinions that are hard to deal with like "Wednesday is my lucky day."

The question now arises whether the definition of E is operational. The answer to this question can only be negative, for undoubtedly no individual is able to draw up a list of *all* his knowledge in explicit terms, any more than he can at any one time list from memory *all* his material possessions (which is by far a simpler task). The negative answer sheds a great deal of light on the frequently entertained hope that decisions based on any expectations whatever might one day become a matter of numerical computations, complicated though these may be. Since individual knowledge, and all the more the entire human knowledge, cannot be effectively listed, to reach a decision in the case of some expectations will remain the appanage of the most primitive electronic brain, the human mind. And as any decision is based *only* on that part of the individual's E he can bring into sharp focus at the proper moment, the rapidity with which a given individual can gather and analyze the elements of his knowledge most relevant to P constitutes one important coordinate which accounts for the difference between good and poor judgment. A discussion of this slippery, nonetheless capital, concept of "judgment" will have to wait until the final section of this paper.

2. Expectation has more than one facet to catch our fancy; these facets are reflected in the various approaches to probability. But let us begin by examining the problem of the structure of expectations. This problem involves two distinct, yet not unrelated, questions: (1) Are expectations *comparable?* and (2) If they are *comparable*, are they also *ordinally measurable?*[13]

3. Expectations are *comparable* if they can be compared in a consistent manner according to some meaningful criterion. In more specific terms this means that: (1) given any two expectations $\mathcal{E}(I, E_1, P_1)$, $\mathcal{E}(I, E_2, P_2)$, the individual I can say whether or not they are "equal," and if they are not, which is "higher," and (2) if $\mathcal{E}(I, E_1, P_1) \geq \mathcal{E}(I, E_2, P_2)$, $\mathcal{E}(I, E_2, P_2) \geq \mathcal{E}(I, E_3, P_3)$, then $\mathcal{E}(I, E_1, P_1) \geq \mathcal{E}(I, E_3, P_3)$.[14] A meaningful criterion of

---

12. To see the futility of attempting to eliminate from E such opinions, one should recall, for instance, that all mathematicians once held that a function continuous everywhere in a closed interval has everywhere a derivative, and the medical men, that the appendix performs no useful function in the human body.

13. The terms used above were introduced in my "Choice, Expectations and Measurability," *Quarterly Journal of Economics*, LXVIII (1954), 520 [in the present volume, p. 201], to replace the mathematical terminology by one more familiar to the economist. For the mathematical terminology, see G. Birkhoff, *Lattice Theory* (New York, 1948), p. 10.

14. The latter is the famous transitivity condition. It should be added that equality cannot prevail in the last relation unless present in the other two.

comparability seems immediately available. The individual I is asked to choose between the outcomes P_1 and P_2, the reward for guessing right being the same in each case with no penalty for guessing wrong. If he chooses P_1, then $\mathcal{E}(I, E_1, P_1) > \mathcal{E}(I, E_2, P_2)$. The rest follows "rationally." But a closer look at our own decisions reveals some troublesome spots in this construction. How many times do we hear around us that "this is a difficult decision to make"? Of course, the difficulty may not necessarily reflect the impossibility of comparing the two expectations *per se*; it may reflect the incomparability of the alternative situations. To be sure, ultimately the individual reaches a decision — he does or does not take an umbrella — but this may at times be the result of intuition, if not of impulse, both of which have nothing to do with the objective elements of $\mathcal{E}$. The problem of comparability is, therefore, far from settled, unless we find another justification for comparability than the fact that no man would die like Buridan's ass.

4. Let us now see what the second question, that of ordinal measurability, involves. Although things may be ordered, it may be impossible to indicate the rank of each element by a real number; there may not be enough numbers for this operation. The twenty steps of a staircase, although ordered, cannot all be assigned a numerical rank if there are, say, only ten numbers available. Something like this may happen also when we try to assign ranking numbers to a comparable set containing an infinity of elements; only the contradiction is not so intuitive. All the points on the continental United States, for instance, can be made comparable by the following criterion: (1) of two points of the same meridian, the one situated north of the other is ranked higher, (2) of two points not situated on the same meridian, the one situated east of the other ranks higher. Yet we cannot assign to each point on the map a unique ranking number.[15]

Writers on probability or utility have for a long time ignored the essential differences between comparability and ordinal measurability, even after "an American reviewer" (whose identity I have not been able to discover) raised the point in discussing Jeffreys' *Scientific Inference*.[16] Typical of this error ("the ordinalist's fallacy") is the argument of De Finetti, who asserts that the doubt "is susceptible of comparison, and, consequently, of graduation." He seems not to suspect the startling effect of his next affirmation that "in this way, although starting from a system of purely qualitative axioms, one arrives at a quantitative measure of

15. The mathematical proof of this follows immediately from Theorem 2 or Ex. 6* in Birkhoff, *Lattice Theory*, p. 32.

16. See Jeffreys, *Theory of Probability*, p. 19. "The American reviewer" supported his objection not by an example from the field of probability, but by a lexicographic ordering, like that used above for the points on the map.

probability." [17] The distinction between comparability alone and ordinal measurability has been generally regarded as a logical nicety without any relevance. Jeffreys, for instance, in his *Theory of Probability* (published after the criticism of "the American reviewer"), mended the leak in his axiomatic setup only by adding an explicit postulate equivalent to that of ordinal measurability, but did not bother to offer the slightest justification for it.

II. EXPECTATION AND THE MATERIAL WORLD

1. An important class of expectations, that referring to the outcome of a random mechanism, can be readily ordered according to a specific coordinate of the mechanism: the probability coefficient. Much simpler than other types of expectations, the probability expectation was the first to attract the attention of theorists. Its continual influence upon the development of thought on expectations, however, is responsible for the tendency of most theorists to reduce all expectations to numerical probability.

According to the Laplacian or Classical School, probability is measured by the relative frequency of favorable cases among all possible cases, *provided these are equally possible*. According to the Frequency School,[18] probability is measured by the frequency limit of favorable cases in an infinite sequence of observations, *provided the limit exists and the sequence satisfies the condition of randomness*. The intensity of the well-known controversy between the two schools is a partial yet impressive indication of the difficulties raised by the general concept of expectation.[19] On the

17. Bruno de Finetti, "La prévision: ses lois logiques, ses sources subjectives," *Annales de l'Institut Henri Poincaré*, vol. 8, fasc. 1 (1937), pp. 4–5. De Morgan also reasoned once in this fashion: "Whenever the terms greater and less can be applied, there twice, thrice, etc., can be conceived, though not perhaps measured by us" (quoted in Keynes, *A Treatise*, p. 21n).

18. John Venn, *The Logic of Chance* (London, 1866), marked the coming of age of the Frequency Theory. It was, however, Richard von Mises who raised it to fame through his fundamental work, *Wahrscheinlichkeit, Statistik und Wahrheit* (Wien, 1928), of which *Probability, Statistics and Truth* (New York, 1939), is an enlarged English translation. Actually, the first hint of a frequency definition appeared in S. D. Poisson's celebrated *Recherches sur la probabilité des jugements* (Paris, 1837): "They [the frequencies] reproduce them [the probabilities] exactly if it were possible to make the series of observations of an infinite length."

19. An example of the acuity reached by the debates between the followers of the two schools is offered by the articles of Fréchet and Mises in *Les fondements du Calcul des Probabilités*, vol. II of *Colloque consacré à la Théorie des Probabilités*, 8 vols. (Paris, 1938–1939), and the discussions summarized by B. de Finetti in *Compte rendu critique du Colloque de Genève sur la Théorie des Probabilités*, which is vol. VIII of *Colloque*.

surface, the controversy turned upon the definition of probability, but a closer examination shows that it stemmed from two questions: (1) the relation between the probability coefficient and the observed frequency ratio, and (2) the definition of randomness.

2. From the early stage, the Classicists related the Laplacian probability to the observed frequency,[20] but for a long time they failed to realize that this relationship poses an additional problem and reasoned on a bivalent interpretation of the Law of Large Numbers. The confusion led one writer to remark that "the experimenters believe it [the Law of Large Numbers] to be a theorem of the mathematicians, and the mathematicians, an experimental fact." [21] Later, under the sustained fire of the Frequentists,[22] the Classicists came out with the answer that the Laplacian coefficient is a theoretical value of which the frequency is an empirical estimation.[23] This implies — it seems quite correctly — that empirical estimation and physical probability form an indissoluble complex in our phenomenological understanding. The only difference between the two concepts is that probability lends itself to a simpler analytical treatment.

For the Classicists, randomness is a special kind of causal relation. As pointed out by Poincaré,[24] this causal relation characterizes the mechanisms whose results can be completely reversed by imperceptible variations in the initial conditions. And, indeed, there is no other sense in which we can speak of randomness.[25] Concretely, randomness can be obtained only through random mechanisms. This is why the "random numbers" have not been constructed according to an abstract rule, but obtained from a "random mechanism." And also why the Chinese game of "matching fingers" poses some extremely subtle problems foreign to

20. Buffon had the patience to toss a coin 4,040 times in order "to test" the theoretical probability of a "head." Cf. De Morgan, *Formal Logic*, p. 185. Laplace thought of obtaining an empirical value of π from a great number of observations in Buffon's Problem of the Needle (P. S. de Laplace, *Oeuvres Complètes*, Paris, 1836, VII, 366).

21. Reported by H. Poincaré, *Calcul des Probabilités* (2nd ed., Paris, 1912), p. 171. The person quoted is identified only as "Lippmann."

22. It is Mises' great merit to have most clearly exposed the lacuna in the thinking of the Classicists. See Mises, *Probability, Statistics and Truth*, pp. 156 ff.

23. Cf. M. Fréchet and M. Halbwachs, *Le Calcul des Probabilités à la portée de tous* (Paris, 1924), p. 9, and M. Fréchet, *Recherches théoriques modernes sur la Théorie des Probabilités* (Paris, 1937), I, 4–10.

24. Poincaré, *Calcul des Probabilités*, pp. 5–6.

25. In addition to the Frequentists, others have attempted to arrive at a logical, i.e., nonintuitive, definition of randomness. The contributions of Alonzo Church are particularly noteworthy. Yet, contrary to H. Reichenbach's assertion (*The Theory of Probability*, Berkeley, 1949, p. 149), the question is far from being solved. Actually, Church, in "On the Concept of a Random Sequence," *Bulletin of the American Mathematical Society*, XLVI (1940), 130–135, emphasizes the fact that a satisfactory formal definition of randomness not only raises insurmountable difficulties, but is confronted *ab initio* with a logical dilemma.

our game of "matching pennies." In fact, Borel has long since warned us that the human mind is incapable of imitating "randomness." [26]

3. The Frequentists began by criticizing the vagueness of "equally possible cases" [27] and ended by claiming that their own definition of probability is "free from contradictions and obscurities of any kind" and does not require the "rather vague idea of the relationship between probability and sequences of events [as does] the so-called modernized classical theory [of] Fréchet." [28] According to Mises, a collective, i.e., a countably infinite sequence of observations, fulfills the randomness condition if the frequency limit remains the same in any subcollective "formed from the original one by an arbitrary place selection." [29] This definition soon proved to be useless, for no collective can fulfill the condition unless all its elements are identical. [30] The impasse was avoided by a new definition, that of "restricted randomness," which requires the frequency limit to be invariant only with respect to a countable infinity of place selections. But by this restriction, the content of what is generally understood by "randomness" is greatly reduced, while the intuitive transparency of the original definition is completely lost.

But where the Frequentists really came to mutilate randomness is in their definition of the probability coefficient as the "limiting value" of the observed frequency in a collective. The "limiting value" of the Frequentists[31] is not identical to "limit" as defined in mathematics. In mathematics a countably infinite sequence has a limit if for any arbitrary, positive ϵ, we can *actually name* a rank N, beyond which all terms differ from the limit by less than ϵ. In the case of the "limiting value," the rank N cannot be named; we can only *affirm its existence*. The Frequentists claim that this is all that matters. [32]

The fundamental property of the collective is usually formulated as follows:

(α) Given an arbitrary positive number ϵ, there *exists* an N such that

26. M. Fréchet, "Exposé et discussion de quelques recherches récentes sur les fondements du Calcul des Probabilités," *Colloque*, II, 36.

27. Cf. Mises, *Probability, Statistics and Truth*, pp. 117–120; Nagel, "Principles" (note 7, above), p. 45.

28. R. von Mises, "On the Foundations of Probability and Statistics," *Annals of Mathematical Statistics*, XII (1941), 192, 196–197.

29. Mises, *Probability, Statistics and Truth*, pp. 87–88. See also his "Quelques remarques sur les fondements du Calcul des Probabilités," *Colloque*, II, 57.

30. A. Wald, "Die Widerspruchsfreiheit des Kollectivbegriffes," *Colloque*, II, 81, and Jean Ville, *Etude critique de la notion de collectif* (Paris, 1939), p. 40.

31. Mises, "On the Foundations of Probability and Statistics," p. 192; Reichenbach, *Theory of Probability*, pp. 69–70.

32. See, for instance, Nagel, "Principles," pp. 52 ff, and Reichenbach, *Theory of Probability*, pp. 338 ff.

all frequencies f_n, $n > N$, satisfy the relation $|f_n - p| < \epsilon$ where p is the probability of the attribute A.[33]

But from this, by simple algebra we can derive the following property:

(β) Given an arbitrary positive ϵ smaller than p and $1 - p$, there *exists* an N such that for all $n > N$, (1) if $f_n \geq p$, the nth observation cannot be followed immediately by a run of A's greater than

$$n(p + \epsilon - f_n)/(1 - \epsilon - p),$$

and (2) if $f_n \leq p$, the nth observation cannot be followed immediately by a run of non-A's greater than $n(f_n - p + \epsilon)/(p - \epsilon)$.[34]

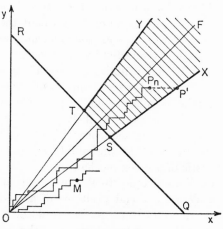

FIGURE 6–1

This can be illustrated by a simple diagram. In Fig. 6–1, x and y are the absolute frequencies of A and non-A respectively in the first n observations ($x + y = n$). A sequence of observations is represented by a zigzag path such as OM. The observed frequency of A at any moment, M, is represented by $1/(1 + m)$ where m is the slope of OM. Assuming that A is a "head" in coin tossing with $p = \frac{1}{2}$, to say that f_n does not differ from p by more than ϵ ($< \frac{1}{2}$) means that P_n is inside the angle XOY, where OX and OY are equally inclined with respect to OF and depend on ϵ.[35] The proposition (α) says there exists a boundary ST ($x + y = N$) such that beyond it, P_n must remain inside the shaded area XSTY. What (β. 1) then says is that after P_n the run of "heads" cannot be greater than P_nP'. For instance, if $N = 100$ for $\epsilon = .03$ and, say, $f_{110} = .518$, then the next two observations *must* contain at least one "tail."

33. Cf. Nagel, "Principles," p. 52.
34. It is easy to see how the formulation of (β) should be modified if $p = 0$ or $p = 1$.
35. Since p is taken equal to $\frac{1}{2}$, the slope of OF is unity.

Now, it is clear that since (α) and (β) have the same assertive value, they must stand or fall together. But (β) imposes a restriction upon the length of the runs at *every* rank higher than N.[36] In the above numerical example, for instance, if the 111th observation happens to be a "head," the next observation must necessarily be a "tail." It is as if the coin would remember at this point that it had already produced enough "heads." But this is contrary to the very essence of probabilistic phenomena, which Bertrand cast in the well-known aphorism, "the coin has no memory." According to the Frequentist analytical setup, however, this aphorism ought to be amended by the continuation that "nevertheless, the coin will develop a memory if it is tossed long enough." That is strongly reminiscent of the views of Azaïs about the compensation of occurrences,[37] or those of Marbe about the uniformity of the world, views that have been emphatically rejected by all schools.[38]

But the Frequency Theory contains also a serious internal contradiction. According to the very theorems of this theory, all distinct sequences of n tossings of a true coin are equally probable (probability $\frac{1}{2}^n$).[39] All infinite sequences also are equally likely to occur. Consequently, any symmetrical coin may produce a sequence with a frequency limit different from $\frac{1}{2}$. But this means that the probability coefficient is no longer an invariant characteristic of a given mechanism.

I foresee one possible objection to the preceding argument: that the collective includes *all possible* end positions of a mechanism.[40] Clearly then, any infinite sequence of coin tossings would be a place selection of the collective, and because of "randomness," the sequence could not possibly yield a frequency limit different from $\frac{1}{2}$. But as we have seen, the modified condition of "randomness" does not and cannot stipulate that *all* place selections should have the same frequency limit. Besides, a collective could not possibly contain all tossings of a coin, for obviously all possible distinct trajectories form a continuum.

36. Criticism of the limit proposition of the frequency theory has been formulated by various writers (e.g., T. C. Fry, *Probability and Its Engineering Uses*, New York, 1928, p. 90; R. B. Lindsay and H. Margenau, *Foundations of Physics*, New York, 1936, pp. 165–167). However, their arguments made use of a precise place in the sequence of observations and, because of this, have been easily refuted by the frequency theorists (e.g., Nagel, "Principles," pp. 36–37; Reichenbach, *Theory of Probability*, pp. 346–347).

37. Pierre Hyacinthe Azaïs, *Des compensations dans les destinées humaines* (3rd edn., Paris, 1818).

38. Mises, *Probability, Statistics and Truth*, pp. 198 ff, devotes a whole section to denouncing Karl Marbe's philosophy.

39. Mises, *Probability, Statistics and Truth, passim*, esp. p. 134.

40. The assumption that a collective contains *all possible* cases does not seem to be explicitly made by Mises, but it does appear in other writings adopting a frequency approach, e.g., R. A. Fisher, "On the Mathematical Foundations of Probability," *Philosophical Transactions*, Series A, CCXXII (1922), 312.

4. This continuum, known as the *phase space* of the mechanism, furnishes the basis of a new definition of the probability coefficient.[41] The definition, based on the theory of measure, is given by the formula

$$(1) \qquad \text{Probability of A} = \frac{\text{Measure of } \Omega_A}{\text{Measure of } \Omega}$$

where Ω is the phase space, and Ω_A the subset formed by the phases where A occurs. This does not answer the problem of how to tell whether all phases are equally probable, for the answer depends on the actual distribution of the initial conditions of the system when at work. Poincaré, however, proved that under very broad conditions the distribution of the operator's thrusts do not influence the outcome of card shuffling or the roulette.[42] This is probably true for other random mechanisms. Be that as it may, formula (1) avoids two main stumbling blocks of the Frequency Theory: first, it does not make the value of the probability coefficient depend upon the order of observations, as does the collective,[43] and second, it allows for a frequency limit different from the probability coefficient defined over Ω.

III. CERTAINTY AND QUASI CERTAINTY

1. It is clear then that the probability coefficient can be uniquely defined as a frequency ratio only in the entire phase-space. A sequence of observations, be it infinite, represents but a mere sample of this space. This does not mean that we can dispense with it. As Borel put it, the countable infinity is the only infinity intuitively known to us and, therefore, the only door through which we can penetrate into the higher infinities.[44] In our case, the infinite sequence of observations constitutes the only link between the empirical basis of our knowledge — the finite number of observations — and the invariant properties of the continuum Ω, which alone are knowledge. But can we trust the intermediary

41. A most remarkable study on the connection between the phase space and probability is that of E. Hopf, "On Causality, Statistics, and Probability," *Journal of Mathematics and Physics*, XIII (1934), 51–102, which contains a detailed analysis of the phase space of a coin and the die box. A similar idea was suggested by Paul Lévy, *Théorie de l'addition des variables aléatoires* (Paris, 1937), p. 7, who described the probability coefficient of a mechanism as a particular integral invariant, something like the moment of inertia.

42. Poincaré, *Calcul des Probabilités*, pp. 148–150, 301–312.

43. This shortcoming of the Frequentist definition was pointed out by Fréchet, "Exposé et discussion," p. 27.

44. Emile Borel, *Leçons sur la théorie des functions* (2nd edn., Paris, 1914), pp. 182–183.

link, knowing that a sequence of observations may lead to any frequency limit or even to none?[45] Fortunately, the infinite sequences that have a frequency limit equal to the frequency ratio in Ω overwhelm in numbers all the others. In the terminology of the theory of measure this is expressed by saying that those other sequences form a set of zero measure; but this does not mean that they could not cross our path at all.[46] This is what the Classicists understood when, in their own language, they used to say that "we cannot affirm that in the long run an event will occur with a frequency proportionate to its probability; but we can affirm that it is more likely to occur with this than with any other precise degree of frequency." [47] The Neoclassicists express the same thing in a slightly different manner: we are *quasi certain* that the frequency ratio in an infinite sequence of observation will equal the probability coefficient.[48]

Quasi certainty is, therefore, the state of the mind created by an outcome whose probability measure is unity but whose occurrence is not logically implied by E. On the other hand, certainty is the state of the mind that corresponds to an outcome whose occurrence is a logical necessity of E. If an "urn" contains an infinite number of elements — like the set of all infinite sequences of tossings of a coin — the probability measure of extracting a certain type of element is obtained by a process of limit. It may very well happen — as for instance with the infinite sequences whose frequency limit equals the theoretical probability — that the limit might be unity without the "urn's" containing only that particular type of element. It is thus seen that even if we tried to confine our analysis to probability as a physical constant, we would nevertheless be led to realize that two entirely different states of the mind are repre-

45. With reference to Fig. 6–1, the diagrammatical image of a sequence of coin tossings may or may not end by remaining inside the angle XOY. The last alternative comprises the paths that end by remaining either inside another such angle corresponding to another value of p, or inside no such angle.

46. The relationship between the frequency limit of a sequence of observations and the frequency ratio of the same attribute in Ω is admirably analyzed in J. L. Doob, "Probability as Measure," *Annals of Mathematical Statistics*, XII (1941), 206–214. It is interesting that in the discussion of Doob's paper, Mises sounds less categorical than usual about the invariance of the frequency limit, since he admits that his own "conception is simpler in its application and closer to reality, while [Doob's] model may be considered more satisfactory from a logical standpoint since it avoids the difficulties connected with the concept of 'all place selections.' " J. L. Doob and R. von Mises, "Discussion on Probability Theory," *ibid.*, p. 216.

47. G. Boole, *Studies in Logic and Probability* (London, 1952), p. 422.

48. It is surprising, therefore, to see Mises, *Probability, Statistics and Truth*, p. 43, affirm that according to the Laplacian theory, "the probability value 1 means that the corresponding event will certainly take place." Mises' theory does not deny that an event may have a probability equal to 1 and still not be certain, as may happen if most but not all elements in the collective are identical. Where Mises differs from the Classicists is in assuming that this cannot be true of a collective of collectives.

sented by the same real number, the unity.[49] This is the first crack in the doctrine of ordinal measurability of expectations.

2. Some writers have regarded the distinction between certainty and quasi certainty as hairsplitting. The usual argument is that quasi impossibility can be described as the chance of selecting a point from a set of zero measure on a line; it is then observed that a set of zero measure exists only in theory because in practice even a single point is replaced by a dot of some width.[50] True, in real life we cannot ascertain the exact position of a point. In fact, the physicist can express all actual measurements only with integers, but this does not mean that physics can throw away the irrational numbers as hairsplitting. Besides, to impugn the distinction between certainty and quasi certainty by arguments invoking actual measurements is as unavailing as trying to solve the Pythagorean controversy by comparing the experimental measure of the diagonal of a square with that of its side.

The negation of the distinction between certainty and quasi certainty means the negation of probability itself. For if we deny this distinction, we implicitly accept the compensating effect of Azaïs-Marbe and therefore ought to join the ranks of those gamblers who, believing in this effect, use all types of schemes to beat the roulette. Let us also suppose that one is asked to guess whether a triangle in a complicated geometric construction is equilateral. The probability measure of an equilateral triangle is zero, and remains zero even after the information that one angle has 60° becomes available. However, the "logical relation" between E and P being increased by this additional information, the initial state of the mind must also be affected thereby. In real situations, therefore, we do distinguish even between two quasi impossibilities.

IV. EXPECTATION AND PURE LOGIC

1. Situations like that just mentioned led theorists to investigate another facet of expectation: the logical connection displayed by some pairs (E, P). The clear-cut cases, those where E either entails or contradicts P, correspond to certainty and impossibility. There are, however, cases where E does not entail P though it contains some of the necessary conditions for P to be true, thus yielding some ground for believing the

49. It should be noted that this is true not only for the extreme values 0 and 1. A quasi-impossible event may be added to any other event without changing the latter's probability.

50. Savage, *Foundations of Statistics* (my note 6, above), p. 39.

prediction. If the prediction is "P = ABCD is a square," the expectation moves in the direction of certainty as we pass from one to the next of the following evidences:

e_0 = "ABCD is a trapezoid," e_1 = "ABCD is a parallelogram," e_2 = "ABCD is a rectangle." [51]

Actually, we need one more step — to know that the dimensions of the rectangle are equal — to reach certainty. This ordering has led some theorists to define probability as the measure of the logical connection between E and P or, as Keynes put it, as the degree of certainty of $\mathcal{E}$.[52] This idea can be traced back to De Morgan,[53] but the first cogent elaborations of it are due to Jeffreys[54] and, especially, to Keynes. However, neither set the probability relation on purely objective foundations. This is clearly expressed by Keynes: "A *definition* of probability is not possible, unless it contents us to define degrees of the probability relation by reference to degrees of rational belief." [55] Only by appealing to this degree of rational belief are we able to compare most probability relations that involve either the same P (as in the example used above), or the same E. But a stronger assumption is needed to make any two probability relations comparable.[56] On a purely logical basis it seems impossible to compare the two expectations:

$\mathcal{E}^1$: e_1 = ABCD is a rectangle; P_1 = ABCD is a square.
$\mathcal{E}^2$: e_2 = ABC has a 60° angle; P_2 = ABC is equilateral.

2. In a theory that, for good reason, has become attached to his name, Carnap endeavored to arrive at an objective numerical measure of any probability relation, independent of any factual knowledge.[57] The cornerstone of this theory is the definition of probability relation as the proportion of the *range* of the evidence contained in that of the prediction.[58]

A few examples may be helpful in showing the generality of the problems that would come under the domain of Carnap's theory, undoubtedly the most ambitious of all theories of probability. The first example is

51. We shall use the notation e to denote that part of E which changes from one case to another.
52. Keynes, *A Treatise on Probability*, p. 15.
53. De Morgan, *Formal Logic*, p. v.
54. D. Wrinch and H. Jeffreys, "On Some Aspects of the Theory of Probability," *Philosophical Magazine*, 6th series, XXXVII (1919), 715–731. Jeffreys' ideas were further developed in his two books already cited.
55. Keynes, *A Treatise*, p. 8; also chap. ii. An almost identical view is found in Jeffreys, *Theory of Probability*, pp. 332–333.
56. Keynes, *A Treatise*, pp. 67, 112.
57. On this point, see also Section VI, 2, below.
58. Carnap, *Logical Foundations* (cited in note 10, above), p. 297. Though in somewhat loose terms, the same idea had been expressed earlier by various authors, in saying that probability relation measures the degree of "truth" of P contained in E. Cf. Keynes, *A Treatise*, p. 15n.

chosen to emphasize the absence of any connection between factual truth and probability relations: The statement "my dog is square" justifies the statement "Caesar was a bird," with a probability of $\frac{2}{11}$.

Here, the number $\frac{2}{11}$ expresses the degree of logical relation between the premise and the conclusion, and nothing else. But we can apply the probability relation to a factually true premise such as: "On the basis of the information provided by my business analyst there is a probability of $\frac{1}{10}$ that the sales of color television sets in 1957 will amount to one million."

On the day when the probability theory is able to set expectations on such a solid, objective basis, many of the problems that still baffle us will be solved. The role of the entrepreneur, for instance, will then be taken over by a self-guiding section of the enterprise — the Probability-Relation Bureau — which will automatically make the "right" decisions, including that of what funds should be allocated to its proper budget. But this day still seems far off; for, as we shall presently see, in addition to its incontestable merits the work of Carnap also had the effect of clearly bringing to the surface the insuperable issues involved in such a project.

3. Although Carnap's theory deals with a language consisting of a countable infinity of "individuals" and a finite number of logically independent "predicates," [59] its essence will appear in sharper relief, yet unmarred, against the background of a simpler structure. Let us then assume that there are only two "individuals," i_1, i_2, and only two predicates $P_1 = $ "square" and $P_2 = $ "white." With the aid of these two predicates we can express only four properties:

$$Q_1 = \text{"square and white,"}$$
$$Q_2 = \text{"square and nonwhite,"}$$
$$Q_3 = \text{"nonsquare and white,"}$$
$$Q_4 = \text{"nonsquare and nonwhite."}$$

These properties are noncontradictory since "square" and "white" (in

TABLE 1. Statements in a language of two predicates and two individuals

	S_1	S_2	S_3	S_4	S_5	S_6	S_7	S_8	S_9	S_{10}	S_{11}	S_{12}	S_{13}	S_{14}	S_{15}	S_{16}
i_1	Q_1	Q_2	Q_3	Q_4	Q_1	Q_2	Q_1	Q_3	Q_1	Q_4	Q_2	Q_3	Q_2	Q_4	Q_3	Q_4
i_2	Q_1	Q_2	Q_3	Q_4	Q_2	Q_1	Q_3	Q_1	Q_4	Q_1	Q_3	Q_2	Q_4	Q_2	Q_4	Q_3
m^*	$\frac{1}{10}$	$\frac{1}{10}$	$\frac{1}{10}$	$\frac{1}{10}$	$\frac{1}{20}$	$\frac{1}{20}$	$\frac{1}{20}$	$\frac{1}{20}$	$\frac{1}{20}$	$\frac{1}{20}$	$\frac{1}{20}$	$\frac{1}{20}$	$\frac{1}{20}$	$\frac{1}{20}$	$\frac{1}{20}$	$\frac{1}{20}$
m^+	$\frac{1}{16}$	$\frac{1}{16}$	$\frac{1}{16}$	$\frac{1}{16}$	$\frac{1}{16}$	$\frac{1}{16}$	$\frac{1}{16}$	$\frac{1}{16}$	$\frac{1}{16}$	$\frac{1}{16}$	$\frac{1}{16}$	$\frac{1}{16}$	$\frac{1}{16}$	$\frac{1}{16}$	$\frac{1}{16}$	$\frac{1}{16}$

English) are logically independent. All possible statements about all individuals are given by Table 1. Thus, S_3 means that "i_1 is Q_3 and i_2 is Q_3,"

59. Carnap, *Logical Foundations*, pp. 58, 73.

and S_9, that "i_1 is Q_1 and i_2 is Q_4." The statements S_i are called by Carnap "state descriptions." All isomorphic descriptions, such as S_5 and S_6, constitute a *structure*. The structure of S_5 and S_6 is "one individual is Q_1 and one individual is Q_2."

The crucial step in Carnap's argument is the introduction of the *regular measure function*, m^*, which is distributed according to the following rules: (1) the measure of all state descriptions is one; (2) all structures have the same measure; (3) all isomorphic state descriptions have the same measure.[60] Each of the 10 structures of Table 1 is assigned the measure $\frac{1}{10}$, which is then equally divided among all isomorphic state descriptions. The probability relation between a prediction P and an evidence E is measured by

$$(2) \qquad C^*(P, E) = \frac{m^*(P, E)}{m^*(E)}$$

where $m^*(P, E)$ is the measure of all state descriptions in which both P and E are true, and $m^*(E)$ is the measure of all state descriptions in which E is true. If, for instance, $P = $ "i_2 is square," $E = $ "i_1 is white," the state descriptions in which E is true are those in which i_1 is either Q_1 or Q_3: S_1, S_3, S_5, S_7, S_8, S_9, S_{12}, S_{15}. Their measure is $\frac{1}{2}$. The state descriptions in which both P and E are true are those where i_1 is either Q_1 or Q_3 *and* i_2 is either Q_1 or Q_2. They are S_1, S_5, S_8, S_{12}, and their measure is $\frac{1}{4}$. Therefore, in this case $C^* = \frac{1}{2}$. In the same way, we find that if the above P is replaced by "i_2 is white," then $C^* = \frac{3}{5}$. In general, the probability relation between $P = $ "i_{n+1} is W" and $E = $ "m out of $i_1, i_2, \ldots, i_n$ are W" is given by the formula

$$(3) \qquad C^* = \frac{m + w}{n + k}$$

where w is the number of Q's for which W is true, and k is the number of all Q's.[61]

4. We can now examine the issues that are so clearly unveiled by Carnap's theory. There is first the restriction that the number of "predicates" be finite. If they formed a countable infinity, formula (2) would give different results for "properties" expressed by an infinite number of "predicates," depending on the lexicographic method used to order the predicates.[62] If the "predicates" formed a continuum, as in any language that includes all real numbers, the difficulty is simply staggering. The second restriction, that the "predicates" should be logically inde-

---

60. *Ibid.*, pp. 294, 563.
61. *Ibid.*, p. 568. Formula (3) generalizes Laplace's first rule of succession.
62. This is related to the impossibility of defining frequence (or probability) in a countable infinity without assuming a given ordering.

pendent, is far more serious. No actual language could possibly satisfy
it, even in a very rough way. However, this is not all. "The man Aristotle
is mortal" remains in the same logical relation to "All men are mortal"
whether or not men are divided into "tall" or "short." The same is not
true in Carnap's inductive logic. Let us assume that we want to find out
the degree of confirmation of "i_2 is a human being" on the evidence that
"i_1 is a human being." If the only other predicate is "good," then according to formula (3) we obtain $C^* = \frac{3}{5}$. As soon as a new predicate, say
"tall," is introduced — and our vocabulary is being thus continuously
amended — the degree of confirmation of the *same* prediction by the
same evidence in the *same* universe of individuals decreases to $\frac{5}{9}$.

It is evident that in reaching for a measure of probability relation we
are confronted with the familiar difficulty that besets the Bayes formula.
Actually, Carnap's definition of the degree of confirmation is nothing but
a transliteration of that formula, while his "regular measure function,"
or "the degree of confirmation by *null evidence*," [63] is nothing but our
old friend, the *a priori* probability. It is also clear that his rules for determining the regular measure involve the Principle of Insufficient Reason.
A discussion of these rules will disclose, for the nth time, the slippery
handles of that principle. Carnap decided that there was no reason for
treating one *structure* differently from another, and no reason for not
treating the *state descriptions* of the *same* structure on the same footing.
But as he remarked,[64] on the Principle of Insufficient Reason one would
be equally justified in placing *all* state descriptions on the same footing,
in which case we would obtain a different regular measure, m^+ (Table 1).
On the basis of this new measure the probability relation between the
prediction "i_2 is square" and the evidence "i_1 is square" is equal to that
between the *same* prediction and the *contrary* evidence "i_1 is not square"
(namely, to $\frac{1}{2}$), which is also the *a priori* probability of the same prediction.[65] Carnap rightly observes that this result is tantamount to denying
that one can learn from *experience*. For this reason Carnap rejects the
measure m^+. To be sure, he admits that the reason is not too strong. However, since his concept of probability relation is conceived as a purely
logical one, in no relation with the real world, the way we learn about
this world has nothing to do with the problem under discussion. From
this viewpoint the objections raised above against m^* seem far more
effective since they remain within the domain of pure logic.[66]

63. *Ibid.*, pp. 289–290, 557–559.
64. *Ibid.*, p. 564.
65. This result is general. *Ibid.*, p. 565.
66. I do not wish thereby to advocate the adoption of m^+ in preference to m^*, but
only to point out the arbitrariness involved in the construction of an objective measure
of the degree of logical connection.

V. EXPECTATION AND SUBJECTIVE BELIEF

1. Most expectations common in real life do not fit even vaguely into any of the patterns discussed thus far. Consider the expectation of a firm that spends large sums on research with the hope of discovering new chemical products. Clearly, the evidence includes the past cases where research has been crowned with success. But how can one compute here the frequency ratio of success? The expectation also involves some logical connection between E and P, since the new product cannot logically contradict all E. But who can appraise the probability relation of this involved logical connection? Yet the management of that firm does act on the basis of such an expectation. This observation led to another approach to probability, that of the Subjectivist School.[67]

2. According to the Subjectivist doctrine, probability is a numerical coefficient that measures the subjective degree of belief in P. This doctrine is the first to have realized that $\mathcal{E}$ involves also the parameter I. Indeed, the Subjectivists emphasize the fact that the numerical probabilities corresponding to $\mathcal{E}(I, E, P)$ and $\mathcal{E}(I', E, P)$ are not necessarily the same, although both I and I' are "reasonable individuals." [68] But while bringing I into the picture, the Subjectivists have thrown a veil over the other aspects of expectation. For they maintain that the numerical coefficient a person attributes to his degree of belief in P is largely a matter of subjective evaluation, not necessarily supported by a compelling argument.[69] It may have any value provided that it corresponds "to a coherent opinion, to an opinion legitimate per se; every individual is free to adopt that [coherent] opinion he prefers, . . . that which he *feels*." [70] The only restriction is that the beliefs of any individual must display an internal consistency that reduces to the addition of probabilities of mutually exclusive events and the multiplication of probabilities of independent events.[71]

A close view of the essence of De Finetti's theory can be obtained through a simple example. An individual knows that in an urn there are

67. Although its seeds are discernible in some Classicist writings, the Subjectivist idea came into full light only with the works of F. P. Ramsey, collected into *The Foundations of Mathematics and Other Logical Essays* (New York, 1950), and De Finetti, "La prévision" (cited in note 17, above).

68. Savage, *Foundations of Statistics*, p. 3.

69. *Ibid.*, p. 65.

70. De Finetti, "La prévision," p. 4. See also Savage, *Foundations*, p. 57.

71. The fact that logic must be used in applying these principles is, of course, well understood here as in any theorizing about "the rational man."

only white, red, and black balls. If he *feels* that the probability of extracting a white ball, p, is greater than that of extracting a red ball, q, but equal to that of extracting a black ball, r, his degrees of belief in each of these alternatives may be represented by any numbers, provided they satisfy the relations $p = r > q$, $p + q + r = 1$. The reasons that prompted the individual to choose these numbers need not necessarily be related to the composition of the urn or to the records of past drawings, and thus need not have a justification in the eyes of someone else. But once these numbers are chosen, the individual (if he is rational) must attribute the probability $p + q$ to the extraction of a nonblack ball, and the probability of $2qr$ to "one red and one black" in two independent extractions. If all uncertain events in the world of that individual were compound extractions from the same urn, the three numbers (p, q, r) would completely describe the particular structure of his probability beliefs. To put it in more general terms, if *all* events could be expressed as Boolian polynomials of some elementary events that need only to be mutually exclusive, the structure of the beliefs of any individual would be completely characterized by the manner in which he would distribute probabilities to these elementary events. This probability distribution is otherwise arbitrary and does not have to reflect any stochastic aspect of the material world. In maintaining that such a theory is fully adequate to deal with rational actions in the face of uncertainty, the Subjectivist is exactly like a geometrician who would claim that any geometry topologically equivalent to that of the material world is all we need to explain our understanding and use of space properties. For, certainly, a theory of probability cannot concern itself only with the internal consistency of the acts of an individual, as maintained by Savage.[72] Ordinary logic would suffice for this. The reasonableness of the individual's decisions involves also the question whether his scale of beliefs reflects the properties of random phenomena. De Finetti himself raised this question: "Among the infinity of scales perfectly acceptable per se, is there a particular one that can be regarded as *objectively right*, though in a sense still unknown?" [73] The controversies between the various schools center precisely upon what is an *objectively right* coefficient, a problem which cannot be settled subjectively by the individual "interrogating himself." [74] The connection between the physical reality and the subjective world requires an additional postulate or principle of some sort. Savage thinks that the Principle of Insufficient Reason partly serves

72. Savage, *Foundations*, pp. 56–57. "Because that theory [of personal probability] is a code of consistency for the person applying it, *not a system of predictions about the world around him*" (*ibid.*, p. 59, italics added).

73. De Finetti, "La prévision," p. 16.

74. Savage, *Foundations*, p. 51.

this purpose.[75] But if the Subjectivist Theory includes this principle, it can hardly be distinguished from the Classical doctrine.[76]

Savage's second claim, that the Subjectivist Theory is susceptible of being adapted so as to include any "acceptable criteria for reasonableness . . . that may have been overlooked," [77] is either trivial or exceptionally strong. For, if it is to be interpreted otherwise than that the theory will have to be altered in the face of new evidence, the statement looks like a definitive judgment on the two basic axioms of probability.

3. In a recent theory, which may be classified as Ultra-Subjectivist, Shackle dispenses almost entirely with the internal consistency of the beliefs.[78] But the most conspicuous innovation of Shackle is the substitution of "surprise" for "belief." It is not difficult to see that this novelty actually draws the analysis away from the main issue. Indeed, in the end potential surprise is nothing but a truncated complement of belief.

Surprise as understood by everybody and as described by Shackle is simply the state of the mind created by knowledge of the actual result of an experiment. But this *ex post* state of the mind is conditioned by the *ex ante* degree of belief. The stronger the belief in a given outcome, the smaller will be the surprise if that outcome actually occurs, and conversely.[79] With one single exception (to be mentioned later), there is no surprise without an *ex ante* degree of belief. Since the surprise at a result varies inversely with the degree of belief in the same result, nothing stands in the way of taking $1 - p$ as an ordinal measure of surprise (where p is the subjective probability of the same result). Now, an ordinal measure of *potential surprise* can be obtained by simply following Shackle's definition of this concept: it is equal to $s = (1 - p) - (1 - \pi) = \pi - p$, where π is the greatest of all degrees of belief corresponding to the various mutually exclusive outcomes into which the individual partitions the event.[80] Of course, the degree of potential surprise is not additive. It is

75. *Ibid.*, pp. 63 ff.

76. The Subjectivist angle is apparent in most writings of the Classicists. Witness the use of probability in Pascal's famous argument about the advantage of believing in God (*Pascal's Pensées*, tr. H. F. Stewart, London, 1950, pp. 119–120). See also the review of various Classicist opinions in Keynes, *A Treatise*, pp. 282 ff.

77. Savage, *Foundations*, p. 67; also pp. 59–60.

78. G. L. S. Shackle, *Expectation in Economics* (Cambridge, Eng., 1949); *Uncertainty in Economics and Other Reflections* (Cambridge, Eng., 1955).

79. Shackle, *Expectation in Economics*, p. 11.

80. *Ibid.*, p. 12. The axiomatic formulation of the theory (*ibid.*, pp. 131–132) is, however, far less clear. Axiom 1, for instance, can hardly deserve its title. Some of the axioms, I am afraid, will prove to contradict the attitude of any rational individual. If I understand it correctly, Axiom 7 implies that the coincidence of two unusual events should not surprise one more than each occurrence taken separately. It may be that the ordinal measure proposed above does not reflect the concept Shackle has in mind. I hope that he will one day offer a more precise formulation of the relationship between degree of potential surprise and degree of belief than the vague wording of Axioms 1 and 3.

true that "the degrees of potential surprise assigned to the members of an exhaustive set of rival hypotheses . . . need not sum to any particular total," [81] not because they "are independent of each other" [82] — and they obviously are not — but because they change with π, i.e., with the partition of the set of all outcomes. Though Shackle sees in the non-additivity of the potential surprise an advantage over the probability approach,[83] the very reason that makes for the nonadditivity seems the most vulnerable part of his thesis. Let us take the case of an urn containing 51 black and 49 red balls. According to Shackle's definition, an individual may be entirely justified in feeling some positive degree of potential surprise if he has bet on black, and a red ball is extracted. But according to the same definition,[84] the same individual should not be surprised at all if, being unable to decide which number will win the Irish Sweepstake, he buys the first ticket offered to him and then wins.[85]

In giving the *raison d'être* of the concept of surprise, Shackle affirms that "you cannot at one and the same time believe positively in (have more than zero confidence in) both (all) of two or more rival (mutually exclusive) hypotheses. Hence we must use a measure of disbelief [i.e., potential surprise]." [86] One would certainly agree with this argument, if we always disbelieved all but one of the rival hypotheses. But since this is not so — and Shackle himself does not assume that it is — we can reverse the question and ask: How is it possible for a person at the same time to have various degrees of disbelief in more than one of the rival hypotheses?

There is, however, one problem for which the *raison d'être* of surprise is evident and for which "the degree of belief is no remedy." We need "surprise" to describe the *ex post* state of the mind that has no *ex ante* correspondent or, in other words, the state of the mind created by an outcome that was not even considered by any of the rival hypotheses. This "surprise" forms the punctuation of history, of new discovery, of learning in general. And although Shackle introduces an *ex ante* correlative for this situation (the *residual* hypothesis), the *pure* surprise remains an expression of which only one of Janus' faces — that looking towards the

81. Shackle, *Uncertainty in Economics*, p. 69.
82. *Ibid.*, p. 69.
83. Shackle, *Expectation in Economics*, p. 40; *Uncertainty in Economics*, pp. 9–10, 27.
84. See *Expectation in Economics*, p. 113, and *Uncertainty in Economics*, pp. 41, 69.
85. This paradox can be avoided if π, for instance, instead of representing the highest degree of belief in a given partition, is taken as that degree of belief which the individual considers as the upper limit of unusual outcomes. Some passages in the works of Shackle seem to suggest that all outcomes of positive surprise are unusual in this sense, and vice versa, e.g., *Uncertainty in Economics*, pp. 39 ff. But then, why not simply use the degree of belief and replace the diagrams of his Fig. III, 1 (*ibid.*, p. 40) by one in which the ordinate represents the degree of belief up to π? (The "inner range" would then become a sort of "confidence interval.")
86. *Ibid.*, pp. 66–67.

past — is capable. Since the theory of decisions is concerned exclusively with the future as viewed from the present, it has little, if any, use for this signal concept.

4. Ordinal measurability is a far more crucial problem for the Subjectivist philosophy than for the other schools, which start from a numerical formula to define probability. Yet, numerous Subjectivists approached this question in a wholesale fashion. An epitome is offered by Shackle's authoritative assertion that "a man's judgment compels him to attach certain degrees of potential surprise to given hypotheses of gains." [87] Most Subjectivists — with the notable exception of Savage — seem not even to suspect that it may be otherwise. The philosophy underlying the views of the Subjectivists was disclosed in Ramsey's remark that it is "conceivable that the degrees of belief could be measured by a psychogalvanometer or some such instrument." [88]

Schemes have not been lacking to show how, by making use only of comparability, numerical ranks can be assigned to expectations. A familiar one can be traced back to the works of Edgeworth and of Pareto on utility.[89] The idea is to assign ½ to the expectation "equally distant" between impossibility and certainty, ¼ and ¾ to those "equally distant" between this middle expectation and impossibility or certainty respectively, and so on. But beneath the surface of this seemingly convincing argument is concealed "the ordinalist's fallacy." [90] We can refer again to the example of the staircase. If there are, say, nine steps and only five numbers available, we can by the procedure described above attribute the numbers 1, 2, 3, 4, 5, to the 1st, 3rd, 5th, 7th and 9th steps, respec-

87. Shackle, *Expectation in Economics*, p. 40. Shackle is probably the only Subjectivist to go so far as to regard the degrees of belief as having an absolute scale valid for interpersonal comparisons (*ibid.*, pp. 32–33; also *Uncertainty in Economics*, p. 32). The idea that complex and largely imponderable variables, such as the state of mind, possess an *absolute scale* is highly surprising in view of the fact that physics has long since discarded the absolute scale of some physical variables. For an admirable discussion of the meaninglessness of the absolute scale, see M. Allais, *Traité d'économie pure* (Paris, 1952), I, 220–239.

88. Ramsey, *Foundations of Mathematics* (note 67, above), p. 161.

89. F. Y. Edgeworth, *Papers Relating to Political Economy* (London, 1925), II, 473; V. Pareto, *Manuel d'économie politique* (Paris, 1909), p. 264. Other writers made use of the same device: e.g., A. L. Bowley, *The Mathematical Groundwork of Economics* (Oxford, 1924), p. 2; O. Lange, "The Determinateness of the Utility Function," *Review of Economic Studies*, I (1933–34), 218–225, and "Notes on the Determinateness of the Utility Function, III," *ibid.*, II (1934–35), 75–77; W. E. Armstrong, "Determinateness of the Utility Function," *Economic Journal*, XLIX (1939), 453–467, and "Uncertainty and the Utility Function," *ibid.*, LVIII (1948), 1–10.

90. Georgescu-Roegen, "Choice, Expectations and Measurability," *Quarterly Journal of Economics*, LXVIII (1954), 522, 527–528 [in the present volume, pp. 202–203, 208]. Of course, this does not refer to those arguments — e.g., Shackle's or Savage's — that used the same or a similar device to construct a particular scale, only after asserting that probability is a number.

tively. There will remain, however, four steps not included in our numerical scale.[91]

5. An arresting way of settling the difficulty was broached by Ramsey in a now famous essay, "Truth and Probability." [92] He proposed to measure the subjective probability by the betting quotient the individual is willing to accept on the given uncertain event,[93] an idea that was not altogether new. But Ramsey was first to realize that in order to obtain the correct measure, the betting quotient has to be expressed in terms of utility, not in terms of money. He was thus compelled to extend his inquiry so as to include the measurement of subjective utility. The result was a theory of remarkable elegance in which the two subjective concepts appeared as an indissoluble complex. But Ramsey's essay constitutes a landmark in the development of the Subjective Theory for still another reason. Ramsey reversed the traditional view that the purpose of the theory of probability is to tell a person what odds to accept in a given bet, and proposed to study the "measurement of belief *qua* basis of action." [94] With this, he laid the first cornerstone of the modern theory of decisions. Subsequent contributions along these lines have polished Ramsey's original argument, but to this day the leading ideas of his essay dominate the thinking of the Subjectivist School.[95]

It would be a mistake, however, to regard Ramsey's contribution as a theory of probability in the true sense of the word. Indeed, his theory does not tell us what probability *is*. It merely shows how the various choices of a person, if interpreted according to Ramsey's assumptions, can lead us to an ordinal measure of subjective probability. Consequently, the lacuna of the Subjectivist Theory (mentioned above, 2), is not in the least affected by Ramsey's theory.

VI. THE MAP OF EXPECTATIONS

1. The fact that expectation has led to so many theories of probability — irreconcilable with each other, although each makes perfect

91. This is what happens if we apply the procedure to an infinite class of expectations (see below, Section VI, 6). A closely related idea used by De Finetti, "La prévision," pp. 5, 16, who arrives at a numerical scale by decomposing certainty into an arbitrarily large number of mutually exclusive and equally probable events, can be shown to suffer from the same fault.

92. Ramsey, *Foundations of Mathematics*, chap. vii. Written in 1926, the essay was posthumously published in 1931. The book cited is the 1950 collection of his essays.

93. *Ibid.*, pp. 172 ff.

94. *Ibid.*, p. 171.

95. Witness Savage's *The Foundations of Statistics*. This stimulating work, by the leading representative of the Subjectivist School in the United States, offers an extremely careful and penetrating reformulation of Ramsey's synthesis, with De Finetti's conception of subjective probability as a general cadre.

sense within its own analytical setup — should have been taken long ago as a sign that the map of expectations does not consist of a single trail between two poles: the absolute impossibility, I, and the absolute certainty, C (Fig. 6–2). These poles have the same role in all theories, because they correspond to two states of the mind having the same meaning for all rational individuals. But what can be said about the other expectations? Is there a unique path between I and C on which any expectation must have a definite place? Are there several such paths, or do expectations fill the whole space between I and C?[96] No understanding of the process by which decisions are reached can be obtained without first answering these questions, at least in some general manner.

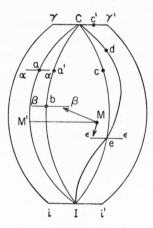

FIGURE 6–2

2. A one-to-one order preserving correspondence can be established between the expectations whose evidence contains the numerical value of the frequency ratio of the outcome P computed over the corresponding phase space Ω,[97] and the points on a line IaC. The expectations envisaged by the Frequency Theory also can be represented in the same way by the points of a line IbC.[98] The complex infinity of quasi certainties and quasi impossibilities are represented by the side alleys opening from these two paths, γC and iI. The correlative situation for expectations such as a or b is shown by the line segments α a, β b.

96. I should make clear that the use of a *diagram* in two dimensions (Fig. 6–2) does not imply that I associate the set of all expectations with a plane map. The true map of expectations is certainly far more intricate.

97. Section II, 4, above.

98. We have seen (Section II, 3) that in the way of identifying this class with the previous one stands the Bertrand-Azaïs controversy. Consequently, although in our behavior we may abstract from this controversy most of the time, for the sake of logic we should keep separate the two images a and a' corresponding to the same value of the two probability coefficients, waiting for a solution of the paradox.

Next comes the class of expectations whose evidence does not contain any frequency ratio, but which display some logical connection between E and P. Although there is no way to render all these expectations comparable,[99] there are cases, like the example discussed previously,[100] where the comparison of such expectations raises no difficulty. They can be thought of as lying on a fictitious line between I and C in the order of their comparison, but without having their position identified by a number determined for each expectation taken by itself. Moreover, the class of all expectations that can be compared with others does not form a class of comparable expectations, because the criterion that justifies their comparison changes in type and in content. We have to think, therefore, of an infinity of such fictitious paths between I and C that connect the images of probability relations that can be compared. One path, IecC, for instance, will be used for those expectations that have the same E', another, IedC, for the expectations having the same P'. Clearly, such paths intersect at the point e corresponding to the expectation $\mathcal{E}(I, E', P')$.

But by placing the probability relations in this fashion on the map, have we not overlooked some elements of the problem that would justify another configuration? Carnap endeavored to establish a bridge between these latter classes of expectations and those of IbC, by arguing that probability$_1$ (probability relation) is equal to probability$_2$ (frequency probability).[101] If this were true, all paths IeC would become one with IbC. It is not difficult, however, to see that Carnap's argument is far from establishing a bridge between the two concepts.[102] The contradiction involved in the equivalence of the two concepts is clearly illustrated by the example, used previously, of an individual who knows that a triangle has a 60° angle, and wonders whether it also is equilateral.[103]

3. Of course, the map is incomplete unless we include expectations such as M, which present at the same time some similarity with more than one pattern already traced on the map. Although such expectations almost completely fill the paths of human life, most theories have set out by concealing them behind claims of a unified concept of measurable

99. In addition to the arguments given in Section IV, it is useful to mention here a very important result reached by Koopman, "Axioms" (cited in note 3, above), p. 289: even the introduction of a postulate similar to that used by De Finetti — that for *any* integer n there is a scale of n equivalent and mutually exclusive assertions — does not make it possible to assign a number to each expectation. There still remains a class of nonappraisable expectations.

100. Section IV, 2, above.

101. Carnap, *Logical Foundations*, pp. 164 ff.

102. G. Bergmann, "Some Comments on Carnap's Logic of Induction," *Philosophy of Science*, XIII (1946), 77, left little unsaid in remarking that "as there is no chapter in mechanics that leads from a (heuristic) principle of indifference to the frequency produced by throwing a die, so there is no philosophical analysis of induction to mediate between probability$_1$ and probability$_2$."

103. Section III, 2, above.

probability. To see whether the human mind can by some power of abstraction reduce the whole map to a single path or, on the contrary, if unable to do so can move within this intricate structure, let us consider the following evidences about urns containing balls identical save for color:

e_1 = in U_1, two thirds of the balls are white and one third are black;

e_2 = the frequency of white in 3,426 independent extractions from U_2 was $\frac{2}{3}$;

e_3 = three independent extractions from U_3 resulted in two white and one black ball;

e_4 = the urn U_4 contains some balls.

Now, let us ask an individual to name a betting quotient for each urn, after warning him that he will have to accept the betting either on white or black, according to the choice of his opponent who knows the composition of the urn. Any rational individual will name $\frac{1}{2}$ as betting quotient for U_1, U_2, U_3, and possibly for U_4, too. Yet he will prefer U_1 to U_2, U_2 to U_3, and U_3 to U_4. The reason for this preference is that e_2 and e_3 contain only an empirical estimation, and that the information of e_3 is less *credible* than that of e_2. It is thus seen that even expectations of a very simple structure have two dimensions: the betting quotient and the *credibility* attached to it. A more realistic example may reveal the presence of other dimensions. We may know, for instance, that the person who reported the evidence e_2 is not as reliable as the one who reported e_1. Other dimensions may stem from the logical connection between the elements of E and P. Moreover, it does not seem likely that all credibility dimensions are readily measurable. But even if they were, we would need more than one psychogalvanometer to detect all variations in expectations, probably one psychogalvanometer for every basic reaction of which the human mind is capable, i.e., we would have to resort to complex numbers to identify each expectation.

4. The Knightian uncertainty is illustrated by a U_4 of a shape, size, color, etc., not recorded in the past experience of E. Knight's thesis is that we should distinguish between (1) *risk expectations*, whose evidence includes the theoretical value of the probability of the predicted event, and (2) *uncertainty expectations*, whose evidence includes no information about the predicted event. These two classes constitute, however, two extreme categories which, strictly speaking, cannot represent expectations common in real life. But exactly as the "infinite" of theoretical concepts becomes "a sufficiently large number" in actual applications, so can we include under risk expectations those expectations of sufficiently large credibility (as in the case of U_2), and under uncertainty expectations those of very small credibility (urn U_3).

5. The position of the advocates of probability as a unifying concept of all expectations is that any uncertainty expectation can be reduced to a risk expectation. And since risk expectations are ordinally measurable, so are *all* expectations. The most noteworthy argument in support of this equivalence is the observation that any individual, if *compelled*, will quote a betting quotient on any outcome. In this way all other dimensions are avoided by excluding them from the question asked of the individual. For if one captain is asked to communicate only the latitude of his ship, his answer will contain only one number. But this is no proof that position on the globe consists of latitude alone. It may be argued, though, that this analogy ignores one important distinction between the location on the globe and expectation, that the betting quotient and credibility are substitutable in the same sense in which two commodities are: less bread but more meat may leave the consumer as well off as before. If this were so, then clearly expectation could be reduced to a unidimensional concept, exactly as the multidimensional basket of commodities is reduced to the unidimensional Paretoan ophelimity. However, the substitutability of consumers' goods rests upon the tacit assumption that all commodities contain something — called utility — in a greater or less degree; substitutability is therefore another name for compensation of utility. The crucial question in expectation then is whether credibility and betting quotient have a common essence so that compensation of this common essence would make sense. This is not an idle question, for important concepts directly pertaining to expectations are known to be essentially different from probability taken in the sense of rational betting quotient: neither the inverse probability nor the likelihood can be reduced to probability.

However we look at it, the betting quotient reflects an estimation of a physical constant describing the composition of the actual urn or of that which in the mind of the individual represents an equivalent model. It would not help the thesis of ordinal measurability to argue that the essence of the betting quotient is not always that of a physical constant because it is the expression of a purely subjective phenomenon, not the result of a process of equivalence. But let us assume, for instance, that an individual names ⅙ as the betting quotient for raining during the weekend. If he is not indifferent between this bet and betting with the same odds on the ace in tossing a true die, it simply means that the same betting quotient covers two nonequivalent expectations. If he is indifferent, it means that the essence of the betting quotient is that of a physical constant. But then there can be no question of compensation between it and credibility. If the evidence provided by a very unreliable barometer is that of an atmospheric pressure of 27.5 inches, is there any sense in saying that a little of the atmospheric pressure can be traded for an additional increment of credibility? If not, then there is no sense either in trading a fraction of the betting quotient for an additional degree of credibility.

6. The well-known rules of succession seem to constitute examples of compensation between the betting coefficient and credibility, although the rules themselves are not concerned at all with this aspect. To clarify this point let us consider the set of all expectations $\mathcal{E}_{m,n}$ defined by

$e_{m,n}$ = "Out of $m + n$ balls drawn from $U_{m,n}$, m were white,"
P = "The next ball drawn from $U_{m,n}$ will be white."

Each $\mathcal{E}_{m,n}$ can be represented by the point (n, m) of the nonnegative quadrant nOm (Fig. 6–3). To each expectation corresponds an observed frequency of "white," $f_{m,n} = m/(m + n)$, with the exception of the pure uncertainty $\mathcal{E}_{0,0}$, for which there is no observed frequency. All points on OA, for instance, have the same frequency, $\frac{7}{8}$. There seems to be no difficulty in the way of taking the number of observations $(c = m + n)$ as an index of credibility for each $\mathcal{E}_{m,n}$. Now, if the drawing of a white ball is the most desirable outcome, any rational individual will prefer the drawing from $U_{3,0}$ rather than $U_{0,3}$. This seems to be the practice of all times, a justification of which was first offered by Laplace.[104] There is, however, another preference ordering among the expectations having the same observed frequency. Thus, N_3 is preferred to N_5, and M_7 to M_3.[105] This principle also seems intuitive, although it can be supported by other arguments. The individual has a stronger belief in the hypothesis that the urn $U_{0,5}$ contains only nonwhite balls than in the same hypothesis for $U_{0,3}$.

Leaving aside $\mathcal{E}_{0,0}$, we see that the set of all $\mathcal{E}_{m,n}$ is comparable if the two criteria involved are used in the order: betting quotient, credibility.[106] The comparability is not disturbed by the inclusion of the risk expectations $\mathcal{E}^p$, corresponding to the urn U^p with a known proportion, p, of white balls. These expectations can be represented by the points at infinity, Ω_p. The set thus completed offers a more instructive ground for discussing ordinal measurability than the elementary examples used in the preceding sections. Clearly, the ranking of all $\mathcal{E}^p$ completely exhausts all real numbers between 0 and 1. Consequently, the expectation $\mathcal{E}_{3,0}$, for instance, must either remain rankless or be given the same rank as one of the $\mathcal{E}^p$'s. In the first alternative, as we pass from one expectation to

104. Known as "Laplace's second rule of succession" but presented by Laplace as the method of the most probable value, this rule says that the "best guess" of the true composition of an urn is the observed frequency, which thus becomes the betting quotient (Laplace, *Oeuvres*, VII, 280 ff).

105. The direction of the preference is reversed as we cross OB.

106. The longitude and latitude in the example in Section I, 4, served an identical purpose.

the "next" beginning with $\mathcal{E}^0$ and ending with $\mathcal{E}^1$, we encounter steps that have no rank.[107] If, on the other hand, the expectations $\mathcal{E}_{m,n}$ are assigned ranks already used for $\mathcal{E}^p$, it means that there is some principle of equivalence among all expectations involved. But such a principle supposes that the betting quotient and credibility are substitutable. All "rules of succession" actually amount to the introduction of some indifference curves in Fig. 6–3. However, the rules proposed thus far (Laplace's and Carnap's) assume that the *a priori* probabilities are known. Consequently, there is no trace of pure uncertainty in their theoretical setup which involves only risk expectations.[108] Such succession rules merely establish an equivalence among risk expectations.

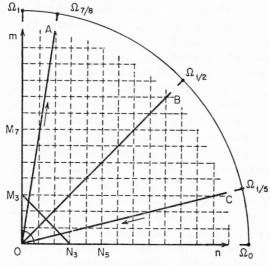

FIGURE 6–3

To be sure, one may see a justification of the principle of substitution in the fact that a person to whom a free prize is promised if a white ball is drawn from a chosen urn may choose $U_{20,4}$ rather than $U_{1,0}$. This choice may reflect, however, only the fact that information of very low credibility is completely ignored if others more credible are available. People seem to prefer dealing with risk expectations to bearing Knightian uncertainties, as two separate classes; this, however, is no proof that inside each class they substitute credibility for the betting quotient.[109]

107. This also exposes the fallacy of the scheme mentioned in Section V, 4, which assigns ranks only to $\mathcal{E}^p$ and overlooks the fact that an infinity of expectations are left rankless.

108. Witness the fact that a betting quotient is assigned to $\mathcal{E}_{0,0}$.

109. Incidentally, none of the known rules of succession displays *uniformly* the invoked substitution.

VII. DECISIONS AND RATIONALITY

1. Even expectations of a geometrical simplicity like $\mathcal{E}_{m,n}$ cannot be represented on our map by a unidimensional trail between I and C. We have to use an additional dimension and represent all expectations $\mathcal{E}_{m,n}$ having the same betting quotient, b, by b β'. One can easily imagine, then, the insuperable difficulties of representing on that map slightly more complicated expectations, let alone those ordinarily confronting us in life. To arrive, therefore, at a decision in the face of uncertainty requires nothing less than to navigate with the help of a map that surpasses any imaginable anfractuosity.

The difficulties of this perplexing task are lessened, however, by some ordinary circumstances of real life. The cases where we need to locate exactly our position on the map are rare, and they usually involve expectations definitely connected with a main passage, IaC or IbC. And when we have to choose among several expectations in order to make a decision, the evidences and predictions of these expectations usually have some elements in common. We are hardly ever called upon to choose in a decision between expectations as disparate as the following two: (1) a rectangle may turn out to be also a square, (2) product A may yield a profit of 12 per cent. The common elements of the expectations envisaged at one and the same time make it possible to establish various hierarchies among these expectations: some, according to the importance of past observations; others, according to the credibility of each piece of evidence; still others, according to the various logical connections between each E and the corresponding P. Since not all these variables are ordinally measurable or even comparable on an objective basis, the process by which they are compared is largely subjective. (This is the moment when the variable I steps well into the forefront.) But once these variables are appraised, the hierarchy in each direction can be established by any rational individual according to that theory of probability connected with the particular facet of expectation: past observations, credibility, logical relation. Thus, expectations, such as M (Fig. 6–2), may have to be ranked once according to the frequency and once according to the logical-relation criterion. If all expectations involved are ranked in the same order, the task is ended. Unfortunately, the various hierarchies are contradictory most of the time. Take a textile manufacturer who contemplates moving one plant to the South. He probably will have no difficulty in ranking the various regions according to the frequency of success of other firms that have moved their plants there. But how is he going to reconcile this ranking with the logic of economic theory, according to which increased demand leads to increased prices of factors of production? Will he decide to move into the region where there are already the greatest number of successful plants or into that where there is no competition for the local factors of production? To make a choice, the manufacturer must weigh the opposing hierarchies, as one weighs the pros and cons in ordinary arguments. But this weighing is largely a subjective, if not somewhat arbitrary, matter.[110] And since the weighing of arguments seems refractory to any numerical analysis, it is hard to visualize how the entire operation may result in the assignment of a number to each expectation involved. The most our acts reflect is preference, and preference is not necessarily measure.

To the argument that no one seems capable of assigning a precise number to each expectation, the advocates of "measurable expectation" have answered that this only reflects the difficulty inherent in accurate measurement, common to all sciences.[111] This analogy overlooks one essential difference. The difficulty of assigning a number to M is not one of accurately locating its position on an ordinally measurable scale — as is the case with physical constants and, therefore, with a or a' — but stems from the fact that the corresponding expectation is not an element of an ordinally measurable set. In one case we look for something that is there, in the other, for something that is not.

2. In some cases the role of the personal variable I acquires an overwhelming importance because of the nature of the problem involved. An excellent illustration is provided by the question whether France will become a monarchy within the next decade. But precisely because the subjective role of the individual in comparing this prediction with its contrary, for instance, is predominant, Savage's opinion that this problem can be analyzed "in terms of mathematical probability" [112] should be viewed with great reserve so long as the actual mathematical treatment is not forthcoming. Decisions regarding expectations of this type involve mainly intuition and good judgment in weighing the evidence, and scarcely any numerical operations. The vast class of nonappraisable expectations, nevertheless, constitutes the only firm ground of the Subjectivist view on probability, and offers a justification for the claim that probability belief unsupported by compelling arguments is the only concept that works where other theories fail. One can hardly question this claim. Where the Subjectivists overstate their case is in claiming also that even if the evidence contains some data relevant from the viewpoint

110. The example of the Chinese sentry, used by Shackle, *Uncertainty in Economics,* pp. 3 ff, tends to prove that some decisions are the result of pure impulses.
111. Cf. Savage, *Foundations of Statistics,* p. 59.
112. *Ibid.,* p. 62.

of other theories, these data are completely useless as a guide in a single decision. The amusing aspect of the ensuing controversy is that, for this thesis, the Subjectivists borrowed their arguments from the Frequentists.

3. The Frequentists have maintained for a long time that "it is utter nonsense" to use the coefficient of mortality in reaching a decision involving the possibility of death of a given person within the next year. For — they argue — at the end of one year from now that person will be either dead or alive, independent of the probability of death at his age;[113] the probability coefficient can be applied only to decisions concerning large groups. Thus, if the probability of A is 0.65, then we can *predict* that A will appear *about 6,500* times in 10,000 observations.[114] But this use of the probability coefficient is not free from the sin imputed to the other use. A group of 10,000 observations is just one *single* drawing from the population of all samples of the same size (or one single observation in the collective of samples of 10,000), and will either contain *about 6,500* A's or not, independent of the probability of A — unless by "about 6,500" is meant any number from 0 to 10,000. Clearly, one cannot quarrel with the statement that where prediction *stricto sensu* is impossible, "knowledge of the past results will not tell us the result of any *single* future trial," [115] provided that it is well understood that stochastically speaking there are only *single* trials.[116] It is, therefore, difficult to see why past frequency-ratios are relevant for the decisions involving composite events, but not for decisions regarding unique events. It may well be that by a unique trial Shackle understands a historically unique trial — like the enterprise of Christopher Columbus, for instance — in which case the question of the relevance of past frequency-ratios becomes vacuous.

I wonder though whether any Subjectivist believes that past frequency-ratios, even when available and sufficiently credible, are entirely irrelevant for the subjective degree of belief or of potential surprise; for how could one explain that Shackle would not be "in the least surprised if the book [read by another passenger on a London tube train] turns out to be in English or in French, [but would be] astonished if it is in Welsh," [117] if not because of the frequency ratios of the books read by Londoners?

113. Mises, *Probability, Statistics and Truth*, p. 23. See the parallel reasoning in Shackle, *Uncertainty in Economics*, pp. 28–29; also his *Expectation in Economics*, p. 110.

114. Mises, *Probability, Statistics and Truth*, p. 61. Shackle, *Uncertainty in Economics*, p. 23, also states that "the frequency-ratios are knowledge, and this knowledge can be used to predict the result of similar series of trials conducted under the same variability of circumstances."

115. Shackle, *Uncertainty in Economics*, p. 8.

116. Even an infinite sequence of observations is a single stochastic trial, unless we fail to distinguish between certainty and quasi certainty (Section III, 1, above).

117. Shackle, *Uncertainty in Economics*, p. 31.

4. In extolling the superiority of the concept of potential surprise, Shackle makes a claim on behalf of it which surpasses all other Subjectivist claims. A person decides to bet on a certain horse A on the ground that the winning of the race by A would not surprise him, whereas he attaches various degrees of potential surprise to a victory by the other horses B, C, etc. According to Shackle, if A wins, the *ex ante* "judgment is vindicated," and if B wins, "the degree of misjudgment is exactly measured and represented by this [corresponding] degree of potential surprise." [118] To see how unsubstantial this argument about a fully vindicated judgment and exact measure of misjudgment is, however, one has only to observe that Shackle's reasoning would apply equally well to any other judgment and any other result of the race.

It is nevertheless true that, lacking an *ex ante* criterion for good judgment, we currently use Shackle's *ex post* criterion to vindicate decisions concerning uncertainty expectations. As we shall see, this practice has another justification than that the criterion applied individually rests on logically sound foundations. Knowledge of the actual result can neither confirm nor infirm a judgment about an uncertain event, unless the actual experiment reveals that an element of E was not taken into consideration. Because we have learned *ex post* that Napoleon underestimated the danger of a severe winter in his Russian campaign, we can say now that he misjudged the situation. But had the winter of 1812 happened to be relatively mild, he might have won the war, in which case his misjudgment probably never would have been discovered. Probably this is why people frequently confuse success with mere luck.

5. In opposition to the view that a judgment can be vindicated by a single result, Marschak proposes to regard a decision as rational if it is based on a policy that succeeds in the long run. [119] This remarkable idea invites two observations. First, Marschak's principle, which is intended as an *ex ante* criterion, can be applied only to the events whose outcomes are, or can be considered as, random drawings from a population of known composition. [120] Indeed, the principle reflects the well-known fact that, if we choose that outcome whose *probability coefficient* is greater than that of any other outcome of a given event, we are *quasi certain* that in an infinite sequence of decisions this policy will result in a greater number

118. *Ibid.*, p. 34; also p. 28, and his *Expectation in Economics*, p. 116.

119. Jacob Marschak, "Why 'Should' Statisticians and Businessmen Maximize 'Moral Expectation'?" *Proceedings of the Second Berkeley Symposium on Mathematical Statistics and Probability* (Berkeley, 1951), p. 504.

120. Marschak fails to emphasize this limitation. The very inclusion of the word businessmen in the title of his paper seems to suggest that he does not distinguish between probability and Knightian uncertainty, with which businessmen are ordinarily confronted. He is more careful in other writings, e.g., "Money and the Theory of Assets," *Econometrica*, VI (1938), 320, 324.

of successes than any other rule.[121] The second observation is that it would not suffice only to *feel* that one outcome has a greater probability-coefficient than all others: the principle assumes that the individual estimates correctly the physical dimensions of the world around him.

But — one may ask — what about the case where the individual is confronted only once in his entire life with a particular urn?[122] The answer is that unless the individual is certain of not being confronted with any other urn in the future, the principle retains its entire validity. For, by always choosing the outcome of highest probability p_i in each decision d_i, we are quasi certain of being successful more often than by following any other rule. The extension of this result is that if we never accept a bet in which the stake is greater than the mathematical expectation measured in the same units as the stake, we are quasi certain not to lose in the long run.

In connection with this reasoning, Shackle raised an important question. If in the sequence of decisions there is a crucial one, whose loss could not possibly be compensated even if the individual is successful in *all* other decisions, does not the above argument err in not distinguishing between successes?[123] Let us think, for instance, of a person who contemplates fleeing from a Communist-controlled country. If he is caught while escaping, any number of successes in other future enterprises could hardly compensate for his failure. There is no denying that impulse may have a large part in the decision to escape. But can that person's decision completely disregard the past frequency-ratios of those caught? Moreover, after deciding to flee can he disregard the fact that there is one particular way of escaping that has proved in the past to be the safest among those he deems appropriate for his own physical and mental qualities? What would be a rational criterion for his choice, assuming that he *has* a choice among several alternatives? There seems to be no other way for a rational individual than to turn to Marschak's "rule of long run success," even if the man is aware of the cruciality of the decision. The view that the rule of long-run success is entirely irrelevant for decision making can be justified, therefore, only if we deny that expectations have any relation with the material world. It is natural for a Subjectivist to hold such a view; but in doing so, he exposes the weakest spot in the Subjectivist Theory.

6. The difficulty of formulating an objective criterion for decisions involving Knightian uncertainty stems precisely from the fact that the

121. We are only quasi certain, for we cannot exclude as *absolutely impossible* the case where a greater number of successes would be obtained by other methods, such as following the advice of a fortune teller.

122. In Shackle's terminology (*Uncertainty in Economics*, p. 5), this is a "non-seriable" trial.

123. *Ibid.*, pp. 5–6.

corresponding expectations are not ordinally measurable, probably not even comparable, and have only a distant and debatable connection with past observations. Under such circumstances there seems to be no other recommendation for dealing with Knightian uncertainty than the common advice: "get all the facts and use good judgment." But what is "good judgment"? The concept seems to resist any attempt at an objective definition that also would be operational *ex ante*. But this difficulty is common to many other human faculties. For instance, how can we decide who is the better marksman of two persons? Certainly not by a single shot, be it in the bull's eye. The only way to answer the question is by the results of a series of shots. Similarly, the result of a single decision cannot vindicate a judgment, but if in a series of decisions one person has been successful more than another, we do hold that he has better judgment. As difficult as it may be to justify this principle from a strictly logical viewpoint, inductively there is much to be said in its support. Indeed, the principle merely interprets a practice of long standing: when people have to entrust statesmen, generals, business executives, trustees of all sorts, with making crucial decisions, instead of themselves handling these situations, they ordinarily look for individuals of "proven" good judgment. Together with gathering, presenting, and analyzing in a logical fashion as many facts as possible, to detect and to use good judgment constitute the only means by which we can respond to living without divine knowledge in an uncertain world.

To many this may sound very discouraging, but the opposite view, that good judgment is an obsolete concept in an era of panlogistic models is patently delusive. Most people refuse to believe that there is no such thing as "the best technique" for dealing with historical variables; and when occasionally one tells them that they have to use their own judgment in extrapolating the trend of an economic phenomenon, they interpret this advice as a polite refusal to share one's philosopher's stone with them. Few social scientists have emphatically denied, as Keynes did, having the same lively hope as Condorcet, or even Edgeworth, "*éclairer les Sciences morales et politiques par le flambeau de l'Algèbre.*" [124] And yet, there is a limit to what we can do with numbers, as there is to what we can do without them.

124. Keynes, *A Treatise on Probability*, p. 316.

PART III

Special Topics of Production

7

Fixed Coefficients of Production
and the Marginal Productivity Theory

The theory of marginal productivity has already been discussed to such an extent that a new paper dealing with the subject seems from the beginning to be superfluous, but, at the same time, the many controversies of the matter plead to the contrary.[1] It is far from my intention to criticize the opinions expressed by many eminent economists; my purpose is rather to see if it is not possible to modify the theory of marginal productivity in such a way as to include in it also the case of constant coefficients of production and to draw the immediate conclusions, in the hope that these may throw some light upon the controversial opinions.

The first main objection to the classical theory of marginal productivity was raised by Pareto, namely, the possible existence of factors of production that are fixed.[2] Pareto himself admits that in this case the marginal productivity theory "cannot be applied without corrections." But he apparently renounced this idea and furnished an entirely different solution of the problem.[3]

In the approach to the problem, there is a great difference between the classical theory and that proposed by Pareto: the marginal productivity theory arises from the consideration of a single producer; Pareto's

NOTE: This paper is reprinted by courtesy of the publisher from *Review of Economic Studies*, III (1935), 40–49. I am adding a postscript to the essay, and I am also correcting an error in the original version (as explained in the postscript).

1. The most recent being raised by H. Schultz and J. R. Hicks: H. Schultz, "Marginal Productivity and the General Pricing Process," *Journal of Political Economy*, XXXVII (1929), 505–551; J. R. Hicks, "Marginal Productivity and the Principle of Variation," *Economica*, XII (1932), 79–88; H. Schultz, "Marginal Productivity and the Lausanne School," *Economica*, XII (1932), 285–296; J. R. Hicks, "A Reply," *ibid.*, 297–300.

2. Vilfredo Pareto, *Cours d'économie politique*, 2 vols. (Lausanne, 1895–1897), vol. II, para. 717.

3. Vilfredo Pareto, *Manuel d'économie politique* (Paris, 1927), pp. 631 ff; "Economie mathématique," *Encyclopédie des Sciences Mathématiques*, tome I, vol. 4, fasc. 4, pp. 591–640.

theory considers the economic universe as a whole. The former is analytical, the latter synthetic.

There is no doubt that this generalization constitutes a great improvement and that the conclusions thus reached are an important scientific advantage. A certain methodological drawback, however, may be found in it. Pareto's theory enables one to reach but one single conclusion, viz., there is a point of equilibrium which, in a static state, the whole system tends toward and finally reaches.[4] His formulae are bound, because of their generality, to be rather complicated, and this complication prevents the economist from finding an economic interpretation, the most usual method in economics being that of analysis and not of synthesis.

In the second place, Pareto's method, because of its generality, does not require any further examination of the nature of the different factors of production and of the way they enter into the productive process. Do we not lose anything by neglecting to investigate these facts that are unimportant details for Pareto's general theory?

I shall, in the present paper, follow the old trail of marginal productivity theory and try to introduce the modifications necessary to incorporate the case of fixed factors of production. These modifications are those to which Pareto alluded in his *Cours*.

Let us consider a given method of production of a product Q in which three factors A, B, Y are used. Let the corresponding production function be[5]

(1) $$q = F(a, b, y).$$

The geometric representation of the relation between q and a, b, y necessitates a four-dimensional space. We can, however, have a geometrical representation by using isoquants in a three-dimensional space. Thus, in this latter space O, ABY, we shall assign to each point the corresponding value of q given by (1).

Let us consider a point $M_2(a_2, b_2, y_2)$, and let $a_1 \geqq a_2$, $b_1 \geqq b_2$ (the signs $=$ being not taken simultaneously) and $y_1 > y_2$. If subject to the conditions just stated

(2) $F(a_1, b_1, y_2) > F(a_2, b_2, y_2)$, $F(a_2, b_2, y_1) > F(a_2, b_2, y_2)$.

4. One might sometimes feel that this conclusion does not go much beyond the hypothesis that everything is connected with everything else. But this would not be logically justified.

5. A suitable way of defining what is meant by a production function would be to show how one might construct it. Taking a definite combination represented by the amounts a, b, y, of the factors considered, we shall experiment, using the available technical knowledge, to obtain the maximum amount of product. This maximum is the value of the function F for that particular combination. Varying the amounts a, b, y, of the *same* factors of production in such a way as to cover all the possible combinations, we get the production function determined throughout.

the factor Y will be called *substitutable* at M_2. If, however,

$$(3a) \qquad\qquad F(a_1, b_1, y_2) = F(a_2, b_2, y_2),$$

$$(3b) \qquad\qquad F(a_2, b_2, y_1) = F(a_2, b_2, y_2),$$

this factor will be called *limitational* at M_2.[6] Of course, there may be regions for which Y is substitutable and others for which it is limitational — at least theoretically. The factor can be treated as such in each of these regions, but for the sake of simplicity we shall suppose that it is limitational throughout.[7]

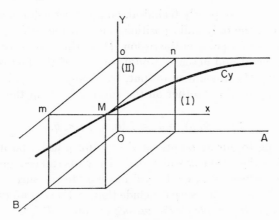

FIGURE 7–1

The locus of the points M for which (3) is satisfied is easily worked out. Let b and y remain constant and a vary. On the line mx (Fig. 7–1) there exists a point $M(a, b, y)$ such that

$$F(a', b, y) = F(a, b, y) \qquad \text{when} \qquad a' \geqq a,$$
$$F(a', b, y) < F(a, b, y) \qquad \text{when} \qquad a' < a.$$

Keeping now only y constant and varying b as well as a, the point $M(a, b, y)$ will describe in the plane mon an isoquant curve (C_y). And if we finally make y vary too, the point M will describe a surface (S), the equation of which is

$$(4) \qquad\qquad F(a, b, y) = \bar{q},$$

6. The terms "substitutable" and "limitational" factors were introduced by Ragnar Frisch, "Einige Punkte einer Preistheorie mit Boden und Arbeit als Produktionsfaktoren," *Zeitschrift für Nationalökonomie,* III (1931), 64.

7. In the case of the indivisibility of a factor we shall have to consider only those points the coordinates of which are an integral multiple of the indivisible unit of the respective factor. The comparisons (2) and (3) should be made only for such points: the fact that F is constant between two such consecutive points has nothing to do with the concept of limitational factor.

where $\bar{q}$ is the maximum output which can be obtained by using the amount y of the factor Y; hence

(5) $$\bar{q} = g(y),$$

and (4) becomes

(6) $$F(a, b, y) = g(y).$$

Needless to say, the derivative g_y of g is always positive. The surface (S) divides the whole space into two regions for which we have respectively

$$(\text{I}) \quad F(a, b, y) = \bar{q}, \qquad (\text{II}) \quad F(a, b, y) < \bar{q}.$$

For reasons that are purely technical, no producer will remain in either of these regions, the only valid position being on the surface (S). Let us consider the case of a point in the region (I). In this case, because of (3a), the producer could and actually would use less of the factors A and B without altering the amount of output, by choosing instead of the combination already used any other on the curve (C_y). If in the region (II), (3b) yields

$$F(a, b, y) = \bar{q}' = F(a, b, y') < \bar{q},$$

where according to our assumptions $y' < y$ (for g has a positive derivative). Therefore the producer will realize in any case an economy by using the amount y' instead of y, i.e. he will move on to the surface (S).

From these considerations we conclude that, *if Y is a limitational factor*, defined as above, *the points on the surface (S) are really the only ones that have to be considered in the pricing process.* The difference between this case and that in which all the factors are substitutable is that in the latter we can move in *the whole space O, ABY*. At the same time, we have seen that a functional relation (5) between the amount of output and the limitational factor exists.

In what follows, we shall write the equation (6) in the form

$$q = f(a, b)$$

which is a direct consequence of (1) and (6). The valid processes of production are therefore described by the system

(7) $$q = f(a, b) = g(y).$$

I shall now proceed to consider briefly the argument of the classical theory of marginal productivity in a form that will make the analogy with what is to follow much easier.

The process by which the amount of the factors used and consequently the prices of these factors are determined, may be decomposed into three steps.

1. The producer will first try to adjust his plant in such a way as to obtain a maximum output for a given cost of production, or what comes

to the same thing, to have a minimum cost for a given output. This step does not depend upon the price of the product, and profits are completely disregarded — we may call it the *production economy*.

2. The second step is that by which the producer will this time adjust his production in order to maximize his profits. We may refer to this step as the *enterprise economy*.

3. Lastly, there is the process by which competition will bring the price of the product to equality with average cost.[8] This step will be called the *community production economy*.

Let us now follow these three steps one by one in the case where all the factors of production are substitutable, and consider for simplicity a productive process in which only two factors a, b are used, the prices of which are p_a, p_b respectively. The conditions for the cost $\mu = ap_a + bp_b$ to be a minimum when $q_0 = f(a, b) = $ const., are

$$\frac{\partial}{\partial a}\left[ap_a + bp_b - \lambda(f - q_0)\right] = 0,$$

$$\frac{\partial}{\partial b}\left[ap_a + bp_b - \lambda(f - q_0)\right] = 0,$$

where λ is a Lagrange multiplier. Hence,

(8) $$\frac{p_a}{f_a} = \frac{p_b}{f_b}.$$

This equation expresses the path which will be followed by the producer

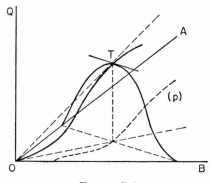

FIGURE 7-2

for a given ratio $p_a : p_b$. We shall refer to it as (p) path (Fig. 7–2). The correspondence between the (p) paths and the ratios $p_a : p_b$ is one to one.

8. This last step is in fact a definition of the status of the community. We are considering only the case of free competition. The results obtained can be, however, adapted also to the case of monopoly, if the necessary modifications are introduced. See Erich Schneider, "Bemerkungen zur Grenzproduktivitätstheorie," *Zeitschrift für Nationalökonomie*, IV (1933), 604.

A pure production economy will follow the (p) path corresponding to the given prices until the average cost is a minimum. The condition is:

$$d\left(\frac{\mu}{q}\right) = \frac{qd\mu - \mu dq}{q^2} = 0.$$

Hence,

(9) $$\frac{\mu}{q} = \frac{d\mu}{dq},$$

and from (8)

(10) $$\frac{p_a}{f_a} = \frac{p_b}{f_b} = \frac{p_a da + p_b db}{f_a da + f_b db} = \frac{d\mu}{dq}.$$

Consequently,

$$\frac{p_a}{f_a} = \frac{p_b}{f_b} = \frac{ap_a + bp_b}{q}.$$

Eliminating p_a, p_b we find

(11) $$af_a + bf_b = f,$$

a relation that represents the locus of the points corresponding to the optimum scale of production when the prices of the factors vary. In other words, the relation (11) represents an invariant of the solution of the problem with respect to the prices p_a, p_b. This is at the same time the equation of the cone with the vertex at the origin and tangent to the production surface. This is shown by the directrix OT (Fig. 7–2). We conclude from the last statement that in a pure production economy the possibility of cooperation prevents the system from working with any scale other than the optimum.[9]

The second step, that of an enterprise economy, will be to find on the (p) path the point for which the maximum profit is reached. Let p be the fixed price of the product; the condition then is

$$d(pq - \mu) = pdq - d\mu = 0.$$

Hence,

$$p = \frac{d\mu}{dq},$$

and from (10),

(12) $$\frac{p_a}{f_a} = \frac{p_b}{f_b} = p.$$

It is clear that the point on the (p) path which satisfies (12), i.e. corresponds to the maximum profit, is not necessarily the same as that corresponding to the optimum scale of production.

9. It is possible that there might be no points in the positive quadrant that satisfy relation (11). The average cost will be then continuously either increasing or decreasing. The scale of production will have a tendency either to be very small or very large.

Now, the production economy leads in the long run to

$$ap_a + bp_b = pq,$$

and from (12),

$$af_a + bf_b = f.^{10}$$

We have proved so far that:

1. The whole product is distributed amongst the factors of production according to their marginal productivity.
2. The optimum scale of production is identical with that obtained in a pure production economy.

Let us now consider the case of a production process involving two substitutable factors A, B and two limitational ones Y, Z. The connection between the output and the amounts of factors used is expressed by the system

$$(13) \qquad q = f(a, b) = g(y) = h(z).$$

We obtain a partial representation of the productive process by considering in a three-dimensional space the surface $q = f(a, b)$.

The first step is to find the minimum of the cost of production $\mu = ap_a + bp_b + yp_y + zp_z$ for a given output. When the output is given, we deduce from (13) that y and z are fixed also. Then the corresponding cost of these last factors, $yp_y + zp_z$, is constant too. The problem resolves itself in finding the minimum of $ap_a + bp_b$; as before, we obtain the condition

$$(14) \qquad \frac{p_a}{f_a} = \frac{p_b}{f_b}.$$

This is a (p) path identical with (8). As we have

$$d\mu = p_a da + p_b db + p_y dy + p_z dz,$$
$$dq = f_a da + f_b db = g_y dy = h_z dz,$$

from (14) we obtain

$$\frac{p_a}{f_a} = \frac{p_b}{f_b} = \frac{d\mu - p_y dy - p_z dz}{dq} = \frac{d\mu}{dq} - \frac{p_y}{g_y} - \frac{p_z}{h_z},$$

or

$$(15) \qquad \frac{p_a}{f_a} + \frac{p_y}{g_y} + \frac{p_z}{h_z} = \frac{p_b}{f_b} + \frac{p_y}{g_y} + \frac{p_z}{h_z} = \frac{d\mu}{dq}.$$

The analogy of this relation with (10) is obvious. The point corresponding to the optimum scale satisfies the relation (9). From (15) we derive

$$(16) \qquad \frac{p_a}{f_a} + \frac{p_y}{g_y} + \frac{p_z}{h_z} = \frac{p_b}{f_b} + \frac{p_y}{g_y} + \frac{p_z}{h_z} = \frac{ap_a + bp_b + yp_y + zp_z}{q}.$$

10. J. R. Hicks, *Theory of Wages* (London, 1932), Appendix.

Hence,

(17)
$$\frac{p_a}{f_a}(af_a + bf_b - q) + \frac{p_y}{g_y}(yg_y - q) + \frac{p_z}{h_z}(yh_z - q) = 0,$$

$$\frac{p_b}{f_b}(af_a + bf_b - q) + \frac{p_y}{g_y}(yg_y - q) + \frac{p_z}{h_z}(yh_z - q) = 0.$$

By comparing the last relation with (11) we see that, as the point corresponding to the optimum scale of production depends in the present case on the prices of limitational factors, there is no longer a locus of these points among all technically possible combinations. Consequently, the prices of these factors varying, any point on the production function might be an optimum scale. This fact constitutes an important difference between the two cases under consideration and it might be looked upon as an explanation of the variability of the scale of production within the same industry.

Proceeding to the second step, we obtain as before

(18)
$$\frac{p_a}{f_a} + \frac{p_y}{g_y} + \frac{p_z}{h_z} = \frac{p_b}{f_b} + \frac{p_y}{g_y} + \frac{p_z}{h_z} = p,$$

and finally, in the long run,

(19)
$$qp = ap_a + bp_b + yp_y + zp_z,$$

which leads us again to relation (16). This proves that also in the case of a productive process involving limitational factors, the optimum scale of production is the same in a pure production economy as in an exchange economy under free competition.

As to the other conclusions reached when all the factors are substitutable, we can no longer make use of the concept of physical marginal product. For no increment of only one factor can bring an increase in the output if it is not accompanied by the appropriate increments of the other factors. It will then be necessary to consider the balance between the increment of the product and the increments of these other factors. This balance is not measurable if we do not introduce prices; in other words, we shall have to give up working with physical increments and consider instead revenue increments. There is, of course, the traditional objection against introducing prices into a theory of the pricing process. However, a sufficient number of arguments have been leveled against this objection to make its standing dubious.[11] But there is something else that has not been mentioned, at least so far as I know. Pareto's

11. The marginal product is a certain part of the total product. But if the same labor is used in several different trades, clearly this concrete conception of the marginal product is no longer possible. The marginal productivity must be the same in every trade, and therefore can only be conceived as a part of the prices of the various products. Gustav Cassel, *The Theory of Social Economy* (New York, 1924), p. 301.

theory of general equilibrium has to introduce prices, first as given and later as unknowns of the problem, to be determined from his general system of equations. It seems then, that if we drop the assumption that all factors of production are substitutable, we shall have either to introduce prices as intermediate unknowns or to go without a theory of distribution.

In order to make this clearer, let us consider the case of a firm using four factors. The producer's equilibrium in a competitive economy might, for instance, lead to one of the following systems:

$$\begin{cases} q = f(a, b, y, z), \\ \dfrac{p_a}{f_a} = \dfrac{p_b}{f_b} = \dfrac{p_y}{f_y} = \dfrac{p_z}{f_z} = \dfrac{ap_a + bp_b + yp_y + zp_z}{q}; \end{cases}$$

$$\begin{cases} q = f(a, b) = g(y) = h(z), \\ \dfrac{p_a}{f_a} + \dfrac{p_y}{g_y} + \dfrac{p_z}{h_z} = \dfrac{p_b}{f_b} + \dfrac{p_y}{g_y} + \dfrac{p_z}{h_z} = \dfrac{ap_a + bp_b + yp_y + zp_z}{q}; \end{cases}$$

according to the assumptions made concerning the nature of the factors used.

Both systems include five relations. But in the first case, the four prices — which represent really only three unknowns, for they are relative prices — are connected by four relations. An algebraic elimination of these unknowns can consequently be performed. In the second case, there are still four prices to be determined, but the number of relations connecting them is diminished from four to two, two other relations having been shifted from the group connecting prices to that connecting physical quantities. The algebraic elimination of p_a, p_b, p_y, p_z is no longer possible. Consequently, the invariance of the solution with respect to prices which was mentioned above, exists only in the case where all factors are substitutable.

Let us return to the production function (13) and let Δa be a small increment in the amount used of the factor A. This increment cannot cause any increase in output unless accompanied by the appropriate increments of the limitational factors, Y and Z. These increments are easily obtained from (13):

	A	Y	Z	Q
Increments	Δa	$\dfrac{f_a}{g_y} \Delta a$	$\dfrac{f_a}{h_z} \Delta a$	$f_a \Delta a$

The marginal expenditure associated with an increment of the factor A will thus be per unit of output

$$\frac{p_a}{f_a} + \frac{p_y}{g_y} + \frac{p_z}{h_z},$$

and similarly for an increment of the factor B

$$\frac{p_b}{f_b} + \frac{p_y}{g_y} + \frac{p_z}{h_z}.$$

Relations (18) express therefore the fact that for the producer's equilibrium the marginal expenditures are equal to the price of the product.

The corresponding element of the marginal product of a factor is the marginal revenue attached to it. The marginal revenue of A is

$$\left(p - \frac{p_y}{g_y} - \frac{p_z}{h_z}\right) f_a \Delta a,$$

and per unit of A

$$r_a = \left(p - \frac{p_y}{g_y} - \frac{p_z}{h_z}\right) f_a;$$

and for B,

$$r_b = \left(p - \frac{p_y}{g_y} - \frac{p_z}{h_z}\right) f_b.$$

Let us now return to relations (18). These may be written as follows

$$p_a = f_a\left(p - \frac{p_y}{g_y} - \frac{p_z}{h_z}\right) = r_a,$$

$$p_b = f_b\left(p - \frac{p_y}{g_y} - \frac{p_z}{h_z}\right) = r_b,$$

and introducing them in (19) we obtain

(20) $$a r_a + b r_b + y p_y + z p_z = p q.$$

This is the fundamental relation we need. It shows that the shares of the factors in the product are determined in a way that combines the classical marginal and net productivity theories, i.e. the part of the revenue going to the substitutable factors is determined according to their marginal revenue, that of the limitational factors is obtained according to their net revenue.

So far, no particular assumptions have been made about the functions f, g, h. Let us suppose that g and h are homogeneous of first degree

$$q = ky = lz$$

(i.e., constant coefficients of production, of which the typical example is provided by some raw materials). In this case relation (17) becomes

$$a f_a + b f_b = f.$$

The form of this relation, but the form only, reminds us of the marginal productivity formula. In any case, it shows that under the assumptions made, there is a locus of the points corresponding to the optimum scale of production.

If f too is homogeneous of first degree, relation (20) is satisfied not only in the long but also in the short run, a consequence that has its perfect analogy in the marginal productivity theory.

The example chosen for the support of our ideas in the present paper is that of a productive process in which there are involved only limitational and substitutable factors divided into two groups. The consideration of a more complicated but similar case is not at all difficult. For instance, the case may be imagined in which there are several groups of substitutable factors, the production being then described by a system such as

$$q = f(a, b) = \varphi(c, d) = g(y) = h(z).$$

The conclusions will be the same as those already obtained. There is, however, still another way of generalizing the concept of limitational factor, by introducing a productive process represented by the system

$$(21) \qquad q = f(a, b, c) = g(a, b, c)$$

from which it is not possible to deduce any relation of the form $q = h(c)$.[12] In this case, or that of a more complicated scheme, there seems to be no way of obtaining results to which we may easily attach an economic meaning. The question arises whether there is any productive process that has to be pictured by a system such as (21). The concepts of substitutable and limitational factors have been introduced by the analysis of some of the actual productive processes; so they have real economic meaning. What about the system (21)? Does economics need so many purely formal mathematical generalizations?

A POSTSCRIPT (1964)

1. In the version of the above paper published in 1935, the characteristic properties of a limitational factor though stated correctly by relations (I) and (II) — upon which my formal argument was based — were incompletely expressed by (3). The slip was called to my attention by

12. The generalization is evident. The productive process will be valid only on a two-dimensional surface

$$f(a, b, c) = g(a, b, c),$$

analogous to (S). The condition that no relation of the form $q = h(c)$ can be deduced from (21) is that all Jacobians

$$\frac{\partial(f, g)}{\partial(a, b)}, \qquad \frac{\partial(f, g)}{\partial(b, c)}, \qquad \frac{\partial(f, g)}{\partial(c, a)},$$

must be different from zero.

N. Kaldor, "Limitational Factors and the Elasticity of Substitution," *Review of Economic Studies*, IV (1937), 162–165. The error has been corrected in the present version — relations (3a) and (3b).

Kaldor's paper, however, is valuable for other reasons. First, he brings to our attention a general property of production factors which before was only tacitly and vaguely implied by other authors (Kaldor, p. 162).[1] Second, he considers the limitational production function corresponding to my equations (21) and observes that its isoquants have the shape of a boat-bow instead of a cylinder with a flat bottom, as is the case for the type defined by (7) and represented diagrammatically by my Fig. 7–1. Lastly, and most important, Kaldor offers a 2 × 2 classification to cover all possible "cases." The classification is based on the opposition between "substitutability" and "fixed coefficients." In a more precise formulation, it says that at a *point* of the production surface we may find one of the following situations: 1) every factor is substitutable with any other factor, 2) some (but not all) factors are such that each has fixed coefficients with some (but not all) factors, 3) some (but not all) factors are such that each has fixed coefficients with any factor, and 4) every factor has fixed coefficients with any factor.[2]

Highly interesting though Kaldor's idea is, the classification does not vindicate his main contention (p. 162), namely, that the analysis of the singularities, such as those connected with limitationality, calls for no new concepts: the traditional notions of "substitutability" and "fixed coefficients" suffice. Kaldor's Fig. 2 notwithstanding, the context leaves no doubt that he does not use "fixed coefficients" in the traditional Walrasian sense. The term is old, but not so the concept Kaldor denotes by it. Only in one place does he refer to it as "a special complementarity" (p. 162). Perhaps "special limitationality" would have been a more appropriate choice in view of the fact that Kaldor's concept becomes identical to limitationality in the case of a two-factor production. Be this as it may, because of Kaldor's use of an old term for a new concept, many readers — like myself — must have been puzzled at first by his classification: how can all factors have fixed coefficients with one factor and not necessarily with each other?[3]

In any science of facts new concepts are usually set up because of the realization that a particular group of facts have an empirically relevant characteristic. It stands to reason also that a new label is necessary to

1. The property is stated formally and more strictly below, definition V.

2. Kaldor, as cited in text, p. 165. A point which escaped Kaldor's attention is that the classification is exhaustive only as point-conditions are concerned. The classification cannot apply to production functions unless amended by other concepts and conditions, such as those included in definition VII below.

3. Actually, "substitutability" too is not used by Kaldor — nor by myself in the above paper — in the strict traditional sense. See note 11 below.

denote unambiguously such a group. But we should observe, first, that only after the new concept has been set up can we proceed to establish a new classification incorporating the new logical division. Second, there is no reason why every category of the new classification should have any empirical content or why further analysis should not reveal the necessity of subdividing some of the categories. The case of the concept of a limitational factor and Kaldor's classification constitute a good example for the first point and, perhaps, for the second point as well.

The concept of a limitational factor was introduced because of the analytical importance of a large class of actual processes, a classical illustration of which is the production of wedding rings. The characteristic of this class is that a necessary relation exists between one *input* (gold) and the *output* (wedding rings).[4] And clearly, such a factor does not fit into either of the traditional categories of substitutability and of fixed coefficients.

It is elementary that there exists a maximum output that can be obtained from a given *supply* of a limitational factor and, hence, a further increase in output requires an increase in that supply. But the fact that this property does not characterize a limitational factor has not always been understood clearly. This explains the confusion involved in some of the verbal definitions of limitationality.[5] Actually, as pointed out by Kaldor,[6] the property belongs to any factor or group of factors. Moreover, in some situations it may acquire appreciable empirical significance, a fact that led me to introduce the concept of limitativeness. On the same occasion, I offered a definition of limitationality which pinpoints the relation between the two concepts and the general property discussed in this paragraph.[7]

2. The purpose of this postscript is to present an analysis of the features associated with a limitational production function in a more stringent way than in my previous papers; and to offer further thoughts on this topic, some suggested by Kaldor's article. There are good reasons, as I shall now argue, why such an endeavor is worthwhile.

4. No doubt, this is the most direct definition of a limitational factor. As shown by equations (7), (13), and those preceding (21), and especially by my rejection of (21), it is this definition I had in mind in my argument. O. Lange, "On the Theory of Socialism," *Review of Economic Studies*, IV (1936), 58n2, reports that Tord Palander proposed that such a factor be called limitational of the first kind.

5. In the same issue with my article, Herbert Zassenhaus, in "Dr. Schneider and the Theory of Production," *Review of Economic Studies*, III (1935), 35–39, finds such a confusion to be the main fault of Erich Schneider's *Theorie der Produktion*. (Incidentally, only by ignoring my equations (I) and (II) and analyzing (3) alone was it possible for Kaldor to argue that I defined a limitational factor only by the property just mentioned.)

6. See note 1 above.

7. See my 1955 paper "Limitationality, Limitativeness, and Economic Equilibrium" included as Essay 10 in this volume, and notes 14 and 15 in the present Postscript.

It goes without saying that interest in fixed coefficients or limitationality arose because of the peculiar problems these situations create
for the marginal-productivity theory of factor pricing. The production
processes belonging to these or other similar categories constitute exceptions to the Classical notion that every factor can always be substituted
for any other factor, a notion tantamount to the assumption that an
increase in the *input* of any factor always yields an increase in output.

Now, as we know from the fierce controversies concerning the causes
of economic maladjustment, if the actual economic world would conform
to the Classical assumption, not only could marginal productivity pricing
always work, if we wished to adopt it, but also no factor could exist
for long in excess supply unless all factors were in excess supply. (And
all factors could be in excess supply only as a result of a fall in general
demand.) However, in the actual world we find, more often than not,
that only some (not all) factors are in excess supply. We may find, for
instance, an excess supply of automobile carburetors because carburetors
are produced independently of motors, bodies, etc., and not always in
the right technical proportion. More important is the almost general and
continuous existence of excess industrial capacity. The main cause of
this phenomenon too is some "technical complementarity" hidden behind
the veil of variable proportions. (I refer to the almost unnoticed fact that,
in general, fund-factors and flow-factors are not mutually substitutable.)
Recently, our attention has been caught by the problem of overpopulation
which illustrates the general situation where output may even decrease
if the entire supply of one factor would be used in production. All these
examples show that the significance of the exceptions to the Classical
assumption of a universal substitutability extends beyond the narrow
field of pricing into the more important area of economic institutions.[8]

3. To deal, therefore, with the exceptions to the Classical position we
must focus our attention upon *supply* instead of *input*. In other words,
we need to define the production function as a relation between output, q,
and the supplies $x_1, x_2, \ldots, x_n$ of the factors $\Phi_1, \Phi_2, \ldots, \Phi_n$, instead of
the customary relation between q and the inputs (y).[9] Let then

$$(1) \qquad q = f(x_1, x_2, \ldots, x_n)$$

mean that q is the maximum output over the domain $0 \leqq (y) \leqq (x)$.[10]

8. As I argued in the two papers reprinted as Essays 10 and 11 in this volume,
"Limitationality, Limitativeness, and Economic Equilibrium" (1955) and "Economic
Theory and Agrarian Economics" (1960).

9. In order not to burden the argument with issues irrelevant for my topic, I shall,
as in the past, confine my analysis to the case of a process with a single product and
a *finite* number of factors, $n > 1$. Mainly for the simplicity of the diction, I shall also
assume that the product and the factors have *continuous cardinal measures*. The
assumption that the production function is continuous becomes then almost a necessity.

10. Notation $(a) \leqq (b)$ means $a_i \leqq b_i$. The exclusion of the case $a_i = b_i$ for every i
is shown by $(a) \leq (b)$.

In this manner, (1) is defined over the entire nonnegative orthant of the factor-space as a single-valued function which fulfills the condition

(2) $If \quad (x') \leq (x), \quad then \quad f(x') \leq f(x).$

It is, however, within reason that the ordinary production function fulfills a stronger condition than (2). To shorten the diction in the discussion of these conditions let us first agree to use u, v, w exclusively for denoting

$$(u)_m = (u_k), \; u_k = 0 \; if \; k \neq m, \; u_k = 1 \; if \; k = m,$$
$$(v)_m = (v_1, v_2, \ldots, v_m, 0, 0, \ldots, 0) \geq 0,$$
$$(w)_m = (w_1, w_2, \ldots, w_m, 0, 0, \ldots, 0), \; w_k > 0, \; 1 \leq k \leq m.$$

As we know, Classical analysis assumed general substitutability of factors. In fact, as we find it stated at times explicitly, that analysis assumed that at least at any $(x) > 0$ every factor has a positive marginal productivity. Or with a term I shall use in this note, it assumed that every factor is always effective: that is, if $(x) > 0$, $\alpha > 0$, and $(x') = (x) + \alpha(u)_k$, then, $f(x') > f(x)$.[11]

The denial of general effectiveness is expressed by the following:

A1. *Any m-dimensional space, say, $E_m^0(x_{m+1} = x_{m+1}^0, x_{m+2} = x_{m+2}^0, \ldots, x_n = x_n^0)$, $m < n$, contains a point $M(x^0)$, $(x^0) < +\infty$, such that $f(x^0) = Q_m^0$ is the absolute maximum of $f(x)$ over E_m^0.*[12]

Let Z_m^0 be the set of all points $(z) \in E_m^0$ for which $f(z) = Q_m^0$. Because of the continuity of $f(x)$, Z_m^0 is closed, that is, it contains all boundary points (ζ) such that $f(\zeta) = Q_m^0$ and $f(x) < Q_m^0$ for any $(x) = (\zeta) - (w)_m \geq 0$. Under the usual assumption of convex isoquants, Z_m^0 would be convex. But A1 restricts the shape of Z_m^0 only by the condition — following from (2) — that if $(z) \in Z_m^0$ and $(z') = (z) + (v)_m$ then $(z') \in Z_m^0$. An extreme example of the shape of Z_m^0 compatible with A1 is the two-factor function defined as follows:

(3) $q = \begin{matrix} \text{Min}[x_1, x_2/2, K] & if & x_1 \leq x_2, \\ \text{Min}[x_1/2, x_2, K] & if & x_1 \geq x_2, \end{matrix}$

where K is a constant.

From A1 it follows that for every k there exist two single-valued functions, $X_k(x)$ and $F_k(x)$, each being independent of x_k and such that

(4) $q < F_k(x) \Leftrightarrow x_k < X_k, \qquad q = F_k \Leftrightarrow x_k \geq X_k.$

11. In its strict sense, substitutability means, for instance, that there exists a_1, a_2, such that $a_1 a_2 < 0$, and $f(x_1 + a_1, x_2 + a_2) = f(x_1, x_2)$. As shown by the factor-combinations on C_y in my Fig. 7–1, this definition of substitutability does not necessarily imply effectiveness. Hence, along C_y the factors A and B are not substitutable in the Classical sense. To eliminate the possibility of confusion, we should use a different term, say, compensativeness, if factors are substitutable but not effective.

12. The reason I do not assume A1 to be necessarily true for $m = n$ is to let (1) also represent a molecular process, i.e., an industry.

Because of the continuity of $f(x)$, F_k is continuous. However, as (3) shows, X_k is not necessarily continuous. To eliminate the peculiarities displayed by (3) — which can hardly have any empirical justification — we shall assume that a production function has also the following property:

A2. $X_k(x)$ *is continuous at* $(x) > 0$. *Moreover, if* $(x') = (x) + \alpha(u)_k$, $x'_k, x_k < X_k$, *then*

$$(5) \qquad\qquad f(x) < f(x') \Leftrightarrow \alpha > 0.[13]$$

DEFINITION I. *Let* $(x) \in E^0_m$. *The group of factors* $(\Phi)_m = (\Phi_1, \Phi_2, \ldots, \Phi_m)$ *is* effective, sufficient, *or* redundant *at* (x) *according to whether* (x) *is an exterior, a boundary, or an interior point of* Z^0_m.

If $m = 1$, then at (x) Φ_1 is effective, sufficient, or redundant according to whether $x_1 <, =, > X_1$. From A1, it is immediate that if $(\Phi)_m$ is redundant at (x), then every Φ_k, $k \leq m$, is redundant at (x). By using A2 we can easily show that the converse of the last proposition is true. However, a sufficient group may contain as many as $m - 1$ redundant factors. More salient is the fact that an effective group need not contain any effective factor, in which case it must contain at least two sufficient factors.

DEFINITION II. *Let* $(\Phi)_m$ *be an effective group at* (x). *If no proper subset of* $(\Phi)_m$ *is effective at* (x) *then* $(\Phi)_m$ *is a* strictly effective *group at* (x).

Clearly, if $m > 1$, then every factor of a strictly effective group is sufficient; if $m = 1$, the factor is effective, and conversely.

A special situation arises if all strictly effective groups at (x) have a common subgroup, say, $(\Phi)_l$, $1 \leq l \leq n$. Any factor Φ_k, $k < l$, is characterized by a property which can be cast into the following

DEFINITION III. *If at* (x) *an increase in* x_j *is a necessary but not sufficient condition for an increase in output, then* Φ_j *is a* limitational *factor at* (x).[14]

Another relevant case is that in which at (x) there is only one strictly effective group, say, $(\Phi)_h$, $1 \leq h$. (Clearly, if $k > h$, Φ_k is not effective at (x).) This case leads to

DEFINITION IV. *If at* (x) *an increase in every* x_j, $1 \leq j \leq h$, *is a necessary and sufficient condition for an increase in output, then* $(\Phi)_h$ *is a* limitative *group at* (x).[15]

13. Example (3) does not satisfy A2. On the other hand, the function $f(x_1, x_2) = x_1 x_2/(x_1 + x_2)$, $f(0, 0) = 0$, satisfies A2 but not A1. Therefore, A1 and A2 are independent. We should also note that even with the addition of A2 the isoquants do not have to be convex.

14. See definition I in my Essay 10 reprinted below, "Limitationality, Limitativeness, and Economic Equilibrium."

15. This represents a natural generalization of definition II of the essay just cited. The above definition also covers the case where locks *and* keys are limitative. Incidentally, if $h = n$, then (Φ) is both a limitational and a limitative group.

The existence of limitational factors is proved by each production function examined in my 1935 paper (see note 4 above). Thus, in (13) there are two limitational-effective groups, (A, X, Z) and (B, X, Z); hence, both X and Z are limitational. However, from A1 and A2 it does not follow that a limitational factor is associated with every production: the point is made plain by Kaldor's Fig. 1. It would be an error, however, to infer from the same diagram that if the marginal productivity of every group of factors becomes zero in the end, then every factor becomes ultimately limitative. The function

$$(6) \qquad q = x_1 x_2 + x_1 + x_2, \qquad q \leq K = \text{constant},$$

immediately settles the point that A1 does not entail regularity, regularity being defined as follows:

DEFINITION V. *If each factor is limitative at some (x) on the isoquant I_0, $f(x) = q_0$, then I_0 is a* regular *isoquant.*[16]

Since a factor-combination (x) involves excess supply if and only if some factor(s) are redundant, it is natural to adopt also the following

DEFINITION VI. *If no factor is redundant at (x), then (x) is a* proper *factor-combination.*

Let us now observe that relations $x_i <, =, > X_i$ divide the factor-space into 3^n L-domains, some of which may be empty. For instance, the domain $L^n = [x_i > X_i$ for every $i]$ is empty if A1 is false for $m = n$; in the case of function (6) there are only three L-domains. Let $L^0 = [x_i < X_i$ for every $i]$. Since in every domain other than L^0, there prevails at least one relation $x_i > X_i$, from (4) it follows that f is defined by the functions F_k everywhere outside L^0. Let $F_0(x)$ be a function such that if $(x) \in L^0$ then $q = F_0$, and if $(x) \notin L^0$ then $q < F_0$. (Clearly, if $(x) \in L^0$, then $F_0(x)$ is an increasing function of *every* x_j.) Any production function can then be written as

$$(7) \qquad q = \text{Min}[\epsilon_0 F_0, F_1, \ldots, F_n]$$

where $\epsilon_0 = 0$ or 1 according to whether L^0 is empty or not.[17] (It is possible to leave out also some $F_j, j \neq 0$, if $L(i) = (x_k < X_k, k \neq i$ and $x_i = X_i)$ is empty.)[18]

16. The assumption — in favor of which there is strong evidence — that all isoquants of a production function are regular, has to be added to A1. For this assumption, in turn, does not imply A1. The point is proved by $q = f(x_1, x_2, x_3)$ such that $q = c$ for $x_i = b$ and $(x_j - b)(x_k - b) \geq 0$, where $b > 0$ is an increasing, continuous function of c, and i, j, k any permutation of 1, 2, 3.

17. Actually, whether or not it satisfies A1, any production function satisfying (2) can be represented by $q = \text{Min}[G_1, G_2, \ldots, G_m]$, $m \leq n + 1$, where at most one G_k is a function of every x_i. This relation may be interpreted as a systematization of the technical constraints $q \leq g_i(x)$, to which any production process is normally subject.

18. Clearly, if $L(i)$ is empty, so is $L'(i) = [x_k < X_k, k \neq i, x_i > X_i]$.

Let $L^0 \neq 0$; then $L^0 \subseteq P$, where P denotes the set of all proper combinations of (1). If $(x) \in L^0$, then the dimension of P at (x) is n, and conversely.[19] The situation that concerns us in this note is that where $L^0 = 0$, that is, where $\epsilon_0 = 0$ in (7). In this case the dimension of P is everywhere smaller than n, and as the following theorem will show, greater than zero.

THEOREM I. *Every isoquant of a function satisfying* (2) *contains at least one proper factor combination.*

PROOF. Let $f(x^0) = q_0$, and let Φ_i, $i \leq m \leq n$, be the redundant factors at (x^0). Let $(x^1) = (x^0) - \alpha(u)_1$, $x_1^1 = X_1^0$; then $f(x^1) = q_0$. At (x^1) Φ_1 is sufficient and no Φ_k, $k > m$, is redundant. For if Φ_k is redundant at (x^1), then for $(x') = (x^1) - \beta(u)_k$ and $0 \leq \beta \leq \beta_0$, we have $f(x') = q_0$ and, by (2), $f(x'') = q_0$ for $(x'') = (x^0) - \beta(u)_k$. But then Φ_k would be redundant at (x^0). Since this creates a contradiction, at (x^1) there are at most $m - 1$ redundant factors. Continuing the algorithm we reach some (x^j), $j \leq m$, at which no factor is redundant, and $f(x^j) = q^0$. Q.E.D.

Let ϵ, σ, λ, τ, be the numbers of the following categories of factors at $(x) \in P$: effective, sufficient, limitational, and sufficient but not limitational. Simple considerations show that if $\epsilon \neq n$, then

$$(8) \qquad \epsilon\lambda = 0, \qquad \lambda \neq n - 1, \qquad \epsilon < r,$$

where r is the dimension of P at (x). We should note that the third condition implies $\sigma \geq 2$. These results lead us to distinguish four general types of situations at $(x) \in P$:

I $(\lambda > 0, \tau, \epsilon = 0)$,   II $(0, \tau > 1, \epsilon)$,   III $(0, n, 0)$,   IV $(0, 0, n)$.[20]

Let $(x) \in P$ and let Φ_i, $1 \leq i \leq \sigma$, be the sufficient factors at (x). Then (x) satisfies

$$(9) \qquad x_i = X_i \text{ for } i \leq \sigma, \qquad x_i < X_i \text{ for } i > \sigma.$$

At (x) we also have

$$(10) \qquad q = F_1 = F_2 = \cdots = F_\sigma < F_i, \qquad\qquad \sigma < i \leq n,$$

and, because of the continuity of F_k, in the neighborhood of (x)

$$(11) \qquad q = \text{Min}[F_1, F_2, \ldots, F_\sigma].$$

19. Even though the concept of dimensionality may seem simple enough, I refer to the definition given in W. Hurewicz and H. Wallman, *Dimension Theory* (Princeton, 1941), chap. iii.

20. The production functions considered in my article are of type I. Kaldor's example of boat-bow isoquants belongs to type II. Type III constitutes the transition between I and II. In Kaldor's classification this type is included together with II in category 2, whereas the pattern $(n, 0, 0)$ is set apart as category 4. The above classification, however, better answers the needs of a correlation between pure analysis and factual evidence.

It is not difficult to construct analytical examples showing that A1–A2 do not warrant that P_0, the intersection of P with I_0, has the same dimension for every I_0.[21]

DEFINITION VII. *If the intersection of every L-domain with every isoquant has the same dimension, then the production function is an L-function. If, in addition, $L^0 = 0$, then the production function is an L^0-function.*

Let $f(x)$ be an L^0-function and Π the domain such that if $(x) \in (\Pi)$, at (x) the number of effective factors, ϵ, is maximum.[22] By reasons based on continuity, it is easily seen that the same factor must be effective over the entire (Π). Therefore, by definition VII, Π has the same dimension r, $\epsilon < r \leq n - 1$, and if $(x) \in (\Pi)$, then (x) satisfies $(9)-(11)$. Moreover, $(x) \in \Pi$, if and only if $x_i = X_i$, $1 \leq i \leq \sigma$.[23] However, the same equivalence does not apply to (10) which is satisfied by some $(x) \notin \Pi$ as well as by $(x) \in \Pi$. Nonetheless, (10) is more useful because it determines output in terms of supplies for the most pertinent class of factor-combinations. I used this form to describe the L^0-functions in my 1935 paper.

To be sure, the structure of the F_k's is far from simple. It is, however, possible to find a system equivalent to (10) and simpler than it. An instructive example is offered by the case of $\lambda = n$. Then, $F_i = \text{Min}[a_1, a_2, \ldots, a_{i-1}, a_{i+1}, \ldots, a_n]$, where a_j is a continuous function of x_j alone, increasing for $x_j < \xi_j$, and constant for $x_j \geq \xi_j$. In this case, (10) and (11) become

(10a) $$q = a_1(x_1) = a_2(x_2) = \cdots = a_n(x_n),$$

(11b) $$q = \text{Min}[a_1, a_2, \ldots, a_n],$$

the last relation being valid over E^n, not only in the neighborhood of (x).

THEOREM II. *If $(x) \in \Pi$, then (x) satisfies a system*

(12) $$q = h_1 = h_2 = \cdots = h_{g+1}, \qquad g = n - r,$$

such that: 1) h_k *is a function of* $(x)^{(k)} = (x_{i_k}, x_{j_k}, \ldots, x_{l_k})$, *where* $(i_k, j_k, \ldots, l_k)$ *is a true subset of* $(1, 2, \ldots, n)$, *and is defined over a domain* H_k *such that* $x^{(k)} \in H_k$ *implies* $(z) \in \Pi$ *with* $z_{d_k} = x_{d_k}$ *for* $d = i, j, \ldots, l$; 2) h_k *is continuous over* H_k *and is an increasing function of every argument;* 3) *at least one* h_k *does not involve* x_i, *for every* $i \leq \sigma$; 4) *if* $(\Phi_a, \Phi_b, \ldots, \Phi_i, \ldots, \Phi_l)$ *is strictly effective, then every* h_k *involves one and only one* Φ_i; 5) *the following*

(13) $$q = \text{Min}[h_1, h_2, \ldots, h_{g+1}]$$

21. For instance, $f(x_1, x_2, x_3)$ defined thus: if any $x_i \leq a$, then $f = \text{Min}[x_1, x_2, x_3]$; if $x_i \geq a$, $i = 1, 2, 3$, then $f = \text{Min}[(x_1 + x_2)/2, (3x_1 - a)/2, (3x_2 - a)/2, x_3]$.

22. (D) denotes the interior of D.

23. It is instructive to observe that, as shown by (6), the existence of a nonempty domain characterized by this system does not imply that f is an L^0-function.

holds over the set H, H being such that if $(x) \in H$, then every $(x)^{(k)} \in H_k$. The system (12) is unique.

PROOF: Observing that (10a) and (11a) prove the theorem for $n = 2$ and $\lambda = n$, we shall proceed by complete induction. We need to consider only two cases: (a) $0 < \epsilon \leq n - 1$, (b) $\epsilon = 0$, $\tau \geq 2$. But let us first observe that because of the continuity of X_i and the uniform dimensionality of Π, a strictly effective group at $(x) \in (\Pi)$ must be strictly effective over (Π).

Let notations be chosen so that Φ_n is effective or sufficient but not limitational according to whether (a) or (b) is the case. Let x_n^0 be such that $\Pi \cap E_{n-1}^0 \neq 0$. Then $f_n^0(x)$ defined over E_{n-1}^0 by $f_n^0 = f$ is an L^0-function of $n - 1$ variables. Let Π_n^0 be the set corresponding to Π for f_n^0. By the inductive assumption, the theorem is true for f_n^0, hence $(x) \in \Pi_n^0$ satisfies

(12a) $$q = h_1^0 = h_2^0 = \cdots = h_{g+1}^0,$$

and over H_n^0

(13a) $$q = \mathrm{Min}[h_1^0, h_2^0, \ldots, h_{g+1}^0],$$

where the superscript zero indicates that h_k^0 may depend on x_n^0. Clearly, $(\Pi \cap E_{n-1}^0) \subseteq \Pi_n^0$. Moreover, if $(x) \in (\Pi_n^0)$ then Φ_n is effective or sufficient at (x) according to whether (a) or (b) is the case. For otherwise X_n would not be a continuous function of $(x_1, x_2, \ldots, x_{n-1})$. Therefore, (12a) and (13a) are valid for any $(x) \in \Pi$. In addition, all h_k^0 are increasing functions of x_n^0 if (a) is true, and at least one h_k^0 is independent of x_n^0 if (b) is the case. It is also seen that (12a) is unique.   Q.E.D.

4. As closing remarks, I wish to point out, first, that from theorem II it follows that if Φ_i, $i \leq \lambda$ are the limitational factors, then (12) is

(14) $$q = h_1(x_1) = h_2(x_2) = \cdots = h_\lambda(x_\lambda) = h_{\lambda+1} = \cdots = h_{g+1},$$

where h_k, $k > \lambda$, is independent of every x_i, $i \leq \lambda$. This confirms the fact that an alternative definition of a limitational factor, say, Φ_j, is that x_j is a function of *output alone over* Π. System (14) shows also that over Π there exist also some relations between every x_j and the other variables. Hence, Palander's suggestion to call a factor limitational of the second kind if its input is a function of other inputs,[24] is not operational. The only remedy would be to equate limitationality of the second kind with strict technical complementarity: lock-and-key. But then, a new label becomes superfluous.

My second remark concerns type III which cannot be represented by a plastic body for the reason that it exists only for more than three

24. For the source, see note 4 above.

factors. (This may, perhaps, explain why Kaldor failed to mention it.) Yet this type is the most relevant of all.

Some years ago, I remarked that the two basic models used in economics to describe a going concern — one using only flow coordinates, the other, only stock coordinates — are unsatisfactory.[25] A going concern is first of all described by what *it is*, i.e., by some stock coordinates, S_1, S_2, ..., S_n; what it does, or rather what *it can do*, by some rate of flow coordinates, say, the input rates, r_1, r_2, ..., r_m. Assuming, for simplicity, complete substitutability between stocks, on the one hand, and inputs, on the other hand, the output rate at the proper factor-combinations satisfies $q = F(S_1, S_2, ..., S_n) = G(r_1, r_2, ..., r_m)$; in addition, $q = \text{Min}[F, G]$. Actually, any production function free from limitationality — as is most often the case — is of type III.[26] For this reason alone, it seems worthwhile to bother with L^0-functions.

25. Georgescu-Roegen, "The Aggregate Linear Production Function and Its Applications to von Neumann's Economic Model," in *Activity Analysis of Production and Allocation*, ed. T. C. Koopmans *et al.* (New York, 1951), pp. 100–101.

26. The numerous and interesting consequences of the above view of a going concern are further explored in the author's forthcoming monograph, *Process, Value, and Development*.

8

Relaxation Phenomena
in Linear Dynamic Models

In a short article published in the first volume of *Econometrica*, Ph. Le Corbeiller pointed out the possibility of using relaxation phenomena as a model for business cycles.[1] However, Le Corbeiller's suggestion has found little echo among economists, and the literature shows only sporadic references to his paper. Paul A. Samuelson, speaking of this possible approach, admits that practically nothing has been done along this line.[2] The only economic problem which could be regarded as having something to do with relaxation is the famous cobweb problem, but this has been developed independently of any relation to the concept of relaxation.

Le Corbeiller's article was inspired by the work of B. van der Pol.[3] Relaxation phenomena occupy an important place in modern physics. The difficulty in dealing with such phenomena is that, although most of them are of a *periodic* nature, this periodicity cannot be described by a sine curve. An example of a relaxation phenomenon is found in a hammer which strikes in a periodic way.[4] Obviously, describing such a movement involves additional difficulties in comparison with the case of the move-

NOTE: This paper is reprinted by courtesy of the publisher from *Activity Analysis of Production and Allocation*, ed. T. C. Koopmans *et al.*, Cowles Commission for Research in Economics Monograph No. 13 (New York: John Wiley & Sons, Inc., 1951), pp. 116–131. Along with the original printing I explained that the results contained in the paper were presented for the first time during two meetings of the staff of the Harvard Economic Research Project in April 1949, and I also said: "The author wishes to acknowledge the many inspiring discussions with Professor W. W. Leontief regarding the topics developed here. It is hardly necessary to add that, for any faults the chapter may contain, the author is solely responsible. The facilities at the Institute of Research and Training in the Social Sciences at Vanderbilt University extended to the author in preparing the final version are gratefully acknowledged."

1. Ph. Le Corbeiller, "Les oscillations de relaxation," *Econometrica*, I (1933), 328–332.

2. Paul A. Samuelson, *Foundations of Economic Analysis* (Cambridge, Mass., 1947), p. 339.

3. Balth. van der Pol, "On 'Relaxation Oscillations,'" *London, Edinburgh and Dublin Philosophical Magazine and Journal of Science*, II (1926), 978–992.

4. For other examples from the field of physics, see Ph. Le Corbeiller, *Les systèmes autoentrenus et les oscillations de relaxation* (Paris, 1931).

ment of a pendulum. The pendulum has a *symmetric* periodicity, the hammer an *asymmetric* one. This is due to the fact that the movement of the hammer can be decomposed into two distinct phases, one before and one after the energy is released through the shock. The two courses of the hammer, toward and away from the object being hit, take place under two *different regimes* which lead to two *different phases* of its movement.

However, not all movements that have an asymmetric periodicity are relaxation phenomena. A ball moving without friction over the surface of a washboard with asymmetric waves does not involve any relaxation in the sense used by the writers mentioned. In that sense, a relaxation phenomenon takes place only when the difference between the "up" and "down" swings is created by a certain *discontinuity in the regime*. Such a discontinuity will introduce a discontinuity in the *speed* of the movement (at least in size or in direction). Therefore the movements related to each phase will be described by a *different function*.[5]

The aim of van der Pol's contribution was to approximate these two different functions by a single analytic function. This was achieved by considering the periodic solutions of the differential equation

$$(1) \qquad \frac{d^2y}{dt^2} + \epsilon(y^2 - 1)\frac{dy}{dt} + y = 0$$

for large values of ϵ. By this procedure the analytical difficulty was solved from the practical point of view. But this veiled the real meaning of relaxation, which is the discontinuity of the regime. Indeed, as has already been pointed out, periodicity in the classical sense is a secondary aspect of the oscillations of relaxation. In economics, where most of the so-called periodic phenomena, such as business cycles, for instance, are treated as periodic phenomena in the classical sense only in order to simplify the problem, the discontinuity aspect retains its full significance. This point of view finds an admirable illustration in a dynamic model presented by Leontief in a paper read during February, 1949, before the staff of the Harvard Economic Research Project. The contribution contained in the present chapter has its origin in the author's attempt to answer one problem raised by Leontief in his paper.

The dynamic model presented by Leontief is an extension of his earlier static model.[6] It is defined by the system

5. From the point of view of the Dirichlet definition of a function, this distinction is not possible. However, the laws of the movement, being derived from certain differential equations of an algebraic, or at least analytical, structure, are, in general, analytic functions. As any analytic function has an individuality of its own, it is a perfectly justified attitude to regard two such functions as distinct.

6. Wassily W. Leontief, "Recent Developments in the Study of Interindustrial Relations," *American Economic Review*, Papers and Proceedings, XXXIX (1949), 218–221.

(F_1)
$$x_1 = a_{21}x_2 + b_{11}\dot{x}_1 + b_{21}\dot{x}_2,$$
$$x_2 = a_{12}x_1 + b_{12}\dot{x}_1 + b_{22}\dot{x}_2,$$

where x_1, x_2 are output flows, $\dot{x}_1$, $\dot{x}_2$ are derivatives with respect to time, and the a's and b's are constants.

The system (F_1) takes account of the fact that production requires both input flows and stocks (or inventories). In this particular formulation, inputs and stocks are supposed to be proportionate to outputs. The ratios between input flows and output flows are referred to as *input coefficients* (or a's), and the ratios between stocks and outputs are the *capital coefficients* (or b's).

The system (F_1) leads to the classical solution

(S_1)
$$x_1 = c_1 e^{\lambda_1 t} + c_2 e^{\lambda_2 t},$$
$$x_2 = c_1 u_1 e^{\lambda_1 t} + c_2 u_2 e^{\lambda_2 t},$$

where c_1 and c_2 are integration constants determined by the initial values of x_1 and x_2. Leontief assumes further that at a *turning point*, defined in terms of $\dot{x}_1$ (the change of the demand for x_1), a discontinuity is introduced in the behavior of the entrepreneurs such as to make $b_{11} = 0$ in (F_1). The next phase of the system will, therefore, be defined by the equations

(F_2)
$$x_1 = a_{21}x_2 + 0 + b_{21}\dot{x}_2,$$
$$x_2 = a_{12}x_1 + b_{12}\dot{x}_1 + b_{22}\dot{x}_2,$$

and will last until the demand for x_1 has reached a certain level determined by the one which existed when the second phase began.

The solution of (F_2) is

(S_2)
$$x_1 = c_1' e^{\mu_1 t} + c_2' e^{\mu_2 t},$$
$$x_2 = c_1' v_1 e^{\mu_1 t} + c_2' v_2 e^{\mu_2 t},$$

where the integration constants, c_1' and c_2', are determined by the splicing conditions, namely, that (S_2) must start from where (S_1) left off. Consequently, c_1', c_2' are indirectly determined by the initial values of x_1, x_2.

When the first phase (F_1) occurs again, if it ever does, the new values of the constants c_1 and c_2 will have to be determined by the new splicing conditions between (S_2) and (S_1), and so forth. It is clear that the problem when handled in this way raises almost insuperable technical difficulties, and one may even ask, as Leontief did, whether it would be possible to predict in *a finite number of steps* the final outcome of the system.[7]

7. An analogy, used by the author elsewhere, may aid in grasping the essence of such a question. It is known, for instance, that the decimal digits of the transcendental number π cannot be determined unless this is done step by step. For numbers such as $\frac{7}{23}$, any decimal digit can be determined in a finite number of steps. The question raised in the text is whether the system has the structure of π, i.e., so that the prediction of any phase requires the *actual* splicing (and knowledge) of the preceding one, or the structure of a rational number, i.e., so that any phase can be computed without necessarily knowing the preceding ones.

It is the purpose of the present chapter to offer a way of answering such a question and to develop a method which can be applied to the study of almost all economic problems where a relaxation phenomenon is present. A more general concept of periodicity, of undoubted importance for economic theory, will also be introduced.[8]

1. Let us think of a system, (Σ), which is subject to two different regimes, R_1 and R_2. Assume that two rules, r_1 and r_2, are also given which govern the switching of the system from regime R_1 to R_2 and vice versa. If from a given initial position, (Σ_0), the system develops under regime R_1,[9] the movement of the system is completely determined up to $t = \infty$.[10]

The evolution of the system may be described by the sequence

(F) $\qquad\qquad\qquad F_1, F_2, F_1, F_2, \ldots,$

where F_1 and F_2 represent the phases corresponding to R_1 and R_2. The sequence (F) may be finite or infinite.

Such a scheme constitutes a special type of periodicity, which we shall refer to as *phase-periodicity*. This is a generalized concept of the classical *point-periodicity*. The economic cycles seem to be better described by phase-periodicity than by point-periodicity, since the relevant aspect of the business cycle is the recurrence of the phases and not the repetition, after a constant time-lag, of the *same* values.

Phase-periodicity, as defined above, implies a relaxation phenomenon every time the phase changes. Moreover, it is more general than the relaxation oscillations considered in physics, the latter usually being only point-periodic.

The first problem which arises in connection with a phase-periodic scheme is that of finding out whether or not the sequence (F) is finite, and further, in case (F) is infinite, to determine whether or not the system has an asymptotic movement. If it has, economists would say that the dynamic model tends toward a unique equilibrium.

Different types of schemes may be conceived. We shall deal with those which are more likely to be met in economic dynamic models.

8. The author wishes to emphasize that the object of this chapter is not to appraise the merits, from the point of view of economic theory, of the models used here to illustrate the analytical method devised to deal with relaxation phenomena in economics. Moreover, in the case of Leontief's model this would have been impossible because Leontief's contribution is not yet available in print.

9. This assumption is absolutely necessary. The choice between R_1 and R_2 as the initial regime is therefore supposed to be made according to some outside criteria if the whole evolution is to be considered as having a beginning. This would no longer be necessary if the system were considered as already in movement, in which case the sequence (F) below has no beginning. In some cases, however, a given initial condition may be compatible with only one of the two regimes, R_1, R_2.

10. Provided, however, that situations do not exist where the rule r_1 (or r_2) becomes self-contradictory.

2. A simple model to illustrate the relaxation phenomena would be one where capital accumulation during the up-swing period of the business cycle would obey a dynamic law different from that governing capital decumulation during the down-swing. Such an approach would, indeed, be more realistic than that of most dynamic models constructed so far, which assume that both phases are governed by a reversible law.[11]

Let us assume that the aggregate dynamic laws governing the accumulation (or decumulation) of capital during the two phases of the cycle are represented, respectively, by the equations

$$(2) \qquad \frac{dC}{dt} = f_1(C), \qquad \frac{dC}{dt} = f_2(C),$$

thus making the change in capital stock a function of the existing stock.

The problem of describing the evolution of the system has a simple and immediate solution if the system of partial differential equations,

$$(3) \qquad \frac{\partial C}{\partial t} = f_1(C), \qquad \frac{\partial C}{\partial \tau} = f_2(C),$$

is integrable. Let

$$(4) \qquad\qquad C = \varphi(t, \tau)$$

be the integral of (3), and t_k and τ_k, the lengths of the kth first phase and of the kth second phase, respectively. These lengths are determined by the rules r_1 and r_2. The movement of the system is described in terms of the single function (4):

$$(5) \qquad
\begin{aligned}
C &= \varphi\left(\sum_1^n t_i + t, \sum_1^n \tau_i\right) \qquad (0 \leqq t \leqq t_{n+1}), \\
C &= \varphi\left(\sum_1^n t_i, \sum_1^{n-1} \tau_i + \tau\right) \qquad (0 \leqq \tau \leqq \tau_n).
\end{aligned}$$

In three-dimensional space function (4) represents a surface and rules r_1 and r_2 will correspond generally to two curves (r_1) and (r_2), as in Fig. 8–1. The movement of the system will be described by a path, $C_0, C_1, C_2, \ldots$, on the surface φ. Therefore the whole problem of phase-periodicity can be followed by considering the path $c_0, c_1, c_2, \ldots$, on the $t0\tau$ plane.

It is seen that the value of C at any moment depends only on the lengths of time obtained by summing time separately for each phase.

11. This point was clearly emphasized by Leontief in the paper read before the staff of the Harvard Economic Project to which I referred earlier in the text. Leontief pointed out the deep significance of the turning point in a business cycle, which goes far beyond the mere formal aspect of changing slope. Use of Leontief's argument is made here solely to illustrate the usefulness of the relaxation concept in economic theory.

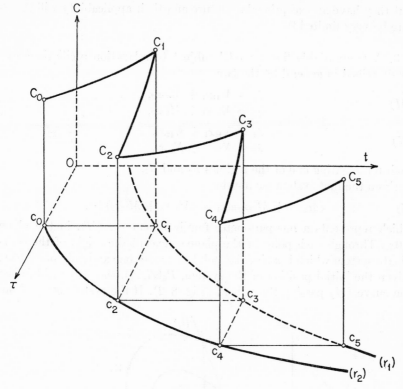

FIGURE 8–1

Therefore the complete evolution of the system can be predicted in a finite number of steps.

It is easily seen that the system will tend toward an equilibrium if either one or the other of the following two conditions is fulfilled:

(a) the curves (r_1) and (r_2) meet at a point E, at a finite distance or at infinity (i.e., they are asymptotic);

(b) the two curves (r_1) and (r_2) do not meet, but, for $t = +\infty$,

(6) $$\lim_{\text{along } r_1} \varphi = \lim_{\text{along } r_2} \varphi.$$

If E is at a finite distance, the equilibrium will be reached over a finite period of time; in all other cases the system will only approach equilibrium.[12]

Economic models can be found which fulfill the integrability condition,

12. Such a conclusion implies evidently that rules r_1 and r_2 never lead to a contradiction. Such would be the case if, in order to go from (r_1) to (r_2) along $c_0, c_1, c_2, \ldots,$ it were necessary to move in the negative direction of time.

but they have an exceptional structure and their applicability will therefore be very limited.[13]

3. A two-variable linear model subject to relaxation oscillations can be described in general by the two systems

$$(M) \qquad \begin{aligned} \dot{x}_1 &= M_{11}x_1 + M_{12}x_2, \\ \dot{x}_2 &= M_{21}x_1 + M_{22}x_2, \end{aligned}$$

$$(N) \qquad \begin{aligned} \dot{x}_1 &= N_{11}x_1 + N_{12}x_2, \\ \dot{x}_2 &= N_{21}x_1 + N_{22}x_2, \end{aligned}$$

each representing one of the regimes R_1 and R_2.

From the first system we obtain

$$(7) \qquad (M_{21}x_1 + M_{22}x_2)dx_1 = (M_{11}x_1 + M_{12}x_2)dx_2,$$

which represents a one-parameter family of curves, (M), in the plane $x_1 0 x_2$. Through each point of the plane — except the origin of the coordinate system which is a singular point — passes one and only one curve. Given the initial position of the system, $P_0(x_1^0, x_2^0)$, the latter will follow the curve (M) passing through P_0 (Fig. 8–2). If the system starts from

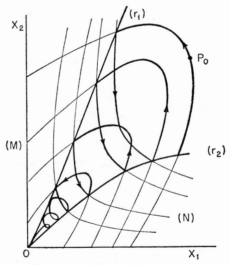

FIGURE 8–2

any point belonging to the *same* curve (M), it will have the same evolution. This is why the curves (M) will be called *isodromes*.

13. As Samuelson, who read the manuscript, further pointed out, in such models the rates of capital accumulation and decumulation at the same level C are in a constant ratio (which obviously must be negative). This follows immediately from the integrability condition of (3): $f_1'(C)/f_1(C) = f_2'(C)/f_2(C)$. Hence $f_1(C) = kf_2(C)$.

It is worth stressing that the evolution of the system described by an isodrome is *time-less* but not *sequence-less*. This does not disturb completely the utility of such a description for economic theory, for the latter deals very often with dynamic formulations of this kind. Indeed, many economic statements tell *what* will happen next but do not consider the more delicate question of predicting also exactly *when*.

Equation (7) is not, however, equivalent to (M), for (7) determines only a family of curves and the *isodromes* must have a *direction*. Therefore (7) must be supplemented by information regarding the direction of movement. This can be derived from (M) and formulated in terms of the sign of dx_1 (or dx_2).[14]

The shape of the isodromes (M) depends, as is known, on the nature of the roots, λ_1, λ_2, of the characteristic equation

$$(8) \qquad \begin{vmatrix} M_{11} - \lambda & M_{12} \\ M_{21} & M_{22} - \lambda \end{vmatrix} = 0.$$

In drawing the shapes of the isodromes it is a great help to keep in mind the properties of the straight lines

$$(9) \qquad M_{11}x_1 + M_{12}x_2 = 0, \qquad M_{21}x_1 + M_{22}x_2 = 0,$$

representing the loci of the points for which $dx_1 = 0$, $dx_2 = 0$, as well as those of the lines

$$(10) \qquad (M_{11} - \lambda_1)x_1 + M_{12}x_2 = 0, \qquad (M_{11} - \lambda_2)x_1 + M_{12}x_2 = 0,$$

representing the loci of the points for which the changes in outputs (dx_1, dx_2) are proportionate to the outputs (x_1, x_2). [The lines (10) are not necessarily real.]

A second family of isodromes is determined by (N). If the curves (r_1) and (r_2) are added to the picture, then, from any initial position, the evolution of the system (x_1, x_2) is perfectly determined. This is described in general by a cobweb-like path (Fig. 8–2) with finite or infinite turning points, depending on the particular shapes of the curves here involved and on the initial position.

A simple illustration of the above scheme is the famous cobweb problem of supply and demand. In this case the isodromes (M) and (N) are given, respectively, by the differential equations

$$(11) \qquad dx = 0, \qquad dy = 0,$$

where x is price and y is quantity. The rules r_1 and r_2 are represented by the demand and supply curves,

$$(12) \qquad x = D(y), \qquad y = S(x).$$

14. The isodromes will not be altered if the right-hand sides of the system (M) are multiplied by the same (arbitrary) function of (x_1, x_2) of constant sign.

The direction of movement on the isodromes (M) is toward (r_1), and on (N) toward (r_2).

4. The Leontief model is a particular case of the model considered in the preceding section.

From (F_1) and (F_2) we obtain

(13)
$$|b|\dot{x}_1 = \beta_{11}x_1 + \beta_{12}x_2,$$
$$|b|\dot{x}_2 = \beta_{21}x_1 + \beta_{22}x_2,$$

where

(14)
$$\beta_{11} = b_{22} + a_{12}b_{21}, \qquad \beta_{12} = -(b_{21} + a_{21}b_{22}),$$
$$\beta_{21} = -(b_{12} + a_{12}b_{11}), \qquad \beta_{22} = b_{11} + a_{21}b_{12}.$$

An alternative formulation of the same system can be made in terms of stocks. The *proper* stock of the commodity G_i in the industry producing G_k is $b_{ki}x_k$. Therefore, X_1 and X_2 being the total stocks,

(15)
$$X_1 = b_{11}x_1 + b_{21}x_2, \qquad X_2 = b_{12}x_1 + b_{22}x_2.$$

Furthermore,

(16)
$$\dot{X}_1 = x_1 - a_{21}x_2, \qquad \dot{X}_2 = x_2 - a_{12}x_1.$$

Eliminating x_1 and x_2, we obtain a system similar to (13), namely

(17)
$$\begin{vmatrix} X_1 & b_{11} & b_{21} \\ X_2 & b_{12} & b_{22} \\ \dot{X}_1 & 1 & -a_{21} \end{vmatrix} = 0, \qquad \begin{vmatrix} X_1 & b_{11} & b_{21} \\ X_2 & b_{12} & b_{22} \\ \dot{X}_2 & -a_{12} & 1 \end{vmatrix} = 0.$$

It is further seen that

(18)
$$|\beta| = |a|\,|b|.$$

We shall not deal with the case $|b| = 0$, where the system (13) is either impossible or indeterminate. But let us consider the special case where $|a| = 0$. This leads to

(19)
$$(b_{11} + b_{12}a_{21})a_{12}dx_1 + (b_{22} + b_{21}a_{12})dx_2 = 0,$$

provided $x_1 - a_{21}x_2 \neq 0$. Therefore the (M) isodromes are parallel straight lines. It is seen that the direction of movement depends on the sign of $|b|$, except in the case where the initial position of the economic model lies on the straight line $x_1 - a_{21}x_2 = 0$ (or $x_2 - a_{12}x_1 = 0$). In this case we have $dx_1 = 0$, $dx_2 = 0$. This means that the straight line $x_1 - a_{21}x_2 = 0$ is the locus of static equilibria. If the system is not in static equilibrium from the beginning, the system will tend toward such an equilibrium along an isodrome (19) if $|b| < 0$, or away from static equilibrium if $|b| > 0$. (Fig. 8–3 represents the case $|b| < 0$.) If $|b| > 0$, we have a typical case of unstable static equilibrium.

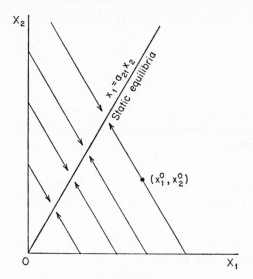

FIGURE 8–3

Furthermore, if the initial position is not one of static equilibrium, the economic system will not be static, nor will it expand at a constant rate of growth. Nevertheless the rate of interest will be zero. This constitutes a peculiar example of the rate of interest being zero, though the stocks are not constant. Only their total value remains constant.

If i is the instantaneous rate of interest and p_1, p_2 the prices of the two commodities, the equality between prices and average cost leads to

(20)
$$(b_{11}p_1 + b_{12}p_2)i + a_{12}p_2 = p_1,$$
$$(b_{21}p_1 + b_{22}p_2)i + a_{21}p_1 = p_2.$$

As prices must be positive, we have

(21)
$$e(i) = \begin{vmatrix} b_{11}i - 1 & b_{12}i + a_{12} \\ b_{21}i + a_{21} & b_{22}i - 1 \end{vmatrix} = 0$$

and

(22) $(b_{11}i - 1)(b_{12}i + a_{12}) < 0,$ $(b_{21}i + a_{21})(b_{22}i - 1) < 0;$

the last two relations being equivalent if taken together with (21). Let

(23) $i_0 = $ the greatest $\left(-\dfrac{a_{ik}}{b_{ik}}\right),$ $i_1 = $ the smallest $\left(\dfrac{1}{b_{ii}}\right).$

As

(24) $e(i_0) > 0,$ $e(i_1) < 0,$

(21) has a root between i_0 and i_1. It is easily seen that this root is the

only one which satisfies (22). This, therefore, is the equilibrium rate of interest. But

$$(25) \qquad e(0) = |a| = \begin{vmatrix} -1 & a_{12} \\ a_{21} & -1 \end{vmatrix}.$$

Hence the rate of interest will be positive or negative according to whether $|a| >$ or < 0. When $|a| = 0$, the rate of interest is zero.

From (20) it follows that the rate of growth of the value of total stocks is equal to the rate of interest,

$$(26) \qquad (p_1 X_1 + p_2 X_2)i = p_1 \dot{X}_1 + p_2 \dot{X}_2.$$

If $i = 0$, the value of the stocks, $p_1 X_1 + p_2 X_2$, remains constant.

Let us now consider the case in which $|a| \neq 0$. The characteristic equation of (13),

$$(27) \qquad \psi(\lambda) = |b|\lambda^2 - (\beta_{11} + \beta_{22})\lambda + |a| = 0,$$

has the real roots λ_1 and λ_2, since

$$(28) \qquad (\beta_{11} + \beta_{22})^2 - 4|a|\,|b| = (\beta_{11} - \beta_{22})^2 + 4\beta_{12}\beta_{21} > 0.$$

There are several alternatives, which shall be examined in turn.

(a) $|b| > 0$, $|a| > 0$. This leads to $\lambda_2 > \lambda_1 > 0$. From (28) it follows that

$$(29) \qquad \beta_{ii} - \lambda_1|b| > 0, \qquad \beta_{ii} - \lambda_2|b| > 0,$$

and consequently the straight line (δ_k),

$$(30) \qquad (\beta_{11} - \lambda_k|b|)x_1 + \beta_{12}x_2 = 0$$

or

$$\beta_{21}x_1 + (\beta_{22} - \lambda_k|b|)x_2 = 0,$$

lies in the positive quadrant only for $k = 1$; it also lies between

$$(31) \qquad \begin{array}{l} (\Delta_1) \quad |b|\dot{x}_1 = \beta_{11}x_1 + \beta_{12}x_2 = 0, \\ (\Delta_2) \quad |b|\dot{x}_2 = \beta_{21}x_1 + \beta_{22}x_2 = 0. \end{array}$$

All the isodromes are tangent to (δ_1) in O. The direction of movement is away from the origin, as in Fig. 8–4(a).

If the initial position belongs to (δ_1),

$$(32) \qquad \dot{x}_1 = \lambda_1 x_1, \qquad \dot{x}_2 = \lambda_1 x_2,$$

and the system will expand at a constant rate of growth. In all other cases, sooner or later, one of the outputs, x_1 or x_2, will expand at the expense of the other until this becomes zero. The dynamic equilibrium is, in Samuelson's sense, unstable.[15]

(b) $|b| < 0$, $|a| < 0$. This leads to $\lambda_2 < \lambda_1 < 0$. It is seen, in a similar

15. Samuelson, *Foundations*, pp. 266 ff.

way, that the shape of the isodromes is the same as those of (a), with the exception that the direction of movement is toward the origin. The economy will end in all cases by contracting toward zero production. The dynamic equilibrium is stable.

(c) $|b| < 0$, $|a| > 0$. This leads to $\lambda_2 < 0 < \lambda_1$. The shape of the isodromes is that shown in Fig. 8-4(b). In all cases the system ends by expanding and tends toward the same equilibrium position represented by a point at infinity on (δ_1). The dynamic equilibrium is stable.

(d) $|b| > 0$, $|a| < 0$. This leads to $\lambda_1 < 0 < \lambda_2$. The shape of the isodromes is similar to that of Fig. 8-4(b), but the sense of the movement is reversed. The dynamic equilibrium is unstable.

Summing up the preceding results shows that (i) the sign of $|b|$ decides the dynamic stability of the system, and (ii) the sign of $|a|$ determines the sign of the rate of interest (i.e., whether the system is contracting or expanding).

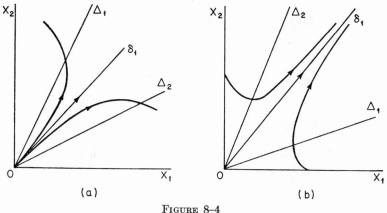

FIGURE 8-4

It is also seen that the existence of a static solution in von Neumann's sense[16] (i.e., where the economy will expand (or contract) proportionately in all sectors) is possible only if the initial position lies on (δ_1).

If the system (F_2) is now considered, it is seen that $|b|$ becomes

$$(33) \qquad b' = -b_{12}b_{21} < 0,$$

and consequently the system will be dynamically stable [cases (b) and (c)]. Furthermore, (13) becomes

$$(34) \qquad \begin{aligned} b'\dot{x}_1 &= \beta_{11}x_1 + \beta_{12}x_2, \\ b'\dot{x}_2 &= -b_{12}x_1 + a_{21}b_{12}x_2. \end{aligned}$$

Consequently Δ_1 in (F_2) is identical with Δ_1 in (F_1).

16. John von Neumann, "A Model of General Economic Equilibrium," *Review of Economic Studies*, XIII (1945-46), 1-9.

Leontief's rules of change from (F_1) to (F_2) and vice versa are formulated, respectively, in terms of $\dot{x}_1$ and of the value of x_1 at the beginning of (F_2). It is easy to see, by using the results of (a)–(d) above, that these rules permit one change at most from (F_1) to (F_2). This is due to the fact that the difference between (F_1) and (F_2) does not affect the sign of the rate of growth since $|a|$ has the same sign in both phases.

5. A better illustration of the phase-periodicity is provided by the Hansen-Samuelson model.[17] For a continuous formulation of this model let y be the flow of national income at the time t; c, consumption; I, private investment; and g, governmental expenditure. The Hansen equations can be written[18]

$$c = \alpha y - \alpha'\dot{y},$$

(35)
$$I = \beta'\dot{c},$$

$$y = I + c + g,$$

with $0 < \alpha \leqq 1$, and α', β' positive.

We obtain, further,

(36)
$$\alpha'\dot{y} = \alpha y - c,$$
$$\beta'\dot{c} = y - c - g,$$

which, through the simple transformations

(37) $t = \dfrac{\alpha'T}{\alpha}, \qquad \beta = \dfrac{\alpha\beta'}{\alpha'}, \qquad y = Y + \dfrac{g}{1-\alpha}, \qquad c = C + \dfrac{\alpha g}{1-\alpha},$

where T is a new time unit, becomes

(38)
$$\alpha\dot{Y} = \alpha Y - C,$$
$$\beta\dot{C} = Y - C.$$

This is a linear dynamic model of type (M), the characteristic equation being

(39) $$\alpha\beta\lambda^2 + (\alpha - \alpha\beta)\lambda + 1 - \alpha = 0.$$

Under the assumption that $\alpha \leqq 1$, the solution of (38) depends on the position of (α, β) with respect to the curve (Fig. 8–5).[19]

17. Paul A. Samuelson, "Interactions Between the Multiplier Analysis and the Principle of Acceleration," *Review of Economic Statistics*, XXI (1939), 75–78.

18. Samuelson's formulation is discontinuous (*ibid.*, p. 76). It must be remarked that Samuelson's formulation, and therefore (35), does not reveal the source of g (*ibid.*, table 1). A model in which g is treated in a more explicit way would constitute a better analytical tool. The incidence of tax shifts are considered in the concluding part of this chapter.

19. Cf. Samuelson, "Interactions," p. 78. The regions considered here differ somewhat from those used by Samuelson.

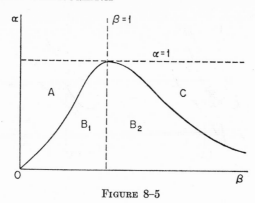

FIGURE 8–5

(40)
$$\alpha = \frac{4\beta}{(1 + \beta)^2}.$$

The shapes of the isodromes are represented in Fig. 8–6 according to the region where (α, β) lies. The result can be summarized as follows:

Region	Stability[20]
A	Perfect
$A \cap B_1$	Perfect
B_1	Perfect
$B_1 \cap B_2$	Unstable (cyclic)
B_2	Unstable
$B_2 \cap C$	Partially stable
C	Partially stable

If we now assume that, soon after income reaches its maximum (i.e., $\dot{Y} = 0$), the propensity to consume, α, increases,[21] or β, "the relation," decreases, or both these things happen simultaneously, there will be a turning point at which a relaxation phenomenon will take place. As an illustration, let us assume that only β decreases from $\beta = 1$ to a point in B_1. The system will become a contracting one instead of a cyclic one. If we assume still further that, as the consumption-income ratio reaches

20. Stability is considered here in Samuelson's sense (i.e., stability is perfect when any displacement, although shifting the system to another isodrome, will not change the limit toward which the system tends). This limit may be regarded as a possible static solution. The term "partially stable" refers to the case where the system will tend toward the same limit only for some (not all) finite displacements. Samuelson, *Foundations*, p. 262.

21. James Duesenberry, "Income-Consumption Relations and Their Implications," in *Income, Employment and Public Policy: Essays in Honor of Alvin H. Hansen* (New York, 1948), pp. 54–81; Franco Modigliani, "Fluctuations in the Saving-Income Ratio: A Problem in Economic Forecasting," in *Studies in Income and Wealth*, vol. XI, National Bureau of Economic Research (New York, 1949), pp. 371–441.

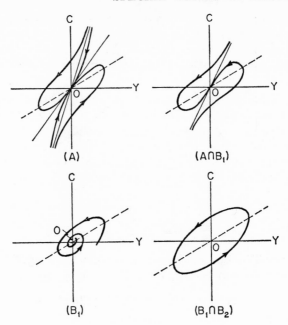

FIGURE 8–6

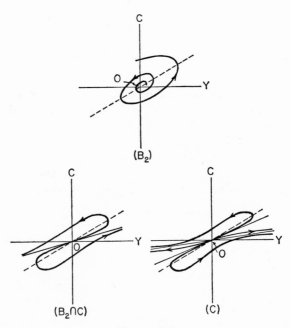

FIGURE 8–7

a certain level, the value of β will recover its former value, the system will follow a cobweb path which will lead, generally but not necessarily, to a contraction towards the origin.

Finally, another illustration of the usefulness of the analysis based on isodromes is provided by the question whether a fiscal policy aimed at decreasing consumption and increasing investment could definitively cure a system having a tendency to consume all its income. The answer is easily obtained by inspecting the shape of the isodromes in Fig. 8–7. If the assumptions underlying the model and also those regarding the invariability of α and β under changing fiscal policies are accepted, it is seen that in the case of C and $B_2 \cap C$ one shift through taxation may be sufficient to cure definitively the tendency of the system toward disinvesting. In all other cases the tax must be applied periodically. Furthermore, in order to insure the shifting of the system to a "higher" isodrome by means of only a small tax, the tax must be applied immediately *before* the income reaches its maximum (i.e., before the isodromes reach the straight line $\alpha Y - C = 0$).

9

Some Properties of a
Generalized Leontief Model

1. The model presented by Leontief in his writings[1] is based on, among other things, the assumption that each commodity can be produced by one method of production only, This is equivalent to assuming that all factors of production are *limitational*. Because of this, the model will be referred to as the Leontief *limitational model* and be denoted by (L).

Samuelson and the author have independently considered the possibility of a Leontief model from which the limitationality restriction could be removed. Such a model will be referred to as a Leontief *generalized model* and be denoted by (GL).[2]

2. A Leontief generalized model is defined by the following assumptions:

NOTE: This paper is now published in full for the first time, in a form that contains some improvements of exposition in comparison with the version that, for lack of space, could not be included in *Activity Analysis of Production and Allocation*, ed. T. C. Koopmans *et al.*, Cowles Commission for Research in Economics Monograph No. 13 (New York, 1951), where only an abstract appeared, pp. 165–173. (See *ibid.*, p. 9.) Along with that abstract I explained that the results contained in it were presented for the first time on March 22, 1949, at a meeting of the staff of the Harvard Economic Research Project, and I added in the same footnote: "The criticism of Professor W. W. Leontief and of other members of the Harvard Economic Research Project showed to the author the path for an ameliorated formulation of the argument related to the consolidation problem. It is hardly necessary to add that, for any faults the chapter may contain, the author alone is responsible. The facilities of the Institute of Research and Training in Social Sciences at Vanderbilt University extended to the author in preparing the final version are gratefully acknowledged."

1. Wassily W. Leontief, "Quantitative Input and Output Relations in the Economic System of the United States," *Review of Economic Statistics*, XVIII (1936), 105–125; "Interrelations of Prices, Output, Savings, and Investment," *ibid.*, XIX (1937), 109–132; *The Structure of the American Economy, 1919–1929* (Cambridge, Mass., 1941).
2. The proof presented by Paul A. Samuelson in chap. vii [that is, chap. vii of *Activity Analysis of Production and Allocation*, as cited in "NOTE," above] is substantially equivalent to Corollary 8.3 and Theorem 9 below. The proofs of 8.3 and 9 do not require, however, the existence of derivatives. Alternative proofs which do not require the existence of derivatives have been given by T. C. Koopmans, *ibid.*, chap. viii and Kenneth J. Arrow, *ibid.*, chap. ix.

ASSUMPTION I: *There are $n + 1$ perfectly defined and homogeneous commodities, $G_1, G_2, \ldots, G_{n+1}$. The commodity G_{n+1} is labor.*

This assumption contains a relevant economic restriction. A heterogeneous product can be replaced by a number of homogeneous commodities so that the above assumption is always fulfilled. This is no longer possible for labor, since the model allows for only one quality of labor.

ASSUMPTION II. *Production takes place in units of production.*

ASSUMPTION III. *The technological information contains no formula for producing several commodities by the same unit of production, but contains at least one production formula for each G_k, $1 \leq k \leq n + 1$.* This assumption makes legitimate the concept of the industry producing G_k.

ASSUMPTION IV. *The production of any G_k, $k \neq n + 1$, requires a positive input of labor.* Labor, therefore, is an *indispensable* factor to industrial production. By Assumption III, it also is the only *primary* factor.

ASSUMPTION V. *The production of G_{n+1} requires at least one positive input.*

ASSUMPTION VI. *Any production formula can always be used simultaneously by any number of production units.*

This stringent assumption constitutes the main characteristic of the Leontief system. It eliminates every kind of interaction between the formulae used not only by units producing the same G_k but also by units producing different commodities.

Because of this assumption, in an industry consisting of a number of production units using *exactly* the same production formula, the output and input flows are proportional to the number of the units. The process represented by such an industry as the number of the production units vary, will be referred to as a *primary* process. Clearly, the process exists only for discrete values of its scale. For analytical convenience, we shall replace it by a continuous pattern — a common artifice in all sciences. (The artifice works quite well as long as the unit does not become prominent but loses its significance in a mass of identical units.) The result is that *a primary process becomes linear in terms of input and output flows for any scale.*

Given a set of primary processes each working at some definite scale, by a *bookkeeping* integration we can represent their ultimate result by a single vector. By Assumption VI, each of the primary processes can work also at a multiple of their initial scales. The same bookkeeping integration will then result in a scalar multiple of the same vector. We have thus come to construct a new linear process, which may or may not be a primary process. Moreover, in the same manner we can integrate any set of processes whether primary or integrated from primary processes. If the result is not a primary process, then the new process will be called a *derived* process. The point is that actual production always takes place through a primary process. On the other hand, a derived process represents only a bookkeeping consolidation of primary processes. By Assumption III, a joint-product process is a derived process. But a one-product process too may be a derived process.

The set of all primary and derived processes form the technological horizon of the given technological information.[3]

In (L) Assumption III is replaced by

ASSUMPTION IIIa: *According to the technological information, each commodity G_k can be produced by only one formula.*

3. We shall restrictively use the notation

$$(1) \qquad P^{(k)}(-a_1^{(k)}, -a_2^{(k)}, \cdots, -a_{k-1}^{(k)}, b_k, -a_{k+1}^{(k)}, \cdots, -a_{n+1}^{(k)})$$

for a primary process of the industry G_k. The a's are input flows, and the b's are output flows. According to Assumptions IV and V

$$(2) \qquad b_k > 0, \qquad a_{n+1}^{(k)} > 0, \qquad a_i^{(k)} \geqq 0 \qquad (i \leqq n),$$

for $k \leqq n$; and

$$(3) \qquad b_{n+1} > 0, \qquad \text{at least one} \qquad a_i^{(n+1)} > 0,$$

for $k = n + 1$. Because of Assumption VI we can always take $b_k = 1$ in (1).

The model including the processes producing all $G_k (k = 1, 2, \ldots, n + 1)$ is a *closed* model. If the processes producing G_{n+1} (labor) are excluded, the model is *open* with respect to labor. Let H and H' be, respectively, the technological horizons of the closed and of the open model.

The process

$$(4) \qquad \pi^{(k)}(0, 0, \ldots, 1, 0, \ldots, 0, -l_k), \qquad (k = 1, 2, \ldots, n),$$

will be called a *completely integrated* process of commodity G^k.[4] If

$$(5) \qquad L_k = \text{greatest lower bound of } l_k$$

for all completely integrated processes $\pi^{(k)}$ belonging to H', the process

$$(6) \qquad \Pi^{(k)}(0, 0, \ldots, 1, 0, \ldots, 0, -L_k)$$

will be referred to as the *most efficient completely integrated process* of commodity G_k.

4. From Assumption IV it follows that, if H' contains a completely

---

3. For further details on linear processes and their integration see T. C. Koopmans, "Analysis of Production as an Efficient Combination of Activities," chap. iii in *Activity Analysis of Production and Allocation*, cited above, and Georgescu-Roegen, "The Aggregate Linear Production Function and Its Applications to von Neumann's Economic Model," *ibid.*, chap. iv. The latter paper also introduces the concepts of technological information and technological horizon.

4. The position of 1 is determined by the superscript k (i.e., $a_i^k = 0$ for $i \neq k, n + 1$).

integrated process $\pi^{(k)}$, then $L_k \geq 0$. According to the same assumption, if $L_k = 0$, then $\Pi^{(k)} \notin H'$ and H' is an *open* cone.[5] *Prima facie*, the relevance of the latter case may seem to be confined to some fine points of analysis. However, this is not so. The case of an open H' is intimately connected with an assumed property of labor in support of which the modern achievements of automation are, quite aptly, invoked. The assumption is that *labor, while an indispensable factor in the production of any commodity, can be substituted by other factors beyond any limit.* Or in other words, an output of any size can be obtained by any industry with as little labor as we may wish, provided that nonlabor inputs are available in unlimited amounts. Since this property presents an obvious analogy with that of a catalyst in a chemical reaction, we propose to say that labor is *catalytic* in industry G_k if it has the mentioned property in that industry.

The following numerical illustration has been constructed mainly for the purpose of providing a convenient background for the mathematical intricacies forced into the analysis of (GL) by the case of an open H'. But, because of the connection between this case and catalytic labor, the same illustration will also clarify some issues raised by this last notion.

Let the technological information consist of the production functions

$$(7) \quad y_1 = (y_2 + \sqrt{y_2^2 + 4\alpha y_2 y_3})/2\alpha, \qquad y_2 = (y_1 + \sqrt{y_1^2 + 4\beta y_1 y_3})/2\beta,$$

for the commodities G_1 and G_2, $(\alpha, \beta > 0)$, and let G_3 be "labor." For $y_3 = 0$ we obtain the processes $S_1(1, -\alpha, 0)$, $S_2(-\beta, 1, 0)$, which cannot possibly belong to H'. However, if (7) are valid for any $y_3 > 0$, labor is catalytic in both industries.

It is convenient to represent the relationships between inputs and outputs by the curves (I_1), (I_2) in the plane $x_1 0 x_2$ (Fig. 9–1). The equations of these curves are obtained from (7) by making $y_3 = 1$ and changing the sign of the input variable in each case:

$$(I_1) \qquad\qquad x_1 = (-x_2 + \sqrt{x_2^2 - 4\alpha x_2})/2\alpha;$$

$$(I_2) \qquad\qquad x_2 = (-x_1 + \sqrt{x_1^2 - 4\beta x_1})/2\alpha.$$

The curves are asymptotic to the straight lines

$$(8) \qquad\qquad \alpha x_1 + x_2 - \alpha = 0, \qquad x_1 + \beta x_2 - \beta = 0.$$

The relevant alternatives are three:

1) $\alpha\beta = 1$. In this case, (8) are parallel, as in Fig. 9–1(b). If $\alpha = \beta = 1$, (8) coincide, as in Fig. 9–1(a). In the latter case, $\Pi^{(1)}(1, 0, -1)$ and $\Pi^{(2)}(0, 1, -1)$, do not belong to H', even though $L_k > 0$. The reason is simple: $\Pi^{(1)}$, for instance, is the limit of $\pi^{(1)}$ as φ_1 and φ_2 (from which $\pi^{(1)}$

5. The term "open" has here the meaning used in point set theory and should not be confused with the expression "open model."

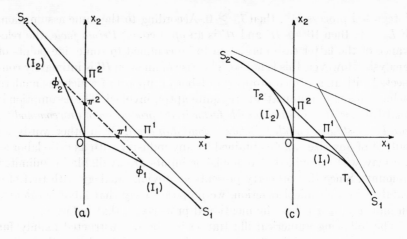

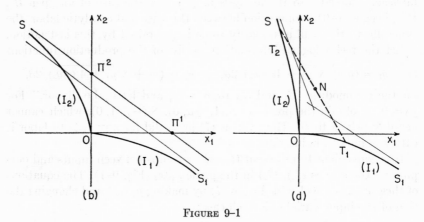

FIGURE 9–1

can be obtained by integration) tend toward infinity along (I_1) and (I_2). By Assumption VI, φ_1 is equivalent to

$$\left[\frac{1}{2}\left(1 + \sqrt{1 + \frac{4\alpha}{|x_2|}}\right), -\alpha, -\frac{\alpha}{|x_2|} \right],$$

which for $|x_2| \to \infty$ tends toward S_1. But S_1 does not belong to H', i.e., it is not *achievable*. Nor is any process on $S_1 S_2$ achievable. However, actual production may tend indefinitely toward S_1 or S_2. This may create the illusion that labor is catalytic for the economy as a whole. Yet we see that labor is not catalytic in the sense that any national net product can be obtained with as little labor as we wish. The same conclusions apply with even greater force if $\alpha, \beta \neq 1$; see Fig. 9–1(b).

2) $\alpha\beta > 1$. In this case the asymptotes meet in such a way that $\Pi^{(1)}$ and

$\Pi^{(2)}$ always belong to H'; see Fig. 9–1(c). Indeed, either process is obtained by integrating two achievable processes, T_1 and T_2. Again, labor is not catalytic for the whole economy. But, since in this case the limit processes can be effectively attained, the illusion mentioned earlier will not last for long.

3) $\alpha\beta < 1$. The relative position of the asymptotes is such that any process, N, in the positive quadrant $x_1 0 x_2$ can be achieved with $y_3 = 1$, for any such process can be decomposed into two achievable processes, T_1, T_2, as in Fig. 9–1(d). This also means that any national product is obtainable with only one unit of labor. Consequently, only if the third alternative prevails is labor catalytic for the economy as a whole. But whether this is the actual case or not depends upon other features of technology than that expressed by "labor is catalytic in every individual process." For, as the preceding illustrations show, while automation may do wonders in each industry, its over-all result may not be of the same nature: more labor may be required to produce an additional automaton than can be saved through its use.

Finally, the illustration pinpoints the reason why Samuelson's formulation of the Substitutability Theorem[6] is inexact and his proof is not quite rigorous (cf. Remark 2 below).

5. The most basic features of (GL) are described by the following two theorems:

THEOREM 1: *A necessary and sufficient condition that any bill of goods be produced by labor alone (i.e., with labor as only net input) is that H' should contain at least one completely integrated process for each commodity* $G_k(k \neq n + 1)$. The proof of this theorem is immediate.

THEOREM 2: *If the conditions of Theorem 1 are fulfilled, and* $(x_1, x_2, \ldots, x_{n+1})$ *is the space of all commodities, then the linear space*

$$(9) \qquad L(x) = \sum_1^n L_k x_k + x_{n+1} = 0$$

is a supporting plane of H'.[7]

PROOF: From the fact that $L(\Pi^{(k)}) = 0$, it easily follows that condition (a) of footnote 7 is fulfilled by some $\pi^{(k)}$ even if $\Pi^{(k)} \notin H'$. For condition (b), let us consider any primary process $P^{(1)} \in H'$ and choose $\pi^{(k)} \in H'$.

---

6. Samuelson, chap. vii in *Activity Analysis*.
7. The linear space $L(x) = 0$ will be said to be a supporting plane of H' if
(a) there are vectors of H' which form with $L(x) = 0$ as small an angle as we want;
(b) one of the open halfspaces $L(x) > 0$, $L(x) < 0$ contains no vector of H'.
If H' is a closed cone, the condition (a) is equivalent to: $L(x) = 0$ should contain at least one element of H'.

The process $P_1 = P^{(1)} + \sum_{2}^{n} {}_k a_k^{(1)} \pi^{(k)} = (1, 0, \ldots, 0, -a_{n+1}^{(1)} - \sum_{2}^{n} {}_k a_k^{(1)} l_k)$ also belongs to H'. From this and the definition of L_k it follows that

(10) $a_{n+1}^{(1)} + \sum_{2}^{n} {}_k a_k^{(1)} l_k \geq L_1.$

From the same definition it follows that we can choose the l_k's so as to have

(11) $a_{n+1}^{(1)} + \sum_{2}^{n} {}_k a_k^{(1)} l_k < a_{n+1}^{(1)} + \sum_{2}^{n} {}_k a_k^{(1)} L_k + \epsilon$

for any given $\epsilon > 0$. Hence, (10) yields

$$L_1 - \sum_{2}^{n} {}_k a_k^{(1)} L_k - a_{n+1}^{(1)} < \epsilon$$

for any ϵ. This shows that $L(P^{(1)}) \leq 0$.

Let now $P \in H'$ be a derived process. By Krein-Milman theorem, P can always be decomposed into at most n primary processes. If these are denoted by $P^{(k)}$, $k = 1, 2, \ldots, n$, we have

(12) $P = \sum_{1}^{n} {}_k \lambda_k P^{(k)}$

with $\lambda \geq 0$.[8] From this it follows that $L(P) \leq 0$, which completes the proof.

6. The following theorems have important applications to both (L) and (GL). However, the relevance of their mathematical substance is not limited to this particular problem.

THEOREM 3: *If the square matrix $[a_{ik}]$ satisfies the conditions*

(13) $a_{ii} > 0,$ $a_{ik} \leq 0,$ $(i \neq k),$

and

(14) $(a_{\cdot k} = \sum_{1}^{n} {}_i a_{ik}) > 0,$

then the system

(15) $\sum_{1}^{n} {}_k a_{ik} \lambda_k = A_i,$ $(i = 1, 2, \ldots, n),$

where

(16) $A \geq 0,$

admits a solution $\lambda \geq 0$.[9]

PROOF: Let us refer to a system fulfilling (13), (14), (16) as an S-system. For $n = 2$ it is immediate that an S-system has a solution $\lambda \geq 0$. Let us

8. I owe to Murray Gerstenhaber the valuable suggestion of invoking the Krein-Milman theorem here in order to shorten an early version of the proof.

9. The signs $\geq$ and $\geqq$ have the standard meanings in vector algebra; see p. 292n10.

make the inductive assumption that the proposition is true for $n' = n - 1 > 2$. Let S_{rs} be the system obtained from (15) by striking out column r and row r of the matrix $[a_{ik}]$, and putting $A_s = 1$, $A_i = 0$ for $i \neq s$. S_{rs}, being an S-system with n' unknowns, has a solution: $\lambda_s^0 > 0$ and $\lambda_i^0 \geq 0$ for $1 \leq i \leq n$, $i \neq r$, s. Let λ^{rs} denote the vector $(\lambda_1^0, \lambda_2^0, \ldots, \lambda_{r-1}^0, 0, \lambda_{r+1}^0, \ldots, \lambda_n^0)$. Since $\lambda^{rs} \geq 0$, by (13) and (14) we have

$$e_{rs} = \sum_1^n {}_k a_{rk}\lambda_k^{rs} \leq 0, \qquad 1 + e_{rs} = \sum_1^n {}_i \sum_1^n {}_k a_{ik}\lambda_k^{rs} = \sum_1^n {}_k a_{\cdot k}\lambda_k^{rs} > 0.$$

Consequently, the system

$$\mu_1 + e_{rs}\mu_2 = 1, \qquad e_{rs}\mu_1 + \mu_2 = 0,$$

has a solution $(\mu) \geq 0$. Therefore,

$$\Lambda^s = \mu_1^0(\lambda^{rs}) + \mu_2^0(\lambda^{sr}) \geq 0.$$

Let now S_s be the system (15) in case $A_s = 1$, $A_i = 0$ for $i \neq s$. It is easily seen that Λ^s is a solution of S_s. The proof is completed by merely observing that

$$\lambda^* = \sum_1^n {}_i A_i(\Lambda^i)$$

satisfies (15) and that $\lambda^* \geq 0$.

COROLLARY 3.1: *If condition* (15) *of Theorem 3 is supplemented by* $A_i > 0$ *for* $i \leq \sigma$, *then* $\lambda_i > 0$ *for* $i \leq \sigma$.

COROLLARY 3.2: *If the square matrix* $[a_{ik}]$ *satisfies* (13) *and* (14), *then* $|a_{ik}| \neq 0$.

PROOF: Let us suppose that $|a_{ik}| = 0$; then, there exists $u \neq 0$ such that

$$\sum_1^n {}_i a_{ik}u_i = 0, \qquad (k = 1, 2, \ldots, n).$$

From (15) it follows that $\sum_1^n {}_i A_i u_i = 0$ for any $A \geq 0$. Hence $u = 0$. The contradiction proves the corollary.

THEOREM 4: *If the square matrix* $[a_{ik}]$ *is nonsingular, and if*

(17) $\qquad\qquad a_{ii} > 0, \qquad\qquad a_{ik} \leq 0, \qquad\qquad (i \neq k),$

and

(18) $\qquad\qquad (a_{\cdot k}) \geq 0,$

then the system

(19) $\qquad\qquad \sum_1^n {}_k a_{ik}\lambda_k = A_i, \qquad\qquad (i = 1, 2, \ldots, n),$

where

(20) $\qquad\qquad A \geq 0,$

admits a solution $\lambda \geq 0$.

PROOF: Let $b_{ii} = a_{ii} + \epsilon$, $b_{ik} = a_{ik}$, $\epsilon > 0$. The system

$$(21) \qquad \sum_{1}^{n} b_{ik} y_k = A_i, \qquad\qquad (i = 1, 2, \ldots, n),$$

being an S-system, has a solution $y \geq 0$. Moreover, because $[a_{ik}]$ is non-singular $b(\epsilon) = |b_{ik}|$ has no zero in the neighborhood of $\epsilon = 0$. From Cramer's formulae it follows that $y_k(\epsilon)$ is a continuous function of ϵ in the same neighborhood. Hence, for $\epsilon \to 0$ we have $\lim y_k = y_k^0 \geq 0$. By taking the limit in (21), it is immediately seen that $y^0 \geq 0$ is the solution of (19).

COROLLARY 4.1: *If, in Theorem 4, (20) is supplemented by $A_i > 0$ for $i \leq \sigma$, then $\lambda_i > 0$, for $i \leq \sigma$.*

THEOREM 5: *If one complete bill of goods, $B^0 > 0$, can be produced by labor alone, any other bill of goods, whether complete or not, can also be produced by labor alone.*

PROOF: Since $P_0(B_1^0, B_2^0, \ldots, B_n^0, -L) \in H'$ and $B_i^0 > 0$, the decomposition of P_0 into primary processes, (12), is

$$(22) \qquad P_0 = \sum_{1}^{n} \lambda_i^0 P^{(i)},$$

with $\lambda^0 > 0$. The theorem is proved if we prove the following proposition:
  If the system

$$(23) \qquad \lambda_i - \sum_{1}^{n} {}' a_i^{(k)} \lambda_k = B_i, \qquad\qquad (i = 1, 2, \ldots, n),$$

where $a_i^{(k)} \geq 0$, has a solution $\lambda^0 \geq 0$ for some $B^0 > 0$, then it has a solution $\lambda \geq 0$ for any $B \geq 0$.[10]
  Because

$$\lambda_i^0 - \sum_{1}^{n} {}' a_i^{(k)} \lambda_k^0 = B_i^0 > 0, \qquad\qquad (i = 1, 2, \ldots, n),$$

the system in μ

$$(23a) \qquad \lambda_k^0 \mu_k - \sum_{1}^{n} {}' (\lambda_k^0 \mu_i) a_i^{(k)} = \lambda_k^0 > 0, \qquad (k = 1, 2, \ldots, n),$$

is an S-system, and hence it has a solution $\mu^0 > 0$. (Corollary 3.1 shows, first, that $\lambda^0 > 0$, and then that $\mu^0 > 0$.) Consequently,

$$\mu_i^0 \lambda_i - \sum_{1}^{n} {}' (\mu_i^0 \lambda_k) a_i^{(k)} = \mu_i^0 B_i, \qquad (i = 1, 2, \ldots, n),$$

is an S-system and, hence, has a solution $\lambda \geq 0$. The fact that this system is equivalent to (23) completes the proof.

  Let $P^{(k)}$, $k = 1, 2, \ldots, n$, be a set of primary processes which satisfy relation (22) of Theorem 5 for some P^0. If the measurement unit of each G_i

  10. The prime sign of Σ' shows that $k = i$ is not covered by the sum.

is changed by a conversion factor $u_i > 0$, in the new standard form of $P^{(k)}$ we have $\mathbf{a}_i^{(k)} = a_i^{(k)}(u_i/u_k)$. And since (23) and (23a) have positive solutions, we can choose $u > 0$ such that one of the systems of inequalities

$$1 - \sum_{1}^{n}{}'_{k}\mathbf{a}_i^{(k)} > 0, \qquad 1 - \sum_{1}^{n}{}'_{k}\mathbf{a}_k^{(i)} > 0, \qquad (i, k = 1, 2, \ldots, n),$$

is satisfied. We can then always assume that units have already been so chosen and, hence, that the given set of $P^{(k)}$'s satisfies any of the following two systems we please:

$$(24a) \qquad\qquad 1 - \sum_{1}^{n}{}'_{k}a_i^{(k)} > 0, \qquad (i = 1, 2, \ldots, n),$$

$$(24b) \qquad\qquad 1 - \sum_{1}^{n}{}'_{i}a_i^{(k)} > 0, \qquad (k = 1, 2, \ldots, n).$$

But we must not forget that *not every set of primary processes can be brought under a form that satisfies one of these systems.*

The following two propositions are easily established.

COROLLARY 3.3: *If $P^{(k)}$, $k = 1, 2, \ldots, n$, satisfies (24a) or (24b), then (23) has a solution $\lambda \geq 0$ for any $B \geq 0$.*

COROLLARY 3.4: *If one of the systems (24) is satisfied, then*

$$(25) \qquad \Delta = \begin{vmatrix} 1 & -a_1^{(2)} & \cdots & -a_1^{(n)} \\ -a_2^{(1)} & 1 & \cdots & -a_2^{(n)} \\ \cdot & \cdot & \cdots & \cdot \\ -a_n^{(1)} & -a_n^{(2)} & \cdots & 1 \end{vmatrix} \neq 0.$$

Let Γ' be the orthogonal projection of H' upon the linear space $x_{n+1} = 0$. In this space, of coordinates $x_1, x_2, \ldots, x_n$, Γ' is a convex cone. Let also Ω_n^+ be the closed positive orthant of the space $(x_1, x_2, \ldots, x_n)$. With these notations and under the assumptions of Theorem 5, we have two corollaries.

COROLLARY 5.1: $\Omega_n^+ \subseteq \Gamma'$. *In particular, if H' is closed, then H' contains a most efficient completely integrated process, $\Pi^{(k)}$, for each G_k.*

COROLLARY 5.2: Γ' *is n-dimensional, or in other words, there exist primary processes satisfying (25).*

THEOREM 6: *If (22) has a solution $\lambda \geq 0$ for a given $B > 0$, then*

$$(26) \qquad \Delta_{j_1, j_2, \ldots, j_s}^{j_1, j_2, \ldots, j_s} > 0, \qquad \Delta_{j_1, j_2, \ldots, j_{s-1}, i}^{j_1, j_2, \ldots, j_{s-1}, k} \geqq 0,$$

where $0 \leq s \leq n - 2$, $k \neq i$. The Δ's with sub- and superscripts, indicating deleted rows and columns respectively, are the classical notations for the cofactors or signed minors of Δ in (25).

PROOF: First, let us observe that changing the units of the G_k's does not change the sign of any minor of Δ. Hence, for the proof we can assume that the elements of Δ satisfy (24a). The reduced system obtained from (23) by eliminating the equations corresponding to $i = j_1, j_2, \ldots, j_s$ with $j_1 < j_2 < \cdots < j_s$ and then making $a_{ik}^{(j)} = 0$ for the same values of j, also satisfies (24a). By Corollary 3.3, the reduced system has a nonnull solution $\lambda_j \geq 0$ for any nonnull set $B_k \geq 0$; by Corollary 3.4, it is determinate. Hence, if we further make all B_k's, except one, equal to zero, Cramer's formulae yield

$$\frac{\Delta_{j_1,j_2,\cdots,j_s,j}^{j_1,j_2,\cdots,j_s,j}}{\Delta_{j_1,j_2,\cdots,j_s}^{j_1,j_2,\cdots,j_s}} > 0, \qquad \frac{\Delta_{j_1,j_2,\cdots,j_s,k}^{j_1,j_2,\cdots,j_s,j}}{\Delta_{j_1,j_2,\cdots,j_s}^{j_1,j_2,\cdots,j_s}} \geq 0,$$

for any j, $k \neq j_i$, $(i = 1, 2, \ldots, s)$. Since

$$\Delta_{1,2,\cdots,n-1}^{1,2,\cdots,n-1} = 1,$$

a familiar chain operation then yields (26).

THEOREM 7: *The necessary and sufficient conditions that the system* (23) *shall admit a solution* $\lambda^0 \geq 0$ *for a given* $B^0 > 0$ *are given by the* $n - 1$ *inequalities*

$$(27) \quad \Delta > 0, \qquad \Delta_1^1 > 0, \qquad \Delta_{1,2}^{1,2} > 0, \qquad \cdots, \qquad \Delta_{1,2,\cdots,n-2}^{1,2,\cdots,n-2} > 0.$$

PROOF: Theorem 6 already shows that conditions (27) are necessary. It remains to prove their sufficiency. This is easily proved for $n = 2$. Let us then make the inductive assumption that sufficiency has been proved for some $n' = n - 1 > 2$. According to this assumption, from (27) and Theorem 5 it follows that the system

$$\mu_i - \sum_2^n {}'_k a_i^{(k)} \mu_k = e_i, \qquad (i = 2, 3, \ldots, n),$$

where $e_r = 1$ and $e_i = 0$, $i \neq r$, and r is arbitrarily chosen, has a non-negative solution $\mu_i = m_i^r \geq 0$. Hence

$$E_r = \sum_2^n {}_k a_1^{(k)} m_k^r \geq 0, \qquad (r = 2, 3, \ldots, n).$$

On the other hand, determinant algebra shows that $E_r = \Delta_r^1/\Delta_1^1$. From (27) it then follows that $\Delta_r^1 \geq 0$ for $r \geq 2$. Therefore, by Cramer's formula we have

$$(28) \qquad \lambda_1^0 = \frac{1}{\Delta} \sum_1^n B_k \Delta_k^1 > 0.$$

Knowing this coordinate of λ^0, we can determine the remaining ones from the reduced system

$$(29) \qquad \lambda_i - \sum_2^n {}'_k a_i^{(k)} \lambda_k = B_i + a_i^{(1)} \lambda_1^0, \qquad (i = 2, 3, \ldots, n).$$

Bearing in mind the definition of m_i^r, we see immediately that this system is satisfied by

(30) $$\lambda_i^0 = \sum_2^n {}_r m_i^r (B_r + a_r^{(1)} \lambda_1^0) > 0, \qquad (i = 2, 3, \ldots, n).$$

Relations (28) and (30) prove the theorem.

COROLLARY 7.1: *A necessary and sufficient condition that a complete bill of goods, $B > 0$, can be produced by labor alone is that a group of primary processes, $P^{(1)}, P^{(2)}, \ldots, P^{(n)}$, belonging to H', can be found such that (27) is fulfilled.*

LEMMA A: *Let*

(31) $$F_i(\lambda) = \sum_1^n {}_k a_{ik} \lambda_k = A_i, \qquad (i = 1, 2, \ldots, n),$$

be a system satisfying (13). If (31) has a solution $\lambda^0 > 0$ for $A = A^0$, $A_i^0 > 0$ for $i \leq t$, $A_i^0 = 0$ for $i > t$, then, with an appropriate choice of notations, it is equivalent to

(32a) $$\sum_i^n {}_k c_{ik} \lambda_k = D_i, \qquad (i = 1, 2, \ldots, m),$$

(32b) $$U_i = 0, \qquad (i = m + 1, \ldots, n),$$

where: 1) $m \geq t$; 2) $c_{ii} > 0$, $c_{ik} = 0$ for $i > k$, $c_{ik} \leq 0$ for $k > i$; 3) $D_i = \sum_1^i {}_k d_{ik} A_k$ with $d_{ii} = 1$, $d_{ik} \geq 0$ for $i \neq k$; 4) $U_i = \sum_{m+1}^n {}_k g_{ik} A_k$ with $g_{ii} = 1$, $g_{ik} \geq 0$ for $i \neq k$ but at least one such $g_{ik} > 0$.

PROOF: The system (31) is equivalent to

(33a) $$F_1 = A_1$$

(33b) $$F_i' = F_i - (a_{i1}/a_{11})F_1 = \sum_2^n {}_k b_{ik} \lambda_k = B_i, \qquad (i = 2, 3, \ldots, n),$$

where $b_{ik} = a_{ik} - (a_{i1}/a_{11})a_{1k}$, $B_i = A_i - (a_{i1}/a_{11})A_1$. From (13) it follows that $b_{ik} \leq 0$ for $i \neq k$. As to the sign of b_{ii} we have to consider two alternatives:

1) $t > 0$. If $a_{i1} = 0$, then $b_{ii} = a_{ii} > 0$. If $a_{i1} \neq 0$, then $B_i^0 > 0$. And since (33b) is satisfied for $\lambda^0 > 0$ and since $b_{ik} \leq 0$, we again have $b_{ii} > 0$. Hence, if notations are appropriately chosen, (33b) satisfies the conditions of the lemma for $B_i^0 > 0$, $2 \leq i \leq t'$, $t' \geq t$, and $B_i^0 = 0$ for $i > t'$.

2) $t = 0$. In this case, $B_i^0 = 0$ for every i. Hence, it is possible to have $b_{ii} = 0$ for some i such that $a_{i1} \neq 0$; but then F_i' must vanish identically. This can happen only if also $a_{1j} = 0$, $a_{ij} = 0$ for $j \neq l$, i. Hence, if F_i' vanishes identically, then no other F_j', $j \neq 0$, can do so. If notations are appropriately chosen, (31) is equivalent to

$$(34a) \qquad\qquad\qquad F_1 = A_1$$

$$(34b) \qquad\qquad F'_i = \sum_2^{n-1} {}_k b_{ik} \lambda_k = B_i, \qquad (i = 2, 3, \ldots, n - 1),$$

$$(34c) \qquad\qquad V_n = A_n - (a_{n1}/a_{11}) A_1 = 0.$$

Again, in (34b) we have $b_{ii} > 0$, $b_{ik} < 0$ for $i \neq k$; (34b) also is satisfied by $\lambda^0 > 0$ and $B_i^0 = 0$, for $2 \leq i \leq n - 1$. Therefore, we can apply to either (33b) or (34b), whichever is the case, the same procedure that led us from (31) to (33) or (34). The algorithm ends when we reach the form (32). This general lemma is thus proved.

THEOREM 5A: (Generalization of Theorem 5): *If one bill of goods, $(B_1^0, B_2^0, \ldots, B_t^0, 0, 0, \ldots, 0)$, $B_t^0 > 0$, $t > 0$, can be produced by labor alone, then any other bill of goods, $(B_1, B_2, \ldots, B_t, 0, \ldots, 0)$, $B_i \geq 0$ can also be produced by labor alone.*

Let us first observe that if in this case the decomposition (12) is such that $\lambda_i > 0$ for $i \leq m$, $m \geq t$, $\lambda_i = 0$ for $i > m$, then $a_k^{(j)} = 0$, for $k > m$, $j \leq m$. Then the proof is immediate by Lemma A.

Let now $t \leq n$ be the maximum number of commodities contained in any bill of goods that can be produced by labor alone; t will be called the *rank of the model*.

COROLLARY 5A.1: *If in Theorem 5A t represents the rank of the model, then H' contains a completely integrated process for every G_k, $k \leq t$, and for no other commodity.*

We are thus led to define an *effective* commodity as one for which H' contains a completely integrated process. From now on we shall assume the notations always chosen so that G_k is an effective commodity if and only if $k \leq t$.

The system (31) has not necessarily a solution $\lambda \geq 0$ for every group of processes $P^{(k)} \in H'$, $k = 1, 2, \ldots, n$. Examples of H' are easily constructed to show that for some group of processes (31) has no solution $\lambda \geq 0$ even if all $B_i = 0$, or that for some groups such a solution exists if and only if $B_j = 0$ for some values of $j \leq t$.

COROLLARY 5A.2: *No bill of goods containing ineffective commodities can be produced by labor alone.*

Given a group of processes $P^{(k)} \in H'$, $k = 1, 2, \ldots, n$, we shall say that $P^{(j)}$, $j \leq t$, is an *effective* process if the system

$$(35) \qquad\qquad\qquad \sum_1^n {}_i \lambda_i P^{(i)} = B$$

has a solution $\lambda \geq 0$ for a $B \geq 0$ such that $B_j > 0$. Clearly, in this case $\lambda_j > 0$. But if $\lambda_j > 0$ only for $B_j = 0$, then $P^{(j)}$ is a *redundant* process. And if $\lambda_j = 0$ for any B for which (35) has a solution $\lambda \geq 0$, then $P^{(j)}$ is an *idle* process. Finally, a group of processes $P^{(k)} \in H'$, containing an effective process for every effective commodity will be called *congruent*.

COROLLARY 5A.3: *The inputs of any effective process of a congruent group are effective commodities.*

PROOF: Let us assume that $a_j^{(k)} > 0$ for some $j > t$ and $k \leq t$. From the fact that (31) has a solution $\lambda^0 \geq 0$, $\lambda_i^0 > 0$ if $i \leq t$, for $B_i^0 > 0$ if $i \leq t$, $B_i^0 = 0$ if $i > t$, it follows that $\lambda_j^0 > 0$. Hence the system

$$(36) \qquad \mu_i - \sum_{t+1}^{n} {}_k a_i^{(k)} \mu_k = \sum_{1}^{t} {}_k a_i^{(k)} \lambda_k^0 = F_i \qquad (i = t+1, \ldots, n),$$

where $F_j > 0$ and $F_i \geq 0$ for $i \neq j$, has a solution $\mu_i = \lambda_i^0 \geq 0$ with $\mu_j > 0$. By Theorem 5A the system derived from (36) by putting $F_j = 1$ and $F_i = 0$ for $i \neq j$, has a solution $\mu_i^0 \geq 0$ with $\mu_j^0 > 0$. Consequently, the process $P(\alpha_1, \alpha_2, \ldots, \alpha_{n+1}) = \sum_{t+1}^{n} {}_i \mu_i^0 P^{(i)}$ belongs to H'. Noting that $\alpha_k \leq 0$ if $k \leq t$, $\alpha_k = 0$ for $k \neq j$, $t < k \leq n$ and $\alpha_j = 1$, we reach the conclusion that $P - \sum_{1}^{t} {}_i \alpha_i \pi^{(i)} = \pi^{(j)}$. As this contradicts Corollary 5A.1, the proposition is proved.

REMARK 1. It is convenient to distinguish the case of an ineffective commodity for which the corresponding process in any congruent group is idle. Such a commodity will be called an *idle commodity*. Let the notations be chosen so that G_k is idle if and only if $s < k \leq n$. For any commodity G_j, $t < j \leq s$, there exists a congruent group such that $P^{(j)}$ is a redundant process. Such a commodity will be called a *redundant commodity*. But again, a process corresponding to a redundant commodity in a group — whether congruent or not — is not necessarily redundant; it may very well be idle.

COROLLARY 5A.4: *If in a congruent group $P^{(k)}$ is redundant and $P^{(j)}$ is idle, then G_j is not an input of $P^{(k)}$.* (The proof is entirely analogous to that of Corollary 5A.3.)

It is immediate that in any group the number of redundant or idle processes must be greater than one. Also the number of idle commodities in a system must be greater than one. Idle commodities can never be produced within the system because, to put it intuitively, together they require at least one input greater than the corresponding output.

From Corollaries 5A.3 and 5A.4, it follows that the matrix of any congruent group has the following standard partition

$$(36) \qquad [\Delta] = \begin{bmatrix} E & S & T \\ 0 & R & U \\ 0 & 0 & I \end{bmatrix},$$

where E, R, I are square matrices. Furthermore,

1) The columns covered by E represent effective processes; E is of order t and satisfies (27).

2) The columns covered by R represent redundant processes; R is singular and its order is at most equal to $s - t$. The system $\mu R' = B$ has a solution $\mu \geq 0$ if and only if $B = 0$.

3) The columns covered by I represent idle processes; its order is at least equal to $n - s$. The system $\mu I' = B$ has no solution $\mu \geq 0$ even if $B = 0$.[11]

Since idle commodities cannot be produced at all, we may ignore them completely and assume $s = n$ in any problem of resource allocation. And since redundant commodities cannot be produced in excess of industrial requirements, the basic problem of resource allocations reduces to that of the solutions $\lambda \geq 0$ of the system

$$(37) \qquad \lambda_i - \sum_1^n {}_k a_i^{(k)} \lambda_k = B_i \geq 0, \qquad (i = 1, 2, \ldots, t),$$

$$(38) \qquad \lambda_i - \sum_1^n {}_k a_i^{(k)} \lambda_k = 0, \qquad (i = t + 1, \ldots, n).$$

From Theorem 5A it easily follows that for any congruent group this system admits a solution $\lambda^* \geq 0$ such that $\lambda_i^* = 0$ for $i \geq t$. In other words,

COROLLARY 5A.5: *Any bill of effective commodities can be produced without using redundant processes.*

Let now $\lambda^0 \geq 0$ be any solution of (37) and (38) such that some $\lambda_i^0 > 0$ for $i > t$. Clearly, $\lambda_i^0 \geq \lambda_i^*$ for $i \leq t$. Hence,

$$(39) \qquad l = \sum_1^n {}_i \lambda_i^0 a_{n+1}^{(t)} > l^* = \sum_1^n {}_i \lambda_i^* a_{n+1}^{(t)},$$

which proves

COROLLARY 5A.6: *The use of redundant processes in producing a bill of goods implies a waste of resources.*

Hence,

THEOREM 8: *The greatest lower bound, L, of the amount of labor necessary to produce a bill of effective goods, $(B_1, B_2, \ldots, B_t, 0, \ldots, 0)$, is given by*

$$(40) \qquad L = \sum_1^t {}_k B_k L_k.$$

COROLLARY 8.1: *If $t = n$, and if H' is a closed cone, the minimum amount of labor necessary to produce a bill of goods, $B \geq 0$, is given by*

11. For R it is easy to show that a chain of relations similar to (27) must begin with equalities and end with inequalities. For I, the conditions are more complicated: the matrix I may be either singular or nonsingular.

$$(41) \qquad\qquad L = \sum_{1}^{n} B_k L_k;$$

i.e., the process $P(B_1, B_2, \ldots, B_n, -L)$, *which belongs to* H', *belongs also to the linear space* (9).

COROLLARY 8.2: *If* $t = n$, *and if* H' *is a closed cone, the process*

$$(42) \qquad P(B_1, B_2, \ldots, B_n, -L) = \sum_{1}^{n} \lambda_k \overline{P}^{(k)} \qquad\qquad (\lambda \geq 0),$$

where $\overline{P}^{(k)}$ *is any primary process belonging to* (9) *and* H'.

PROOF: Let $\overline{P}^{(k)}$ belong to (9), i.e.,

$$(43) \qquad L_k - \sum_{i}' \overline{a}_i^{(k)} L_i = \overline{a}_{n+1}^{(k)} > 0, \qquad (k = 1, 2, \ldots, n).$$

The system

$$L_i(\lambda_i - \sum_{1}^{n}{}' \overline{a}_i^{(k)} \lambda_k) = L_i B_i$$

is an S-system. Therefore, it has a solution $\lambda \geq 0$. On the other hand, (43) yields

$$\sum_{1}^{n} \overline{a}_{n+1}^{(k)} \lambda_k = \sum_{1}^{n} B_k L_k = L,$$

which completes the proof.

For any given k, there may be more than one $\overline{P}^{(k)}$. But as far as the amount of resources needed to produce P is concerned, it is immaterial which $\overline{P}^{(k)}$ is used. Hence,

COROLLARY 8.3: *If* $t = n$, *and if* H' *is a closed cone, the processes* $\overline{P}^{(k)}$ *to be used for producing a bill of goods, B, with the minimum labor are independent of the bill of goods (i.e., of consumers' demand).*

7. The results presented in the preceding sections concern mainly the technological side of economics. If we retain the assumptions that $t = n$ and that H' is a *closed* cone, and assume also the classical conditions for long-run competitive equilibrium for each industry G_k, we have several results.

THEOREM 9: *Any long-run competitive equilibrium process used by the industry G_k is a $\overline{P}^{(k)}$.*

PROOF: If all processes are linear, the principle of maximum profit becomes inoperant unless we fix the scale arbitrarily.[12] Let us then choose the scale corresponding to one unit of labor input in each process. Let V_k be the set formed by the intersections of all elementary processes, $P^{(k)} \in H'$, with the plane $x_{n+1} + 1 = 0$. Let Γ be the intersection of H'

12. See, however, Remark 3, below.

with the same plane; Γ is convex, closed, and *bounded*. It also represents the convex hull of the union of all V_k.

By Assumption III, a firm in the industry G_k can use only formulae corresponding to a $P^{(k)}$. Given a set of prices $(p_1, p_2, \ldots, p_{n+1})$, the profit of such a firm is

$$(44) \qquad w(P^{(k)}) = p_k b_k - \sum_1^n {}'_i p_i a_i^{(k)} - p_{n+1}.$$

Because V_k is closed and bounded, $w(P^{(k)})$ effectively reaches its maximum, W_k, for at least one $P_0^{(k)} \in V_k$. Hence, $W_k = w(P_0^{(k)}) \geq w(P^{(k)})$ for any $P^{(k)} \in V_k$. Therefore,

$$(45) \qquad \sum_1^n {}_i p_i x_i - p_{n+1} - W_k = 0$$

is a supporting plane of the convex hull of V_k.

As long as W_k has not the same value for all industries, resources will keep moving from the low-profit to high-profit enterprises, thus causing prices as well as profits to change. Let p^* be the equilibrium prices and W^* the equilibrium profit; from (45) it follows that

$$(46) \qquad T(x) = \sum_1^n {}_i p_i^* x_i + (p_{n+1}^* + W^*)x_{n+1} = 0$$

is a supporting plane of H' such that it contains the equilibrium process(es) for every G_k and, moreover, $T(P) \leq 0$ for every $P \in H'$. For obvious reasons, this plane must coincide with (9). Hence, any equilibrium process is a $\bar{P}^{(k)}$, and the theorem is proved.

REMARK 2. A simple look at Fig. 9–1(b), especially, suffices to convince us that the assumption of a closed H' is vital for the conclusions of the last three propositions.

However, the assumption $t = n$ is not indispensable for the last two propositions. This is straightforward for Corollary 8.3. For the case of Theorem 9, let us observe that at any one time the prices and the maximum profits satisfy the system

$$(47a) \qquad X_k\left(b_k p_k - \sum_1^t {}'_i a_i^{(k)} p_i\right) = X_k(p_{n+1} + W_k),$$
$$(k = 1, 2, \ldots, t),$$

$$(47b) \qquad X_k\left(b_k p_k - \sum_{t+1}^n {}'_i a_i^{(k)} p_i\right) = X_k\left(p_{n+1} + W_k + \sum_1^t {}_i a_i^{(k)} p_i\right),$$
$$(k = t + 1, \ldots, n),$$

where X_k is the gross output of industry G_k. Because of the fundamental property of the ineffective commodities, (47) cannot be satisfied by $p \geq 0$ with $p_k > 0$ for some $k > t$ unless either every right-hand side in (47b) is zero or some are negative. Let us assume that not all $X_k = 0$ for $k > t$.

Then, in either of the two alternatives just mentioned, there exists $j > t$ such that

(48) $\qquad X_j > 0, \qquad W_j \leq -p_{n+1} - \overset{t}{\underset{1}{\Sigma}}_i a_i^{(j)} p_i \leq -p_{n+1}.$

But (47) yields

$$\overset{n}{\underset{1}{\Sigma}}_k \frac{X_k}{b_k} (p_{n+1} + W_k) = \overset{t}{\underset{1}{\Sigma}}_k p_k B_k > 0.$$

Hence, there exists at least one industry G_h, $h \neq j$, such that $W_h > -p_{n+1}$. Therefore, as long as (48) prevails, resources will move out of the industry G_j in some other industries such as G_h. Consequently, the competitive conditions — as these have been interpreted in Theorem 9 — will ultimately eliminate any industry producing an ineffective commodity.

The preceding analysis proves

COROLLARY 9.1: *If H' is closed, then in a generalized open model the long-run competitive equilibrium brings about the optimum allocation of resources (labor).*

REMARK 3. From the identity of (9) and (46) it follows that

(49) $\qquad p_k^* = L_k(p_{n+1}^* + W^*),$

which shows that *the equilibrium price of each product, G_k, is proportional to the amount of labor necessary to produce one unit of that product with labor as the only input.* We have an equality between labor value and product value if and only if $W^* = 0$. But if processes are linear, the competitive conditions by themselves do not determine the distribution of value added between wages and profits; hence, nothing warrants yet $W^* = 0$.[13] Only by invoking an over-all increase in the scale of production and the resulting stronger competition among entrepreneurs in the labor market can we account for $W^* = 0$. If this procedure is accepted, then (49) becomes

(50) $\qquad p_k^* = L_k p_{n+1}^*,$

as we find it in the labor theories of value.

If we now adopt for each G_k as unit of measurement the quantity produced by one unit of labor in the process $\Pi^{(k)}$, it follows that $L_k = 1$. If labor is taken as *numéraire*, (50) becomes $p_k = 1$, which means that the new unit of G_k is the *dollar's worth*. Relation (41) yields

$$L = \overset{n}{\underset{1}{\Sigma}}_i B_i.$$

13. Actually, in the proof of Theorem 9, we should have connected long-run competitive equilibrium with the uniformity of the return to the dollar, or equivalently with that of the rate of profit (cf. Georgescu-Roegen as cited in note 3, above). Had we done so, instead of a single plane (45) we would have had several *parallel* planes which need not coincide. Indeed, in this alternative the equilibrium profit for the particular scale chosen need not be the same in every industry, and the proportionality between p^* and L_k no longer holds in general.

This relation shows that the natural unit of measurement to be adopted whenever consolidation of industries is contemplated for practical investigations is the dollar's worth of product. This offers a justification for the procedure adopted by Leontief.[14]

8. A normal continuation for the preceding analysis would be to take up the closed (GL). However, there seems to be very little to be said about such a model. The following represents some summary thoughts.

Let U_{n+1} be the convex hull of all $P^{(n+1)}$ known to be feasible. If $P \in U_{n+1}$ and $L(P) = 0$, then P is a *possible* equilibrium. If $P \in U_{n+1}$ and $L(P) < 0$, then by adopting P for the relation between labor and real income, the economy will contract down to nil: not enough labor would be forthcoming to produce the necessary bill of goods. If $P \in U_{n+1}$ and $L(P) > 0$, then the adoption of P leads to an excess supply of labor; the economy may either expand indefinitely, or, alternatively, unemployment may appear. But these results are to be expected in a system from which all interaction is eliminated by an assumption such as VI.

A POSTSCRIPT (1964)

The above paper was initially written in connection with the conference on "linear programming" held at the Cowles Commission, University of Chicago, during June 1949, thanks to the vision and initiative of T. C. Koopmans. At that time, the theorems proved in my paper — like most theorems established by other participants — were quite novel in mathematical economics for the simple reason that the topics covered by the conference themselves were novel. Actually, from the reaction of several first-rate mathematicians who participated in the conference, the theorems seemed then novel even to them. But as an increasing number of economists subsequently came to recognize the tremendous importance of the Leontief system and of linear systems in general, numerous mathematicians too were attracted by the special mathematics associated with the newly opened domains in economics. Through this growing interest in linear structures, it came to light that several relevant theorems pertaining to these new topics had already been established in the mathematical literature. If they were still little known outside a small circle of specialists, it was, no doubt, because of their limited area of application. But by the mid-1950's the name of Frobenius became as familiar to mathematical economists as that of Lagrange had once been. They were in-

14. Leontief, *The Structure of the American Economy* (cited in my note 1), chap. iii, pp. 14 ff.

formed of the existence of three memoirs — a truly pioneering work, by now a classical reference — in which Frobenius presented some extremely interesting results concerning matrices with positive and nonnegative elements.[1] (I should, perhaps, add that mathematicians as a body also discovered Frobenius through the same train of events.)

As was normal, the new trend in mathematical economics induced many authors to write expository articles intended to familiarize economists and econometricians with the results obtained by Frobenius and a few others after him, and to relate these results to the theorems independently proved by some mathematical economists who — like myself — were driven to do so by their own limited knowledge of mathematics. Unfortunately, in some of these papers — as well as in some casual commentaries inserted in articles of a different nature — one finds an appreciable amount of confusion that can be only detrimental to the still uninitiated. The situation prompted this postscript.

Perhaps, the best illustration in point is provided by the otherwise very instructive article of Max A. Woodbury, "Properties of Leontief-Type Input-Output Matrices."[2] On page 358, he asserts that the "result given by Hawkins and Simon (1949) and [its] equivalent formulation by Georgescu-Roegen (1950) . . . were proved by Frobenius [III] and Ostrowski (1937)." The first reference is to a proposition enunciated in 1949 by Hawkins and Simon,[3] the second, to my Theorem 7.[4]

Now, the Hawkins-Simon proposition states that given the system

$$(1) \qquad \sum_{1}^{n} {}_k a_{ik}\lambda_k = A_i, \qquad (i = 1, 2, \ldots, n),$$

where: 1) $a_{ii} > 0$, $a_{ik} < 0$ for $i \neq k$, 2) the matrix $[a_{ik}]$ *is nonsingular*, and 3) $A_i > 0$, the necessary and sufficient condition for λ satisfying (1) to be positive is that *all* principal minors of $|a_{ik}|$ shall be positive (Hawkins and Simon, pp. 246–247). But contrary to Woodbury's assertion none of the Frobenius memoirs includes this proposition: Frobenius' aim was not the study of linear, nonhomogeneous systems of equations. The most one can say is that *the sufficiency part* of the proposition is an immediate corollary of a theorem of Frobenius (I, pp. 471–472) which, reinterpreted,

1. G. Frobenius, "Über Matrizen aus positiven Elementen," Parts I and II, *Sitzungsberichte der königlich preussischen Akademie der Wissenschaften*, 1908, Bd. I, 471–476 and 1909, Bd. I, 514–518; "Über Matrizen aus nicht negativen Elementen," *ibid.*, 1912, Bd. 1, 456–477. (These memoirs will be cited as Frobenius I, II, and III respectively.)

2. In *Economic Activity Analysis*, ed. Oskar Morgenstern (New York, 1954), pp. 341–363.

3. D. Hawkins and H. A. Simon, "Note: Some Conditions of Macroeconomic Stability," *Econometrica*, XVII (1949), 245–248.

4. Georgescu-Roegen, "Leontief's System in the Light of Recent Results," *Review of Economics and Statistics*, XXXII (1950), 214–222, where Theorem 7 appeared for the first time in print.

states that if $[a_{ik}]$ fulfills 1) and 2), and if the principal minors of $|a_{ik}|$ are positive, then the cofactor of every a_{ik} is positive. It is true that Ostrowski uses a stronger form of the same theorem (Frobenius, III, pp. 457–458) to prove the sufficiency part of a proposition analogous to that of Hawkins and Simon for less restrictive conditions.[5] On the other hand, even Ostrowski's paper contains no theorem corresponding to *the necessity part* of the Hawkins-Simon proposition. Woodbury's evidence does not prove, therefore, that on this point they had been preceded by others. Besides, one should not fail to recognize the difficulty of the proof, even if condition 3) is granted.[6]

I shall now turn to the statement that my Theorem 7 is an equivalent formulation of that of Hawkins and Simon.[7] Of course, any two propositions, A and B, stating the necessary and sufficient conditions for the same fact are equivalent in the strict sense in which this term is used in Logic. That is, $A \Leftrightarrow B$. But if one does not want to ignore aspects that greatly matter in mathematics, one must observe the essential difference between "the necessary and sufficient condition for $n > 10$ is that $n = 10 + m, m > 0$" and "the necessary and sufficient condition for $n > 10$ is that $n = 1 + c, c > 9; n = 2 + d, d > 8; \ldots; n = 10 + m, m > 0$." The elementary but vital distinction between these two *equivalent* propositions is that the former is the *stronger*. Or as economists would say, the former achieves the same goal in a more economical manner. That is precisely the sort of difference between the Hawkins-Simon proposition and my Theorem 7. In the case of a system of n equations, one has to compute $2^n - n - 1$ determinants if one uses Hawkins-Simon's proposition, but only $n - 1$ if one uses Theorem 7. The difference is still greater if one takes into account the order of the determinants to be computed. Moreover, the inequalities (27) of Theorem 7 constitute a set of *independent* conditions such that if one is left out the remaining set fails to represent the sufficiency condition. Or in terms of computation cost, they ensure the minimum cost.

5. A. Ostrowski, "Über die Determinanten mit überwiegender Hauptdiagonale," *Commentarii Mathematici Helvetici*, X (1937), p. 71, *Satz* II.

6. Hawkins and Simon failed to notice that their proof does not need that premise, which comes out as a necessary consequence. I wish also to point out that — a strange coincidence — their proof is strictly accurate only for the necessity part. Besides, their argument (p. 245n3) that the proposition is immediately extended for $a_{ik} \leq 0$, $i \neq k$, "because of the continuity of solutions of these equations with respect to variations of these coefficients" is fallacious: on the same ground their proposition should be true for $a_{ii} \geq 0$. But all these blemishes do not diminish in the least their merit of being the first economists to tackle a problem of special importance, of having enunciated a valuable theorem — one part of which was novel — and of having proved that part in a very simple manner.

7. It is, no doubt, because of the frequency with which this statement is found in literature, that as careful a writer as Michio Morishima refers to my theorem as "the Hawkins-Simon theorem." See his *Equilibrium, Stability, and Growth* (Oxford, 1964), p. 15.

The method of proving Theorem 7 is so elementary that I would not be surprised at all to learn that the theorem had been proved on some not yet known paper. But the fact remains that nothing to approach it is found in the works of Frobenius, Ostrowski, or of other (known) authors prior to 1950, at least. The following two theorems are interesting by themselves, but they also show the bearing of Theorem 7 upon the results reached by Ostrowski or Frobenius in making some of them *stronger*.

In his paper, quoted above, Ostrowski proposes to study the properties of the *M-determinant* which he defines by the following properties: 1) $m_{ii} \geq 0$, $m_{ik} \leq 0$ for $i \neq k$, 2) *all* principal minors of any order are nonnegative.[8] Theorem 7 immediately shows that for $m_{ii} > 0$ and positive principal minors, Ostrowski's second condition is *superabundant*. Actually, an even stronger result can be easily established:

THEOREM A: *If M is a determinant of order n with $m_{ii} \geq 0$, $m_{ik} \leq 0$ for $i \neq k$, and if*

$$(2) \qquad M \geq 0, \qquad M_1^1 \geq 0, \ldots, M_{1,2,\ldots,n-2}^{1,2,\ldots,n-2} \geq 0,$$

then M is an M-determinant, that is, all its principal minors are nonnegative.[9]

As Debreu and Herstein remarked, Theorem 7 suggests a stronger form of a proposition they proved and which contains the converse of Frobenius' main theorem.[10] Frobenius proved that if A is a nonnegative matrix and r is the maximal root of the characteristic equation

$$A(s) = |sE - A| = 0,$$

then all principal minors from order 1 to n, of $A(s)$ are positive for $s > r$ (Frobenius, III, p. 457). The proposition suggested by Debreu and Herstein, in a more general formulation, is

THEOREM B: *If*

$$(3) \qquad A(s) \geq 0, \qquad A_1^1(s) \geq 0, \ldots, A_{1,2,\ldots,n-1}^{1,2,\ldots,n-1}(s) \geq 0,$$

then $s > r$ or $s = r$ according to whether no equality sign prevails in (3) or $A(s) = 0$.[11]

8. Ostrowski, p. 69.

9. I need only mention that the proof is obtained by induction, uses the Cauchy expansion of M, and invokes some of Frobenius' theorems (III, pp. 457, 462).

10. G. Debreu and I. N. Herstein, "Nonnegative Square Matrices," *Econometrica*, XXI (1953), 603n6.

11. The proof of this theorem too is made by induction and uses the Cauchy expansion of $A(s)$. It invokes the direct theorem of Frobenius mentioned a few lines earlier.

Limitationality, Limitativeness,
and Economic Equilibrium

1. In the intellectual atmosphere of the second half of the last century, characterized by the belief in the unlimited power of quantitative tools and by the unrestricted hopes that, with their perfection, man would ultimately be decomposed into a series of mechanical gears, the Walrasian formulation of general economic equilibrium was naturally received with unsurpassed enthusiasm and regarded as a new message confirming the prevailing faith. As time went by and Walras' famous system begot no other quantitative results than the initial equality between the number of unknowns and of equations, the reaction moved economists to the other extreme, overemphasizing the sterility of this equality and belittling — or ignoring — the other teachings of the Walrasian theory. "What if the system has no solution?" the more mathematically minded economists began to ask. It was felt that with the answer to this new problem a new light would descend upon the scene. After a long time, the mathematical genius of Abraham Wald provided the first answer to the problem, although his proof used K. Schlesinger's formulation, a simplified version of the Walrasian framework.[1] After an interval of two decades, the combined sagacity of Kenneth J. Arrow and Gerard Debreu finally broke through another rampart, and proved the existence of a solution under conditions more general than those of Schlesinger.[2] However, a closer search for the economic relevance of both these valuable contributions

NOTE: A discussion version of this essay was published in vol. I of *Proceedings of the Second Symposium in Linear Programming*, Washington, D.C., January 27–29, 1955, Office of Scientific Research, United States Air Force, pp. 295–330.

    1. K. Schlesinger, "Über die Produktionsgleichungen der ökonomischen Wertlehre," *Ergebnisse eines Mathematischen Kolloquiums*, VI (1933–1934), 10–11. Abraham Wald, "Über die eindeutige positive Lösbarkeit der neuen Produktionsgleichungen," *ibid.*, pp. 12–18, in translation, "On the Systems of Equations of Mathematical Economics," *Econometrica*, XIX (1951), 368–403.
    2. Kenneth J. Arrow and Gerard Debreu, "Existence of an Equilibrium for a Competitive Economy," *Econometrica*, XXII (1954), 265–290.

will show that — paradoxical as it may seem — they opened an avenue which is more likely to harm the value of the Walrasian system for economics as a social science than to elevate it. Two different aspects of the results attained combine to elicit the main weakness of the Walrasian system. What is relevant in Wald's results is not the uniqueness of the solution, but that the vector price is not restricted by any condition other than to be non-negative. Consequently, the "equilibrium" price of any factor of production may be zero. There is nothing that would exclude zero prices for the factors of production involving human cost, like labor or waiting. Since labor must have a positive share in the national product, a solution in which the wage rate is zero — although mathematically valid — is purely "imaginary" from the point of view of economic reality. Wald's result is therefore analogous to the proposition of analytical geometry stating that any line intersects a circle, at times in real, at others in imaginary points. *Prima facie*, Arrow-Debreu's results seem invulnerable from this point of view, since in their scheme labor always gets a share sufficient to maintain life. But this amazing result is due to one particular premise, which is as idle from the economic point of view as Wald's results may be "imaginary." Indeed, Arrow-Debreu assume that all individuals concerned are endowed *ab initio* with an adequate real income over their entire life span. In this way, the adequacy of labor's share is not a result of the economic mechanism, but an analytically lucrative assumption. To use the same analogy as before, the straight line is in this case bound to intersect the circle in at least one real point, simply because *ex hypothesi* it already has one such point in common with the circle.[3]

The obvious implication of the above remarks is that the Walrasian system may not always be an adequate description of reality, and that, in those cases where its analytical solution is "imaginary," the real solution must be the result of a different system. One certainly would not expect a society that cannot live according to the Walrasian distribution theory — i.e., according to the marginal productivity theory — to commit suicide rather than adopt another system.

Clearly, the relevance of the preceding discussion depends entirely upon whether or not economic structures exist where the marginal productivity theory does not work. To prove the existence of such structures is an almost impossible task, but not more so than to prove *directly* the existence of the other type. In both cases we have to rely on indirect evidence pieced together by a mental process. Actually, the experiment

3. In a way, the Schlesinger-Wald model is superior to that of Arrow-Debreu, because it is free from this *petitio principii* formulation. This is why the former allows one to lay bare the fundamental difficulty immediately.

in vivo is more likely to be inconclusive in the case where the marginal productivity theory works than in the other.[4] Without underestimating the weight of these caveats, one can find actual examples of economies where the marginal productivity theory fails to provide a workable distribution of the national income. They are the so-called overpopulated countries, which in recent years have become the object of increasing interest.

It is the purpose of the present paper to analyze the economic equilibrium of structures designed to represent overpopulated economies and to pinpoint the reasons why the mechanism of the marginal productivity theory cannot in these cases ensure the maximum national product.[5] The various cases of departure from the marginal productivity theory will be illustrated first for isolated economies; in the last sections, foreign trade will also be considered briefly. Since, as we shall see, the entire problem turns on the relative proportion between some primary factors of production and the labor force, some new definitions related to the concept of limitationality must first be made explicit.

2. The existence of limitational factors of production and the special problems in connection with the equilibrium of the firm raised by such factors have long since been mentioned in economic literature.[6] Yet, before the spectacular ascent of the Leontief model and its generalizations little attention was paid to the bearing of limitationality upon general equilibrium. Even in the Leontief model or the linear programming, only one aspect of limitationality is directly involved. Indeed, limitationality has often been interpreted to mean one of two different — but not entirely unrelated — things. The usual interpretation of limitationality is that illustrated by Fig. 10–1(a), to be exact, by a position such as A. However,

4. See, for instance, the fruitful controversy between Richard A. Lester and Fritz Machlup: Richard A. Lester, "Shortcomings of Marginal Analysis for Wage-Employment Problems," *American Economic Review*, XXXVI (1946), 63–82; Fritz Machlup, "Marginal Analysis and Empirical Research," *ibid.*, XXXVI (1946), 519–554; Richard A. Lester and Fritz Machlup, "Communications," *ibid.*, XXXVII (1947), 135–154.

5. There can be little quarrel with quantifying man's economic objective and presenting it as a maximum under restraint of a certain function, $F(I, L)$, of I, real income, and L, leisure — at least for some purposes. It seems, therefore, that the maximum national product should not be brought alone into consideration. Such an objection — perfectly justified in the case where the average income is well above the minimum of subsistence — is unavailing in the case of an overpopulated country. It must not be forgotten that the function $F(I, L)$ has no sense unless (I, L) is at least equal to the subsistence minimum (I_0, L_0). Moreover, $F(I, L)$ is not likely to vary with L in the vicinity of (I_0, L_0).

6. E.g., Erich Schneider, *Theorie der Produktion* (Vienna, 1934); Georgescu-Roegen, "Fixed Coefficients of Production and the Marginal Productivity Theory," *Review of Economic Studies*, III (1935), 40–49 [in the present volume, Essay 7, above]. The term "limitational" was introduced by Ragnar Frisch in 1931; see note 6 in my article just cited.

limitationality has been interpreted also as describing the situation of B in Fig. 10–1(b).[7] The following definitions serve the purpose of eliminating the ambiguity.

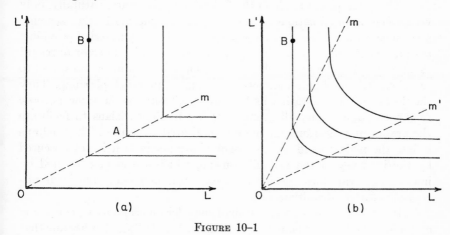

FIGURE 10–1

DEFINITION I. *A factor of production is* limitational *if an increase in its input is a necessary but not a sufficient condition for an increase in output.*

According to this definition, both L and L' are limitational at A in Fig. 10–1(a). It is convenient to refer to a production function displaying the property described by Fig. 10–1(a) as a limitational production function. Some limitational production functions are characterized by the fact that for each output there is a combination where all factors are limitational. If the production function is also homogeneous, then it is completely defined by a single "process," like Om in Fig. 10–1(a).

DEFINITION II. *A factor of production is* limitative *if an increase in its input is a both necessary and sufficient condition for an increase in output.*

L is limitative at B in Fig. 10–1(a and b).

In any production function, there is a domain of limitativeness for each factor of production. Limitativeness is, therefore, a common aspect of all production functions, whereas limitationality imposes upon the production function a special structure.

Limitationality, being essentially a technical property, pertains especially to the "engineering" side of economics. Limitativeness may, however, raise delicate problems in which both social and economic aspects are entangled. The economists who have struggled with the formidable problems of overpopulated countries have in fact been concerned with

7. Cf. Herbert Zassenhaus, "Dr. Schneider and the Theory of Production," *Review of Economic Studies*, III (1935), 35–39.

the particular economic structure where some of the primary factors of production — other than labor — are limitative.

3. The assumptions made in the following analysis are, naturally, only very rough approximations of the real world. Like most other schemes used in economics, those of this paper serve only the purpose of providing the argument with a convenient pedestal. It would be an error to regard them as playing the same role as the models of natural sciences, and to use them *tale quale* for indirect measurement or exact planning. They should be considered rather as "metaphors," serving the same purpose as the two-dimensional diagrams used by mathematicians to facilitate the argument concerning far more complicated structures. They help us isolate the main thread of the economic argument so it is not obscured by the intricacy of the real phenomena, which are in fact affected by innumerable variables — some nonmeasurable and many not even susceptible of exact definitions.

In the following argument, we shall consider an economy C, consisting of P individuals living on a given territory. We shall further assume that in C the amount of every primary factor of production available per unit of time is limited.[8] Let these amounts be denoted by $l = \kappa P$ for labor (κ = constant), l' for land, l'' for mineral resources, etc.[9] We shall also make the following assumptions:[10]

A. The technical information of the economy C consists exclusively of techniques by which only one of the commodities $X_1, X_2, \ldots$ can be produced.

B. The elementary techniques are such that each industry X_i works under constant real cost,[11] or in other words, the gross output of X_i is a homogeneous function of degree one: $x_i = f_i(x_{k,i}; l_i, l'_i, \ldots)$ where $x_{k,i}$ ($k \neq i$) is the input of X_k; l_i, of L; l'_i, of L'; etc.

C. All economic goods are homogeneous.

D. The market conditions are such that equilibrium price equals average cost.

4. To make clear an important point on which the analysis of this paper rests, we shall consider first a very simple scheme: a country which

---

8. Primary factors of production means those factors which either cannot be produced (land, mineral resources, seashore, etc.) or whose supply is largely governed by noneconomic motives (labor force).

9. The assumption of a constant maximum flow of mineral resources raises some difficulty. This difficulty becomes, moreover, extremely serious if the assumption is used in parallel with that of constant returns to scale of industry. Since this study makes use of both assumptions, it has at least this major imperfection.

10. Cf. Georgescu-Roegen, "Leontief's System in the Light of Recent Results," *Review of Economics and Statistics*, XXXII (1950), 214–222.

11. This is the well-known Wicksellian assumption, which is not inconsistent with the existence of an optimum plant scale.

produces only one commodity (X) with the aid of labor (L) and land (L'). If the available amounts of primary factors of production are represented by T (Fig. 10–2), land is a limitative factor. The problem is to find out the size of the national income and its distribution. If the landlords maximize the rent-profit, then equilibrium will be at a point such as I, and the net national product (x^1) will be distributed according to the famous theorem of marginal productivity theory: $x^1 = wl^1 + w'l'$ with $l^1 < l$. The shares of the factors will be determined by the slope of the tangent JJ' to the isoquant (x^1). This is all right, provided $l^1w \geq \beta P'$, where β is the minimum economic standard,[12] and P' the total population of the labor class. Social patterns — like those observed in all countries which are not overpopulated — will easily develop to maintain social equilibrium, too. It will be found, for instance, that only certain members of the family — the husband, or only the men above a certain age — could go to work without losing "social prestige." They will provide the entire family with its living. Very likely, before such equilibrium is reached the underbidding of the laborers has lowered the wage rate until $l^1w = \beta P'$.

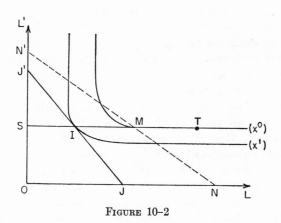

FIGURE 10–2

5. Let us, however, consider the situation where no equilibrium compatible with the marginal productivity theory exists such that $l^1w \geq \beta P'$. Classical and neoclassical theory dismisses such a possibility; to justify this position, it has been advanced that P' will decline until the incompatibility is removed. In this way, the marginal productivity and, implicitly, the principle of maximizing profit have been saved. At the same time, other alternatives have been eliminated from the attention of the

12. This is not equal to the minimum of subsistence. As is well known, β varies with the social patterns and is greatly influenced by economic development.

economists. But these alternatives are found to prevail in the real world
as much as the patterns based on the maximization of profit. They are
readily observable in all overpopulated countries, particularly in the
regions where capitalistic patterns have not yet fully penetrated.

If x^1, when distributed according to the marginal productivity theory,
is not sufficient to provide the working class with the minimum economic
standard, it does not mean that the community under consideration has
no other means to solve the problem. It is obvious that the economy under
consideration can increase its national product (which in this case represents also income) above x^1, namely, up to x^0. It might happen that
x^0 distributed in another fashion would suffice to provide labor with an
adequate share, with no loss, or a very small one, in rent as compared
with what this is for x^1.[13] But in order that the national product should
reach its maximum, its distribution must be made in disregard of the
marginal productivity theory. It is seen, therefore, that the marginal
productivity theory is at its best a translation of the economic aspects
of *some* social patterns compatible with a "land of plenty."

Ample evidence exists to point out that in overpopulated countries
the social patterns differ from those of the more advanced communities.
They provide other rules than the principle of maximizing profit. The
general impression gathered by all Western visitors from their contact
with the overpopulated and poor countries of the East is that their
economies are very badly managed and unpardonably inefficient. The
Western commentators were unanimous in pointing out that laborers
get paid for performing tasks which manifestly are not worth their wages.
(A classical and eloquent example — to mention only one — is that of
the gleaners, who are paid in real terms more than they gather.) Clearly,
from the point of view of marginal productivity such patent facts are
indisputable symptoms of "waste." But none of the outside observers
seems to have realized that only in this way — and not through adherence
to more "efficient" methods — can the maximum national product be
reached. On the other hand, we may understand the internal logic of
those Europeans who, still preserving some traces of those social patterns
where value is not determined according to the Neoclassical formula,
accuse the people of the United States of excessive materialism. Their
value judgment poorly expresses an undeniable fact: under the conditions
assumed in this section, the principle of maximizing profit would lead

13. The condition for this is that $x^0 = w_0 l^0 + w_0' l' = \beta P' + w_0' l'$, where $w_0 l^0$, $w_0' l'$
are the shares of labor and land, respectively, and l^0 is the amount of labor at M.
Nothing prevents the share of land from being the same at M and I, i.e., $w_0' = w'$.
However, *in this case*, the line NN' which represents the distribution of x^1 has a smaller
slope than JJ'. Indeed, since the average marginal productivity of labor at I is smaller
than w, we have $x^0 - x^1 < w(l^0 - l)$. This yields immediately $w_0 < w$.

indeed to "inefficiency," since it would prevent the net product from reaching its maximum.

Finally, we should remark that if the production function of X is limitational, there is no difference between x^0 and x^1, that is, the net national product is the same in both economic systems. But then, neither can the marginal productivity theory by itself offer an explanation of how the net national product is distributed among the factors of production. I assume that for such a situation a one-hundred-per-cent marginalist will have to resort to some rather extra-economic motives, essentially similar to those determining distribution in overpopulated countries.

6. The preceding remarks apply equally well to schemes which reflect a little more of the intricacy of economic reality. We may consider first an economy producing only two commodities X_1, X_2 with the aid of labor and land. We shall begin by assuming that both elementary production functions are limitational. Each production is then described by a set of input coefficients:[14]

	Inputs of			
	X_1	X_2	L	L'
For X_1		a_{21}	b_1	b_1'
For X_2	a_{12}		b_2	b_2'

With given amounts of L and L', the set of feasible net national products is represented by

$$(1) \qquad p_1 x_1 + p_2 x_2 \leq l, \qquad p_1' x_1 + p_1' x_2 \leq l',$$

$$(2) \qquad x_1, x_2 \geq 0.$$

The inequalities (1) are the familiar relations encountered in the Leontief system. Each is obtained by assuming in turn that L' and L are free goods. As is known, p_1, p_2, for instance, represent the equilibrium prices under the assumption that L' is a free good, which reduces the above system to a Leontief one with L as the only primary factor of production.[15] The scheme considered here is a particular instance of Koopmans' model,

14. The well-known condition guaranteeing a positive net national product is that $1 - a_{12}a_{21} > 0$. It can easily be shown that, in this case, the units of X_1 and X_2 can be chosen so that $a_{12}, a_{21} < 1$. A more convenient diagrammatical representation is obtained if it is assumed that the units have been chosen in this manner.

15. Georgescu-Roegen, "Some Properties of a Generalized Leontief Model," in *Activity Analysis of Production and Allocation*, ed. T. C. Koopmans, *et al.* (New York, 1951), pp. 171–172. [In the full version of that paper, printed in the present volume, the reference is at pp. 330–333.]

which offers a more realistic approach than that of Leontief.[16] We may assume, in the first instance, that the straight lines

(1a) (S_1S_2) $p_1x_1 + p_2x_2 = l,$ $(S_1'S_2')$ $p_1'x_1 + p_2'x_2 = l',$

meet inside the positive quadrant (Fig. 10–3). For an isolated economy,

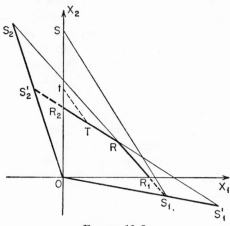

FIGURE 10–3

the opportunity cost curve is R_1RR_2. Labor is fully employed on RR_1, and land, on RR_2. From the point of view of the net national product, labor and land are limitational at R.

In the situation of Fig. 10–3, industry X_1 utilizes more labor per unit of land than industry X_2.[17] Hence

(3) $$\frac{b_1}{b_2} > \frac{b_1'}{b_2'}.$$

We may speak of X_1 as a *labor-intensive industry*, and of X_2 as a *land-intensive* one. The equilibrium prices must satisfy the conditions

(4) $\pi_1 - a_{21}\pi_2 = wb_1 + w'b_1',$
                 $\pi_2 - a_{12}\pi_1 = wb_2 + w'b_2',$

where $w > 0$, $w' \geq 0$. Because

(5) $p_1 - a_{21}p_2 = b_1,$ $p_1' - a_{21}p_2' = b_1',$
             $p_2 - a_{12}p_1 = b_2,$ $p_2' - a_{12}p_1' = b_2',$

16. T. C. Koopmans, "Analysis of Production as an Efficient Combination of Activities," in *Activity Analysis*, p. 37. See also his "Maximization and Substitution in Linear Models of Production," in *Input-Output Relations*, Proceedings of a Conference on Inter-Industrial Relations Held at Driebergen, Holland, ed. The Netherlands Economic Institute (Leyden, 1953), pp. 106–107.

17. This is immediate by observing that in S_1 the amount of land used is less than in S_2, whereas the same amount of labor is used in S_1 and S_2.

the system (4) yields:

(6) $$\pi_1 = wp_1 + w'p_1', \qquad \pi_2 = wp_2 + w'p_2'.$$

Therefore, the equilibrium prices must satisfy the condition

(7) $$\frac{p_1'}{p_2'} < \frac{\pi_1}{\pi_2} \leq \frac{p_1}{p_2}.$$

Hence, equilibrium must be either at R, if $w' > 0$, or on RR_1 if $w' = 0$. If the demand for X_1 and X_2 is satisfied by a point on RR_1, then everything fits perfectly the theory of marginal productivity, i.e., the price ratio equals the technical marginal rate of product substitution. If equilibrium is at R_1, again we find that price ratio is equal to the marginal rate of substitution in demand. But if demand is such that no equilibrium can exist on RR_1, then marginal productivity fails again to explain how equilibrium is reached. If, for instance, x_2 must be at least equal to ξ_2, then equilibrium must be at $T(\xi_1, \xi_2)$, with labor not fully employed.[18] The equilibrium price ratio represented by the slope of Tt must be such as to ensure the minimum economic standard to the working class. But such an equilibrium cannot be maintained unless we assume, as we did in the preceding section, that wages and employment are determined by a mechanism involving some special patterns of behavior. In their absence, or in the face of their dissolution, equilibrium could still be maintained at T by some price regulation, including by necessity minimum *real* wages. On the other hand, we may see in this scheme the real reason why in all overpopulated countries, the landlords are reluctant to encourage the landless to become accustomed to higher economic standards: this would cause total rent to decrease.

The situation with equilibrium at T illustrates a country which may be referred to as *relatively* overpopulated. Although technically the entire labor force could be absorbed in production, this is prevented by the required minimum production of food. As is known, Marxists have usually argued that overpopulation is only a myth, at best supported by false symptoms caused by wrong allocation of resources. The scheme just considered offers a clear-cut example of what they may have in mind when they speak about overpopulation as being only a misleading appearance, but at the same time we see that very likely the pseudo overpopulation cannot be remedied from within, unless we want forcibly to feed people ocarinas instead of bread.

Clearly, equilibrium at T cannot be stable unless the social patterns underlying it remain stable, too. Serious difficulties may occur if changes in social patterns are taking place. Thus, if X_1 is a commodity mainly consumed by the landlords, it may happen that by following more closely

---

18. X_2 may, for instance, represent "wheat," and X_1, "hemp."

the formula of Neoclassical theory the upper class will devote more resources to the production of luxuries at the expense of the basic commodities consumed by the lower class. The occurrence of such phenomena has been observed in almost every underdeveloped country as a consequence of the increasing influence of Western patterns upon the local mores.

7. Let us now consider the case where S_1S_2 and $S_1'S_2'$ do not intersect inside the positive quadrant and S_1', S_2' are nearer the origin than S_1, S_2 respectively. Most of the remarks made in the last part of the preceding section apply with even greater force in this case. This structure provides, however, an illustration of an overpopulated country which will remain so, independent of the distribution of its resources. We have here a counterexample to the Marxist thesis alluded to above.

8. Many other variants of the scheme considered in the two preceding sections lead to similar results. One such variant seems to deserve special attention, because it may illustrate one particular aspect not present in the scheme used above: specific primary factors of production. Thus, let us assume that land is used only in the production of X_1, and mineral resources only in that of X_2. If the elementary production functions are limitational, with given amounts of land, l', and of mineral resources, l'', and a sufficiently large labor force, the opportunity cost curve is represented by R_2RR_1 (Fig. 10–4).[19] The straight lines RR_1 and RR_2 are parallel to OS_2'' and OS_1':

$$(8) \qquad (RR_1) \quad x_1 + a_{12}x_2 = v_1, \qquad (RR_2) \quad x_2 + a_{21}x_1 = v_2.$$

Equilibrium prices must satisfy the conditions

$$(9) \qquad \begin{aligned} \pi_1 - a_{21}\pi_2 &= wb_1 + w'b_1', \\ \pi_2 - a_{12}\pi_1 &= wb_2 + w''b_2'', \end{aligned}$$

with $w > 0$, w', $w'' \geq 0$. It follows that

$$(10) \qquad a_{21} < \frac{\pi_1}{\pi_2} < \frac{1}{a_{12}}.$$

According to the last relation, equilibrium can exist only at R, if it is to conform to the marginal productivity.[20] Then, land and mineral resources are fully employed. But the distribution of the national product

19. We leave aside the case where the utilization to full capacity of one of the primary factors of production is not feasible. The lines (8) are therefore assumed to intersect inside the positive quadrant.

20. This proposition is valid for the generalized scheme of several commodities. See Mathematical Note A, at the end of this paper.

remains indeterminate if only purely economic considerations are invoked. Here again, distribution, if it exists at all, must be determined by extra-economic factors.[21] The need for X_1 could be so great that mineral resources may remain underemployed with a zero, or next to zero, rent. But then we may again be confronted with the situation when at equilibrium, the price ratio does not equal the technical rate of substitution. This may happen even in the case where the labor force would not be sufficient to man both industries to full capacity; then a situation similar to that of Section 4 is obtained.

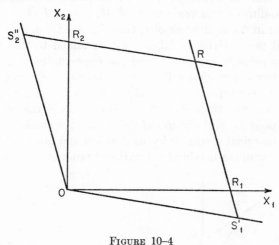

FIGURE 10–4

9. So far, in the schemes involving more than one commodity, we have assumed that the elementary production functions are limitational. We shall now reconsider the scheme of Section 7 with this restriction removed, and formulate the condition used there regarding the labor force in a more precise fashion. The labor force will be assumed so large that from the point of view of the economy *as a whole* land remains under all circumstances a limitative factor. If we define the opportunity cost curve as that which represents the maximum net production of X_1 for any given X_2, then it is easy to show that the theorem proved by Samuelson and the writer for a generalized Leontief model[22] holds also in the present scheme. That is:

The opportunity cost curve is a straight line $p_1' x_1 + p_2' x_2 = l'$, and the

21. On economic considerations, π_1/π_2 can be determined by the demand marginal rate of substitution provided this rate at R satisfies (10). But even in this case (9) fails to determine w, w', w''.

22. In *Activity Analysis* (as cited in note 15, above), Paul A. Samuelson, chap. vii; Georgescu-Roegen, chap. x ["Some Properties of a Generalized Leontief Model," the full version of which is Essay 9 in the present volume].

optimum combinations of factors in each elementary process of production do not change with the structure of the net national product.[23]

Let S_1' and S_2' be the optimum combinations, each using all the land available. In S_1', S_2', the amount of labor used is represented by l_1, l_2, respectively, with $l_2 < l_1 < l$. X_1 is the labor-intensive industry. The opportunity cost curve is R_1R_2 (Fig. 10–5). To be sure, at S_1' and S_2' labor is employed to the point where its marginal productivity is zero. Thus, R_1R_2 represents in a sense a purely technological maximum which can be reached with the available amounts of primary factors. Obviously, it is the two-dimensional equivalent of M, not of I (Fig. 10–2). If we are interested in the equivalent of I, then the condition that wages equal the marginal productivity of labor must be added to the system. Less labor will be actually employed, and the opportunity cost curve based upon this latter assumption will no longer be a straight line, but a curve r_1r_2. This curve corresponds to I, and exactly as I, it represents a *smaller* net national product than R_1R_2. Here, exactly as in the scheme of Section 4, it would be entirely *wasteful* for an overpopulated country to stick to the marginal productivity mechanism and not to abandon it for the sake of reaching a maximal net national product.

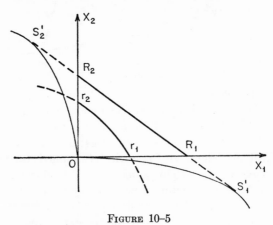

FIGURE 10–5

If the processes OS_1', OS_2' are identified by their input coefficients, with the notations used above the equilibrium prices must satisfy the system (4), where, however, w, $w' > 0$. And if, as assumed, X_1 is the labor-intensive industry, we obtain

(11)
$$\frac{p_1'}{p_2'} < \frac{\pi_1}{\pi_2}.$$

23. For proof see Mathematical Note B at the end of this article. The theorem is not valid if there is more than one limitative factor of production. The number of commodities, however, may have any value whatever.

This means that the price ratio cannot possibly equal the marginal rate of substitution. Because of (11), there exists a tendency for the industrial structure to move towards R_1, i.e., for resources to move into the labor-intensive industry; thus, if no minimum requirement exists for X_2, then equilibrium is established at R_1.

10. Another remark warranted by the preceding analysis is that the opportunity cost curves — of which the theories of general equilibrium, international trade, or positive welfare make an extensive use — are not purely technological entities as is often tacitly implied. The opportunity curve may be either R_1R_2 or r_1r_2 depending upon the size of the labor employed. It is obvious from the uses made of them, that the opportunity cost curves usually found in economic literature are not established on the assumption of maximum utilization of labor. Indeed, the opportunity cost curve can be constructed either by assuming that the entire labor force must be used, or that it is used only up to the point where labor's marginal productivity is sufficient to provide the working class with the minimum economic standard.

Secondly, we may point out that the fact that the mechanism described by the marginal productivity theory does not ensure a maximum net national product is one of the main reasons why its free working has to be suspended in times of emergency — like wars, for instance.

Finally, it should be remarked that to base one's argument on an opportunity cost curve constructed on the basis of a given value of the marginal productivity of labor is equivalent to assuming that the real wage rate is not an unknown, but an exogenous element of the economic system. In general, there is nothing wrong with such a view, but the marginal productivity theory can hardly adopt it.

11. The preceding analysis can be extended to include the case where C has access to foreign markets. There are points connected with this problem that do not necessitate other comments than those usually found in the literature, but there are others that need to be explored further.

Let us assume, for instance, that a country with a productive structure such as that of Section 6 is in contact with a foreign market in which the prices of X_1, X_2 are P_1, P_2 respectively.[24] Several cases must be considered.

a) $p_1'/p_2' \leq P_1/P_2 < p_1/p_2$. Replacing π_i by P_i in (6), we obtain a system which determines w, w', with possible equilibrium at R (Fig. 10–3). Assuming that at R there is an excess demand for X_2, some X_2 will be imported and some X_1 exported. If, however, the share of labor as

24. It is assumed here that these prices are given and country C's trade cannot influence them.

determined by (6) is insufficient, equilibrium cannot possibly be at R, unless the internal price ratio is brought to exceed P_1/P_2. This can be achieved in many ways, such as the introduction of an export bounty for X_1 paid out from the receipts of a sales tax on X_2.[25]

b) $P_1/P_2 \geq p_1/p_2$. If equality prevails in (b), the equilibrium of the industrial structure may be anywhere on RS_1, with land a free good. However, if $P_1/P_2 > p_1/p_2$, wages are determined by $P_1 - a_{21}P_2 = wb_1$ and, since $P_2 - a_{12}P_1 < wb_2$, no X_2 can be produced in the country. The structural equilibrium is, therefore, at S_1; the opportunity of foreign trade is represented by S_1S (Fig. 10–3).[26]

c) $P_1/P_2 < p_1'/p_2'$. Because of (7), the internal price exceeds p_1'/p_2' for any w and w'; therefore only X_2 can be produced in the country, with equilibrium at S_2'. The distribution of the net national product remains in this case, too, indeterminate if only economic factors *stricto sensu* are taken into account. Indeed, the only condition, $P_2 - a_{12}P_1 = b_2w + b_2'w'$, cannot determine both w and w'.

Clearly, unemployment is greatest at S_2'. The country as a whole may be better off by producing only X_2 and importing X_1 than without foreign trade, but increasing unemployment without a substantial increase in the average wage rate is most likely to cause dissatisfaction. Only if the advantage of the foreign trade is indeed tremendous can social discontent be avoided. Thus, England was able to turn almost all arable land into pasture because her exceedingly favorable terms of trade increased the British average income to the point where the surplus of labor could be turned into leisure by the introduction of the British weekend.[27]

12. The analysis of the preceding section can easily be extended to the structure of Section 8. One additional remark seems to be in order. The industrial structure is at S_2'' only if $P_1/P_2 < a_{21}$. This implies that some other country can turn X_2 into X_1 at a higher rate of conversion, which ultimately means that with respect to X_1, land or labor — or probably both — have a higher efficiency in that country than in C. This confirms a well-known result of the theory of international trade.

25. This device was actually used by Rumania in the 1930's; an export bounty for wheat was supported by a sales tax on several manufactured products.

26. This result shows that the usual analyses of foreign trade ignore that, for their purpose, the opportunity cost curve is not identical to that under isolation. The latter must be extended outside the positive quadrant as far as is justified. Thus, in the model used here, for problems of foreign trade the opportunity cost curve is S_1RS_2', and not R_1RR_2 (Fig. 10–3). Only in this way can one take account of those industries which exist because part or all of a certain input is imported.

27. As late as 1927, Sir Max Muspratt admitted before the Geneva International Economic Conference that England's favorable terms of trade led to "shorter hours of labor, and more pleasure and amusement among the mass of the people."

In addition to this, it is clear that at S_2'' land becomes idle and, consequently, a practically free good. If it happens that the terms of trade threaten the income of the landowners, it is only natural that they should campaign for the introduction of an import duty on X_1. Such a campaign is facilitated by the argument that the move from R to S_2'' will decrease employment. Although this is true — and also impressive — the argument is specious since an import duty under the circumstances not only decreases the general welfare but, if effective, may also increase the share of the land without affecting at all the industrial structure. Something like this happened during the 1930's when agricultural protectionism became the rule in Western Europe as the result of the progressive opening of the international price scissors.

13. Another interesting impact of foreign trade upon an economy of an overpopulated country can be illustrated with the aid of the structure analyzed in Section 9. If such an isolated country comes in full contact with an international market, there will be not only a change in its industrial structure, but also a change in its social patterns, i.e., in the regulating principles of economic transactions. Certainly, the country as a whole will greatly benefit from the change, but the contact with the more advanced countries will gradually disrupt the old forms and replace them by the economic patterns of the richer countries, of the lands of plenty. Landlords and traders will gradually adopt the principle of maximizing profit, with the result that unemployment will increase and the share of the labor will decrease. The general picture of all Eastern European countries since the middle of the nineteenth century — with their increasing modernization and their spreading misery among the uprooted masses — is too familiar to need emphasis. It is so much the more an incontrovertible illustration of the argument offered in this study. Familiar also is the fight — so often misunderstood — of the Agrarians who have deplored the rapidity with which the old forms of the peasant countries were disrupted under the impact of capitalism, and who have advocated holding onto them until other patterns could be adopted without making only the masses of the in-between generations pay for it.

Lastly, one may wonder if the Soviet regime, which forcibly and intentionally destroyed the old forms, did not replace them by a regulating economic principle entirely different from that of the capitalistic system, just because it intuitively felt that the latter does not lead to the maximum net national product. This may offer a partial, but not negligible, explanation of the frequently called "spectacular" rise of the net national product in the USSR over the last thirty years.

The foregoing analysis sheds sufficient light upon the main controversial issues surrounding the idea of land reform — a solution persistently advocated by Agrarians. The liquidation of the large estates aimed precisely at the replacement of their capitalist patterns by those of the familial farmsteads, where the family labor can be used to the point where its marginal productivity is zero. In this way, farm income could reach its maximum compatible with the available labor. The troubles of Agrarians came, however, from another direction. In most countries where extensive land reforms have been introduced, the resulting average farm was appreciably smaller than the technically optimum size. What was gained on one hand did not always compensate what was lost on the other.[28]

MATHEMATICAL NOTES

A. For n commodities X_i, the system (8) is replaced by

$$(12) \qquad \Sigma A_{ik} x_i = v_k, \qquad (k = 1, 2, \ldots, n),$$

where A_{ik} are the cofactors of the Leontief determinant

$$(13) \qquad A = |a_{ik}| = \begin{vmatrix} 1 & -a_{12} & \cdots & -a_{1,n} \\ -a_{21} & 1 & \cdots & -a_{2,n} \\ \cdot & \cdot & \cdots & \cdot \\ -a_{n,1} & -a_{n,2} & \cdots & 1 \end{vmatrix},$$

and $v > 0$. The system (12) is determinate because $A \neq 0$. Let x^0 be its solution; then (12) can be written

$$(14) \qquad \Sigma_i A_{ik}(x_i - x_i^0) = 0.$$

Let us recall that the system

$$(15) \qquad \Sigma_i a_{ik} \pi_i = u_k, \qquad (k = 1, 2, \ldots, n),$$

where $u > 0$, has a positive solution $\pi > 0$:

$$(16) \qquad \pi_i = \frac{\Sigma_k u_k A_{ik}}{A}.$$

THEOREM I. *For any* $u > 0$, *the hyperplane* $\Sigma_i \pi_i(x_i - x_i^0) = 0$, *is a supporting plane of the convex set* Γ

$$(17) \qquad \Sigma_i A_{ik}(x_i - x_i^0) \leq 0, \qquad (k = 1, 2, \ldots, n).$$

This is immediate by observing that for any $y \in \Gamma$, $y \neq x^0$,

$$(18) \qquad \Sigma_i \pi_i(y_i - x_i^0) = \frac{1}{A} \Sigma_k u_k \Sigma_i A_{ik}(y_i - x_i^0) < 0$$

28. Several ways to remedy this have been advocated. Unfortunately, Agrarians, who patently had been little aware of all the economic implications of their platforms, came to realize the difficulty and to think of these remedies only *post partum*.

THEOREM II. *No supporting hyperplane of* Γ *at a point* $x^1 \neq x^0$ *can satisfy conditions* (15).

Let $\Sigma_i \pi_i'(x_i - x_i^1) = 0$ be such a plane. Since any z such that $z - x^0 < 0$ obviously belongs to Γ, and since we can choose z such that $z - x^1 < 0$ as well, it follows that $\Sigma_i \pi_i'(y_i - x_i^1) \leq 0$ for any $y \in \Gamma$. According to the preceding theorem, $\Sigma_i \pi_i'(y_i - x_i^0) \leq 0$. Writing the last two inequalities for $y = \lambda x^0 + (1 - \lambda)x^1$, which for $0 < \lambda < 1$ belongs to Γ, we obtain

$$\lambda \Sigma \pi'(x_i^0 - x_i^1) \leq 0, \qquad (1 - \lambda)\Sigma \pi (x_i^1 - x_i^0) \leq 0.$$

Hence, $\Sigma \pi'(x_i^1 - x_i^0) = 0$. This contradicts (18), and the theorem is thus proved.

Considering a generalization of the structure described in Section 8, one can give the following economic interpretation to the theorems just proved:

(1) If $x^0 > 0$, i.e., if it is feasible to use all natural resources to full capacity, the equilibrium compatible with the marginal productivity theory is necessarily at x^0, with distribution not determined by purely economic factors.

(2) If $x^0 \not> 0$, the equilibrium price ratios cannot equal the marginal rates of substitution.

B. The proof follows the line adopted by Samuelson in the case of a generalized Leontief model;[29] for the sake of simplicity, the case of only two industries and only one superabundant factor, L, is considered. The net national product is given by

(19) $\qquad y_1 = f_1(x_{21}, l_1, l_1') - x_{12}, \qquad y_2 = f_2(x_{12}, l_2, l_2') - x_{21},$

with $l_1 + l_2 = l$, $l_1' + l_2' = l'$. The production functions, f_1 and f_2, are assumed to be homogeneous of the first degree. Maximizing y_1, with y_2 constant, we obtain the conditions:

(20)
$$\frac{\partial f_1}{\partial x_{21}} - \lambda = 0, \qquad -1 + \lambda \frac{\partial f_2}{\partial x_{12}} = 0,$$

$$\frac{\partial f_1}{\partial l_1} + \mu = 0, \qquad \lambda \frac{\partial f_2}{\partial l_2} + \mu = 0,$$

$$\frac{\partial f_1}{\partial l_1'} + \nu = 0, \qquad \lambda \frac{\partial f_2}{\partial l_2'} + \nu = 0.$$

29. *Activity Analysis*, chap. vii. As I have shown in "Some Properties of a Generalized Leontief Model" [the full version of which is printed in the present volume], for the Substitutability Theorem we must assume also that, first, the rank of the model is not zero, and, second, that the technological horizon is a closed cone. Both these assumptions are granted here.

These yield

$$(21) \qquad \frac{\partial f_1}{\partial x_{21}} \cdot \frac{\partial f_2}{\partial x_{12}} = 1,$$

$$(22) \qquad \frac{\partial f_1}{\partial l_1} = \frac{\partial f_1}{\partial x_{21}} \cdot \frac{\partial f_2}{\partial l_2},$$

$$(23) \qquad \frac{\partial f_1}{\partial l_1'} = \frac{\partial f_1}{\partial x_{21}} \cdot \frac{\partial f_2}{\partial l_2'}.$$

If l is large enough, then (22) is replaced by

$$(24) \qquad \frac{\partial f_1}{\partial l_1} = 0, \qquad \frac{\partial f_2}{\partial l_2} = 0.$$

In this case, the equations (21), (23) and (24), being homogeneous of degree zero, determine $\frac{x_{21}}{l_1'}, \frac{l_1}{l_1'}, \frac{x_{12}}{l_2'}, \frac{l_2}{l_2'}$ independent of any other condition. This proves that the factor proportions to be used in each industry in order to obtain a maximal net product do not depend upon the composition of the latter or the size of the labor employed.

Furthermore, λ, μ, ν are constants ($\mu = 0$). Using the basic property of homogeneous functions, (20) yields

$$(25) \qquad f_1 - \lambda x_{21} + \nu l_1' = 0, \qquad -x_{12} + \lambda f_2 + \nu l_2' = 0,$$

and further

$$(26) \qquad y_1 + \lambda y_2 + \nu l' = 0.$$

This is readily seen to be equivalent to

$$(27) \qquad p_1' y_1 + p_2' y_2 = l'.$$

If l is not sufficiently large, then conditions (21), (22), (23) fail to determine the factor proportions and λ, μ, ν are no longer constants. Moreover, the opportunity curve is concave towards the origin,[30] not necessarily a straight line.[31]

30. Georgescu-Roegen, "The Aggregate Linear Production Function and Its Applications to von Neumann's Economic Model," *Activity Analysis*, p. 107.

31. After this paper was read at the Linear Programming Symposium on January 28, 1955, I learned of Masao Fukuoka's article "Full Employment and Constant Coefficients of Production," which appeared in the February 1955 issue of *Quarterly Journal of Economics* and which deals with some interesting Keynesian topics within a similar framework. I felt that I would not have done justice to Fukuoka's analysis by inserting hurried comments in my paper, which — save for a few minor editorial changes — appears here [in the *Proceedings* of the Symposium] in the version circulated in mimeographed form during the Symposium.

PART IV

Economic Development

11

Economic Theory
and Agrarian Economics

According to some recent studies, more than 1.3 billion people still live in a self-subsistence economy, that is, as peasants. Most of these also live on the verge of starvation. Asia and Africa, which together represent more than 60 per cent of the world's population, produce only a little more than 30 per cent of the world's agricultural output. Conservative estimates show that if basic nutritional needs for the entire population of the world are to be met, it is necessary that the food production be increased by at least 30 per cent.[1] Neither the overwhelming numerical importance of peasant economy nor the scarcity of food is a new economic development peculiar to our own time.

In spite of all this, agrarian economics — by which I mean the economics of an overpopulated agricultural economy and not merely agricultural economics — has had a very unfortunate history. Noncapitalist economies simply presented no interest for Classical economists. Marxists, on the other side, tackled the problem with their characteristic impetuosity, but proceeded from preconceived ideas about the laws of a peasant economy. A less-known school of thought — Agrarianism — aimed at studying a peasant economy and only this. An overt scorn for quantitative theoretical analysis, however, prevented the Agrarians from constructing a proper theory of their particular object of study, and consequently from making themselves understood outside their own circle. There remain the Standard economists (as a recent practice calls the members of the modern economic school for which neither Neoclassical nor General Equilibrium suffices as a single label). Of late, as economic development has become tied up with precarious international politics, Standard economists have been almost compelled to come to grips with the problem of underdevel-

NOTE: This paper is reprinted by courtesy of the publisher from *Oxford Economic Papers*, XII (1960), 1–40.

1. The above data are found in W. S. and E. S. Woytinsky, *World Population and Production* (New York, 1953), pp. 307, 435, and *passim*.

oped economies, and hence with noncapitalist economics. But in their approach they have generally committed the same type of error as Marxists.

Thus, the agrarian economy has to this day remained a reality without a theory. And the topical interest of a sound economic policy in countries with a peasant overpopulation calls for such a theory as at no other time in history. But one cannot aspire to present a theory of a reality as complex as the peasant economy within the space of an article. My far more modest aim is to point out the basic features that differentiate an overpopulated agricultural economy from an advanced economy. I have endeavored to present the argument in terms of the familiar analytical tools of Standard theory or others akin to these. The brief historical critique which prefaces the theoretical analysis is intended to place the latter in a better perspective, particularly as concerns policy implications.

I. THEORY, REALITY, AND POLICY

1. *Theory and reality.* Theory is in the first and last place a logical file of our factual knowledge pertaining to a certain phenomenological domain.[2] Only mathematics is concerned with the properties of "any object whatever," for which reason since Aristotle's time it has been generally placed in a special category by itself. To each theory, therefore, there must correspond a specific domain of reality. In any science, the problem of precisely circumscribing this domain faces well-known difficulties. Where physics ends and chemistry begins, and where economics ends and ethics begins, are certainly thorny questions, although not equally so. Here, however, I want to discuss a quite pedestrian query pertaining to the problem of the proper domain of a theory. And this query is: Can an economic theory which successfully describes the capitalistic system, for instance, be used to analyze another economic system, say feudalism, successfully?

Let us observe that a similar question hardly ever comes up in the physical sciences, for no evidence exists to make physicists believe that matter behaves differently today from yesterday. In contrast, we find that human societies vary with both time and locality. To be sure, one school of thought still argues that these variations are only different instances of a unique archetype and that consequently all social phenomena can be encompassed by a single theory. This is not the place to show in detail where the weakness of the various attempts in this direction lies. Suffice it to mention here that when the theories constructed by these attempts

2. That is not to deny that theory may serve other purposes, but these are byproducts of its essential nature.

do not fail in other respects, they are nothing but a collection of generalities of no operational value whatever. As Kautsky once judiciously remarked, "Marx designed to investigate in his 'Capital' the capitalistic mode of production [and not] the forms of production which are common to all people, as such an investigation could, for the most part, only result in commonplaces." [3] For an economic theory to be operational at all, i.e. to be capable of serving as a guide for policy, it must concern itself with a specific type of economy, not with several types at the same time.

What particular reality is described by a given theory can be ascertained only from that theory's axiomatic foundation. Thus, Standard theory describes the economic process of a society in which the individual behaves *strictly* hedonistically, where the entrepreneur seeks to maximize his cash-profit, and where any commodity can be exchanged on the market at uniform prices and none exchanged otherwise. On the other hand Marxist theory refers to an economy characterized by class monopoly of the means of production, money-making entrepreneurs, markets with uniform prices for all commodities, and complete independence of economic from demographic factors.[4] Taken as abstractions of varying degree, both these axiomatic bases undoubtedly represent the most characteristic traits of the capitalist system.[5] Moreover, far from being absolutely contradictory, they are complementary, in the sense of Bohr's Principle of Complementarity.[6] This is precisely why one may speak of Marx as "the flower of Classical economics." [7]

A far more important observation is that the theoretical foundations of both Standard and Marxist theories consist of cultural or, if you wish, institutional traits. Actually, the same must be true of any economic theory. For what characterizes an economic system is its institutions, not the technology it uses. Were this not so, we would have no basis for distinguishing between Communism and Capitalism, while, on the other hand, we should regard capitalism of today and capitalism of, say, fifty years ago as essentially different systems.

As soon as we realize that for economic theory an economic system is

3. Karl Kautsky, *The Economic Doctrines of Karl Marx* (New York, 1936), p. 1.

4. I refer to the fact that the assumption of a permanent reserve army simply means that at the subsistence wage rate the supply of labor is "unlimited" both in the short and in the long run, whereas Classical economics held that this was true only in the long run. *Infra*, note 51.

5. I have left the surplus value proposition out of the Marxist axioms because this proposition — as I shall argue later — belongs to feudalism, not to capitalism.

6. This principle by which Bohr overcame the impasse created by the modern discoveries in physics states that reality "cannot be comprehended in a single picture" and that "only the totality of the phenomena exhausts the possible information about objects." Niels Bohr, *Atomic Physics and Human Knowledge* (New York, 1938), pp. 40 and *passim*.

7. Terence McCarthy in the preface to K. Marx, *A History of Economic Theories* (New York, 1952), p. xi.

characterized exclusively by institutional traits, it becomes obvious that neither Marxist nor Standard theory is valid as a whole for the analysis of a noncapitalistic economy, i.e. of the economy of a society in which part or all of the capitalist institutions are absent. A proposition of either theory may eventually be valid for a noncapitalistic economy, but its validity must be established *de novo* in each case, either by factual evidence or by logical derivation from the corresponding axiomatic foundation. Even the analytical concepts developed by these theories cannot be used indiscriminately in the description of other economies. Among the few that are of general applicability there is the concept of a production function together with all its derived notions. But this is due to the purely physical nature of that concept. Most economic concepts, on the contrary, are hard to transplant. "Social class" seems the only exception, obviously because it is inseparable from "society" itself (save the society of Robinson Crusoe and probably that of the dawn of the human species). This is not to say that Marxist and Standard theories do not provide us with useful patterns for asking the right kind of questions and for seeking the relevant constituents of any economic reality. They are, after all, the only elaborate economic theories ever developed.

All this may seem exceedingly elementary. Yet this is not what Standard and (especially) Marxist theorists have generally done when confronted with the problem of formulating policies for the agrarian over-populated countries. And, as the saying goes, "economics is what economists do."

2. *A reality without theory.* As has often been remarked, economists of all epochs have been compelled by the social environment to be far more opportunistic than their colleagues in other scientific fields, with the result that their attention has been concentrated upon the economic problems of their own time.[8] And as the transition of economic science from the purely descriptive (i.e. taxonomic) to the theoretical stage coincided with the period during which in Western Europe feudalism was rapidly yielding to capitalism, it was only natural that capitalism should become the object of study of the first theoretical economists. That may explain only why most Western economists have been interested in developing the theory of the capitalist system, but not why none attempted a theory of a noncapitalist economy. The only explanation of this omission is the insuperable difficulty in getting at the cultural roots of a society other than that to which one actually belongs. And, as we have hinted, an in-

8. The point finds an eloquent illustration in the vogue that the problem of economic development has recently acquired among Western economists: we have reached the point where the development of underdeveloped nations is as much an economic problem of the West as of these other nations.

tuitive knowledge of the basic cultural traits of a community is indispensable for laying out the basis of its economic theory.

By its very nature, a peasant village is the milieu least fit for modern scientific activity. The modern scientist had therefore to make the town his headquarters. But, from there, he could not possibly observe the life of a peasant community. London, for instance, offers indeed "a favorable view . . . for the observation of bourgeois society" — a circumstance immensely appreciated by Marx[9] — but not even a pinhole through which to look at a peasant economy. Even if, unlike Marx, an economist was born in a village, he had to come to town for his education. He thus became a true townee himself, in the process losing most, if not all, *Verstehen* of the peasant society. It was natural, therefore, that to Marx as well as to other Western economists (to those coming from a peasantless country, especially) the peasant should seem "a mysterious, strange, often even disquieting creature." [10] Yet none showed Marx's unlimited contempt for the peasantry. For him, the peasantry was just a bag of potatoes, not a social class. In the *Communist Manifesto* he denounced "the idiocy of rural life" to the four corners of the world. But these Marxist hyperboles apart, there is, as we shall presently see, a spotless rationale behind Marx's attitude towards the peasant.

The difference between the philosophy of the industrial town and of the agricultural countryside has often attracted the attention of sociologists and poets alike.[11] But few have realized that this difference is not like going to another church, and that it involves every concrete act concerning production and distribution as well as social justice. Undoubtedly the basis of this difference is the fact that the living Nature imposes a different type of restriction upon *homo agricola* than the inert matter upon *homo faber*.

To begin with, no parallelism exists between the law of the scale of production in agriculture and that in industry. One may grow wheat in a pot or raise chickens in a tiny backyard, but no hobbyist can build an automobile with only the tools of his workshop. Why then should the optimum scale for agriculture be that of a giant open-air factory? In the second place, the role of the time factor is entirely different in the two activities. By mechanical devices we can shorten the time for weaving an ell of cloth, but we have as yet been unable to shorten the gestation period in animal husbandry or (to any significant degree) the period for maturity

9. K. Marx, *A Contribution to the Critique of Political Economy* (Chicago, 1904), preface, p. 14.

10. Karl Kautsky, *La question agraire* (Paris, 1900), p. 3.

11. In the Western literature, Oswald Spengler is probably the best-known author for placing a great historical value upon this difference. See especially his *The Decline of the West* (New York, 1928), vol. II, chap. iv.

in plants. Moreover, agricultural activity is bound to an unflinching rhythm, while in manufacture we can very well do tomorrow what we have chosen not to do today. Finally, there is a difference between the two sectors which touches the root of the much discussed law of decreasing returns (in the evolutionary sense). For industrial uses man has been able to harness one source of energy after another, from the wind to the atom, but for the type of energy that is needed by life itself he is still wholly dependent on the most "primitive" source, the animals and plants around him. These brief observations are sufficient to pinpoint not only why the philosophy of the man engaged in agriculture differs from that of the townee but also why agriculture and industry still cannot be subsumed under the same law. Whether future scientific discoveries may bring life to the denominator of inert matter is, for the time being, a highly controversial — and no less speculative — topic.

Probably the greatest error of Marx was his failure to recognize the simple fact that agriculture and industry obey different laws; as a result, he proclaimed that the law of concentration applies equally well to industry and agriculture.[12] To repeat, Marx had no opportunity to observe a peasant economy. Nor is there anything in his vast literary activity to indicate that he ever studied a noncapitalist agriculture.[13] The analysis of rent in *Capital* is based entirely on capitalist production even during Marx's brief excursion into peasant agriculture.[14]

Probably no other theoretical aberration has been refuted by historical developments as promptly and as categorically as the Marxist law of concentration in agriculture. During the second half of the nineteenth century one census after another revealed that in agriculture concentration was continuously decreasing while the peasants instead of being proletarianized became landowners in increasing numbers. In Kautsky's own lamenting words, "the capitalists were on the increase, not the proletarians." The indictment was all the more unappealable since this phenomenon was taking place in capitalist countries without any planned intervention. That convinced everyone save the ultra-orthodox Marxists that the concentration law is false.

3. *Policy and factitious theory.* The aftermath of "the sorest experience of Marxist doctrine" — as Veblen labeled the refutation of the concentration law[15] — can be best appraised in the light of the Hegelian tenet which is the cornerstone of the Marxist doctrine. To recall, according to that tenet it is beyond man's power to change the course of history. This

12. K. Marx, *Capital*, vol. I (Chicago, 1906), chap. xiv, sec. 10.
13. Kautsky, *La question agraire*, p. xii. Also F. Engels in the preface to the third volume of *Capital* (Chicago, 1909), p. 16.
14. Marx, *Capital*, vol. III (Chicago, 1909), chap. xlvii, sec. 5.
15. Thorstein Veblen, *The Place of Science in Modern Civilization* (New York, 1919), pp. 450 ff.

is why Marx argued that socialism is to come as the natural product of the evolution of the relations of production, not because the interests of the working class would in any sense be superior to or more important than those of capitalists. Marx even scoffed at those who wanted to base a socialist platform on such "unscientific" arguments as greater social justice. But, always according to the Marxist Hegelianism, man can speed up the historical process so as to shorten the periods of growing pains. A right policy must be based on the acceptance of the inexorable outcome. Because of the belief in the concentration law in agriculture, Socialists were advised to more than welcome any measure that would tend to proletarianize the peasants so that the advent of socialism would be hastened. But since the peasant did not want to hear of proletarianization, Socialist parties found themselves rejected everywhere by the peasant masses. Failures on the electoral front, combined with the mounting evidence against the Marxist theory, brought about the internal crisis known as the Agrarian Question. At the Frankfurt (1894) and Breslau (1895) congresses, the Question almost wrecked the unity of the party.[16] Even though officially this unity was then saved, the Question continued to make life difficult for Marxism. In the end Marx himself was obviously disturbed by the overwhelming evidence and the mounting criticism, for in the last two years of his life he painfully sought to amend his theory, but not so as to jeopardize the political movement which he had set in motion and to which he was attached from first to last.[17] But his desire was unrealizable, because contradictory. After Marx's death the party made great efforts to cover up the Agrarian Question. They vacillated between Leninist opportunism, according to which they proclaimed loudly that no one intends to destroy the peasant, and various dialectic circumvolutions aimed at proving that there *is* concentration although in an

16. For the Agrarian Question one may consult Kautsky, *La question agraire*, and G. Gatti, *Le socialisme et l'agriculture* (Paris, 1902). The Kautsky work is important because it appeared (German edition, *Die Agrarfrage*, Stuttgart, 1899) only a few years after the Breslau congress, where Kautsky had a decisive role in defeating the "deviationist" motion. Gatti, on the other hand, was a prominent Socialist who ultimately embraced the non-Marxist view on agriculture.

17. Marx's public concession, though somewhat veiled, is found in the preface to the 1882 Russian edition of the *Communist Manifesto* — see K. Marx and F. Engels, *Correspondence, 1846–1895*, ed. Dona Torr (New York, 1935), p. 355. A clearer expression of the deviation from "the Marxist line" came in a letter Marx wrote in 1881 to Vera Zasulich in answer to a definite question regarding the necessity of speeding up the proletarianization of the Russian peasant. The letter, however, was published by the Marx-Engels Institute only in 1924, when the struggle between Russian Marxists and their adversaries was long since over. (D. Mitrany, *Marx against the Peasant*, University of North Carolina Press, 1951, pp. 31–33, was the first to draw the attention of the English-speaking reader to this letter.) We know also that in his last years Marx decided to learn Russian (and apparently even Turkish) to have access to the original sources concerning the agrarian problems of Eastern Europe (*Correspondence*, p. 353). For more than one reason, it was too late.

entirely new sense.[18] The Agrarian Question was thus kept on a low flame until Stalin decided to solve it by proclaiming a holy war against the peasants, a war with which neo-Marxism has since become almost synonymous.

It is hard not to see in this momentous decision the ultimate product of Marx's scorn for the peasant. Indeed, this scorn constituted a lasting ferment for the thinking of Marxist leaders. Quite early, none other than Engels spoke of the necessity for the proletariat "to crush a general peasant uprising." [19]

Be this as it may, the Stalinist war, which by its number of victims surpasses all other wars known to history, could not have found sufficient momentum in the cultural opposition between the urban and the rural sectors. Nor could this war feed on "sacking the rich," for precisely in the regions where Stalinism has till now spread, the capitalist-bourgeois class was paper-thin, and the rich peasant quite a rarity. The war must have had other springs.

That the interests of the town conflict with those of the countryside is by now a well-established fact. However, it is not always realized that the price-scissors do not tell the whole story. For this story, we must observe that food is indispensable, whereas the need for industrial products is secondary, if not superfluous. To obtain foodstuff from the agricultural sector, and moreover to obtain it *cheaply*, constitutes a real problem for the industrial community. In the ultimate analysis, "cheap bread" is a cry directed against the tiller of the soil rather than against the capitalist partner of the industrial worker. In some circumstances this conflict may become very spiny. And it is permanently spiny in the overpopulated countries where the income of the masses allows only the satisfaction of the most elementary needs and where the population of the town is unduly swollen by a rural exodus. That has been the situation in all countries — with one or two exceptions — where Stalinism has come to power. And it is in this situation that the war against the peasant found its needed spring.[20]

18. An epitome of these endeavors is offered by Kautsky, *La question agraire*. He argued that, although the concentration law is not true as to the size of the holdings, it is true as to the global ownership, with more landowners having important outside sources of income. Then he threw everything overboard by arguing that peasant agriculture must disappear in any case because the optimum scale of production is that of latifundia.

19. Quoted in Mitrany, *Marx against the Peasant*, p. 219.

20. The conflict between the interests of the agricultural and industrial sectors exists also in the advanced economies, including the United States. Cf. J. D. Black, "Discussion," *Proceedings of the Fifth International Conference of Agricultural Economists* (London, 1939), pp. 86–87. The only difference is that in these economies the conflict is attenuated by the high income, and therefore it can be resolved by such methods as the Agricultural Price Support Program. Overpopulation is the necessary condition for the conflict to become a social *vis viva*.

Clearly, the Stalinist formula constitutes a solution (at least a temporary one) of the conflict between the industrial and agricultural sectors. But the solution is based on the primacy of the interests of the industrial and bureaucratic sections of the society, not on some evolutionary law regarding the inexorable proletarianization of the peasants.[21] Consequently according to the very essence of Marxism the Stalinist formula cannot claim to be "scientific."[22]

Marx was, however, aware of the conflict between the industrial and agricultural divisions of society. He once remarked quite *en passant* that "the whole economic history of society is summed up in the movement of this antithesis [the division between the city and the countryside]." [23] This remark is extremely important. It shows that Marx, for once, recognized the existence of an antithesis which — as we argued in the preceding section — seems rooted in the permanent conditions of the human species, and which should therefore outweigh any antithesis peculiar to a particular economic system. Unfortunately, Marx did not explore this point further to explain how he would have envisaged the *scientific* (in the Hegelian sense) solution of that antithesis.

4. *Policy without theory.* In the first half of the nineteenth century, while the West grew intensively preoccupied with the lot of the industrial masses, Russia witnessed the rise of a social movement concerned solely with the peasant. Time and again, essentially different economic conditions imposed entirely different preoccupations. It was not, therefore, because of the much discussed intellectual isolation of Russia that the founders of this new ideology borrowed nothing from Western economic theories. They simply drew the logical consequences from the fact that these theories were molded on a different economic reality. But as their intellectual inheritance contained nothing regarding the economics of a peasant community, the new social reformers had to start from scratch. They soon discovered that their personal social background could not help them in grasping the problems in which they were interested, and as a consequence decided to go "to the people." This slogan earned them the

21. A London tailor, J. G. Eccarius, *Eines Arbeiters Widerlegung der national-ökonomischen Lehren John Stuart Mills* (Zürich, 1868), was the first to argue that to guarantee "cheap bread" to the industrial worker the peasant must be placed under the dictatorship of the proletariat. The book, it is said, enjoyed great prestige among Marxists during the 1870's (see Mitrany, *Marx against the Peasant*, p. 15). That Eccarius's view has become the basis of Communist agrarian policy is beyond question: "general collectivization of the peasants is indeed a means of . . . securing the supply of food [for the towns]" (V. Lenin, *Selected Works*, Moscow, 1934–1939, XII, 13). The fact is that Narodnikism is an older school of thought than Marxism.

22. As we shall see, neither can it be justified on positive welfare grounds; *infra*, footnotes 83 and 85.

23. Marx, *Capital*, vol. I, chap. xiv, sec. 4, p. 387.

Russian name of *Narodniki*, but outside Russia they became generally known as Populists.[24]

As Marxism began to acquire a basis of its own in Russia, the incompatibility between Marxist theory and the Russian reality gave rise to a fierce and more lasting conflict between *Narodniki* and Marxists than that between the orthodox Marxists and the Agrarian Socialists in the West. Some *Narodniki* did become attracted by Marxism, primarily because its programmatic implications and social dialectics appealed to their revolutionary spirit. But as it was impossible to fit the peculiarities of an agrarian economy into the Marxist frame, most of these succumbed as hetero-Marxists. The great majority of the *Narodniki*, however, refused to be lured into denying the specific traits of that economy. And thus the Agrarian ideology came to be identified with a double negation: not Capitalism, not Socialism. It is precisely this double negation that has been called into question by Western economists, whether Marxist or not.

David Mitrany observes that Marx's view on peasant agriculture combines "the townsman's contempt for all things rural and the economist's disapproval of small scale production." [25] But this is true for most Western social scientists. Add to this, especially, their usual disdain for any idea that is not presented through a mathematical model, and you have the explanation for the misunderstanding of the Agrarians by the West.[26] Indeed the *Narodniki*, like the Agrarians of latter days, have not only failed to construct a theory of the peasant economy — as the others have done for capitalism — but they have distinguished themselves by a lack of interest in, almost a spurn for, analytical preoccupations. They relied exclusively on the intuitive approach, on the *Verstehen* of the peasant's *Weltanschauung*, much as the German historical school advocated (although there was hardly any direct contact between the two schools). Populism, like Marxism, represented not only an economic doctrine but a faith as well. And this faith "fed on a strong sentimental undercurrent,

24. Alexander Herzen, who in 1847 went into exile because of his political activity, is generally regarded as "the founder of Russian 'Socialism,' or 'Narodnikism,' " as Lenin put it. Quoted in Marx and Engels, *Correspondence*, p. 285.

25. Mitrany, *Marx against the Peasant*, p. 6.

26. In this respect, it is highly instructive to compare, for instance, the analysis of Populism by L. H. Roberts, *Rumania* (New Haven, 1951), pp. 142 ff, with that by Rosa Luxemburg, *The Accumulation of Capital* (London, 1959), pp. 271–291. Although Rosa Luxemburg was "a more genuine Marxist than any other member of the German movement" (Paul M. Sweezy, *The Theory of Capitalist Development*, New York, 1942, p. 207), her analysis is far more objective than Roberts's.

On the *Narodniki*, one may fruitfully consult also J. Delewski, "Les idées des 'narodniki' russes," *Revue d'économie politique*, XXXV (1921), 432–462, and above all Mitrany, *Marx against the Peasant*, chap. iv. The memoirs of the "grandmother" of the Russian revolution, Katerina Breshkovskaia, *Hidden Springs of the Russian Revolution* (Stanford, 1931), are interesting as personal history.

on the emotional piety and rustic ties" of its believers.[27] All this laid Populism open to the accusation of romanticism.

The particular circumstances in which *Narodnikism* began its career may account for much of its peculiar spirit. But the lack of any true theorizing in the Populist doctrine was due more to the unusual difficulty of casting the peasant's economic conduct into a schema than to anything else. For this we have the testimony of one of the most praiseworthy Russian Agrarians, Alexander Tschajanov, who gave to one of his works the symptomatic title: *Die Lehre von der bäuerlichen Wirtschaft: Versuch einer Theorie der Familienwirtschaft im Landbau* (Berlin, 1932). In the concluding remarks of this book, in which he submits only the various activities of agricultural production to quantitative analysis, Tschajanov confesses his dissatisfaction over the fact that we still do not possess a theory of the economic behavior of the peasant. He significantly observes that the relation between Classical economics and an economic theory of a peasant community seems to be similar to that between Euclidian and non-Euclidian geometry. Yet he ends with the admission that an abstract theory of agrarian economics cannot easily be constructed.[28]

Whatever the explanation for the outlook of the Agrarians, there is no more dramatic example of the disaster that awaits him who in formulating an economic policy disregards theoretical analysis, than the well-known fate of the Agrarian parties of Eastern Europe.

II. OVERPOPULATION: A RE-EXAMINATION

1. *The facts analyzed.* The Agrarians have at all times sensed that the plague of most underdeveloped agrarian economies is overpopulation and that consequently the problem of a peasant economy is to a large extent a population problem.[29] It is natural, therefore, for us to see whether an analysis of overpopulation would not offer a lead to the solution of the Agrarian riddle.

Whoever speaks of "excess" is naturally expected to define it in terms of a point of reference which in some way must represent a normal if not an optimum situation. But to define "normal" or "optimum" is not easy, especially if one faces a quibbling relativist. Such a relativist may argue, for instance, that the excess capacity of a monopolistic industry is a fiction because all capacity could be used if monopoly were removed and a new

27. Mitrany, *Marx against the Peasant*, p. 40.
28. Tschajanov, *Die Lehre* (as cited in text), p. 130.
29. Cf. Tschajanov, p. 131, for instance.

system of distribution were introduced. A wholly analogous position is adopted by Marx in arguing that overpopulation exists only relative to "the average needs of the self-expansion of capital." [30] Be this as it may, we must recognize that the concept of overpopulation presents unusual difficulties. Normal (or optimum) population implies the concept of normal (or optimum) life. And even if the latter were not such an elusive concept, we would still find it impossible to choose a "normal" valid for all times and localities. To avoid the trivial conclusion that every population is normal for the time and place in which it lives, it is necessary to adopt some criterion of normality. This criterion may be dynamic or static, depending on the problem at hand.[31]

Ever since statistical data have been used for comparative purposes, it has become obvious that some agricultural countries presented symptoms suggesting the existence of some sort of overpopulation. It has been remarked that, given the following data for two prominently agricultural economies,

	Denmark	Yugoslavia
Inhabitants per square kilometer of arable land	36.6	157.4
Wheat yield in quintals per hectare	22.9	11.0

even if Yugoslavia could raise her agricultural yield to the Danish level, the average Yugoslav would still have only one quarter as much food as the average Dane. This observation has supplied the basis for a crude concept of relative overpopulation upon which are based the measures of overpopulation in terms of some crop basket as a standard.[32] As has generally been admitted, the concept of relative overpopulation thus defined is ambiguous and the procedure for its measure debatable.[33] The chief drawback of this approach, however, is that it sidetracked the analysis from the right direction. Indeed, a difference in the *per capita* national product (or a sector of it) may be a *symptom* of the difference between two economic systems, but by no means an *intrinsic coordinate* of that difference. Otherwise we should regard the economic system of Belgium as different from that of the United States. But the belief that the difference between an agrarian and a capitalist economy is a matter of degree only, not of essence, is still very frequent.

30. Marx, *Capital*, vol. I, chap. xxv, sec. 3, p. 695.

31. Marx, for instance, argued that the developed means of communication in the United States at the middle of the last century made that country more densely populated than India. *Ibid.*, vol. I, chap. xiv, sec. 4, p. 387.

32. Cf. W. E. Moore, *Economic Demography of Eastern and Southern Europe* (Geneva, 1945), chap. iii.

33. *Ibid.*, pp. 55 ff.

And yet the elements for the solution of the problem were not out of reach. In the 1930's, studies originating in several countries with large peasantries revealed the astounding fact that a substantial proportion of the population could disappear without the slightest decrease in the national product.[34] The closeness of the independent estimates of the *superfluous* population for each case shows that we are confronted with a real quantitative phenomenon.[35] If additional proof of this is needed, one may invoke some relevant "experiments" history carried out *in vivo*. For two years after the beginning of hostilities in 1914, agricultural production in Russia was maintained at the prewar level, although no less than 40 per cent of the *able-bodied* male peasants were in the army.[36] The same phenomenon occurred in Rumania during World War II. Whenever agricultural production collapsed in Eastern Europe during the World Wars it was solely because of the extreme requisition of draft animals, the difficulty of replacing worn-out implements, and, of course, the disturbances caused by the movement of armies. Even the disappearance of some ten million Ukrainian peasants during the so-called liquidation of kulaks, although accompanied also by a radical disturbance of the entire economy, had only an ephemeral influence on agricultural output.[37]

Now, to say simply that part of the population could disappear without causing any decrease in output is not sufficient for a theoretical characterization of overpopulation. The national product of the United States *could* easily be maintained at the same level even though a large proportion of the population were to disappear. The *differentia specifica* between the two situations is that in the latter the national product could be increased if people simply chose to have less leisure, while in the former, not.[38] This difference reveals that the situation where the marginal pro-

34. References to the earliest studies for Poland and Bulgaria are in Doreen Warriner, *Economics of Peasant Farming* (London, 1939), pp. 68–69.

35. For Rumania, one study (*Enciclopedia României*, Bucharest, 1939, III, 60) estimated the percentage of superfluous peasant population at 48, another, at 45 (V. Madgearu, *Evoluția economiei românești după războiul mondial*, Bucharest, 1940, p. 49). The first estimate was derived from national statistical data, but the second was checked by *direct observation* in extensive field work covering sixty villages chosen at random. Moore, *Economic Demography*, pp. 63–64, using national data, arrived at a percentage of 51.4.

36. Leonard E. Hubbard, *The Economics of Soviet Agriculture* (London, 1939), pp. 59, 65.

37. *Ibid.*, p. 117. Harvey Leibenstein, in a recent paper, "The Theory of Underemployment in Backward Economies," *Journal of Political Economy*, LXV (1957), 103, alludes to some experiences in the Soviet orbit when industrialization would have caused a shortage of labor in the agricultural sector, but fails to say precisely which events he refers to. My guess is that they exhibited only the familiar kind of *spurious* shortage caused by wholesale dislocations of persons, if they did not reflect either peasant resistance or administrative inefficiency.

38. Marx, in *Capital*, vol. I, chap. xxv, sec. 3, p. 698, asserts that if the population of England would be reduced in the same proportion for all categories, the remaining population "would be absolutely insufficient" to maintain the same level of output

ductivity of labor equals zero is the starting point in searching for a definition of overpopulation. And the existence of countries where the actual marginal productivity of labor is zero for all practical purposes has been admitted by all keen observers of peasant economies.[39]

All this clearly conflicts with Professor Schultz's categorical statement that there is "no evidence for any poor country *anywhere* that would even suggest that a transfer of some small fraction, say 5 per cent, of the existing labor force out of agriculture, with other things equal, could be made without reducing its [agricultural] production." [40] Nothing is farther from my thought than to challenge the fact that the concrete cases cited by him prove that in several Latin American countries the agricultural production did fall off after some labor had been transferred to other activities.[41] But that is not sufficient to justify his well-known position, namely that the overpopulation theory of underdevelopment "as a 'theory' . . . fails in that the expected consequences are not those that one observes." [42] The situation of most Latin American countries is not identical with that of the East European or Asiatic countries, although they all have this in common: they are underdeveloped. Although overpopulation is always accompanied by underdevelopment, it is neither a necessary nor the only cause of it. The underdevelopment of Latin American countries may have other bases than overpopulation.[43] Overpopulation, therefore, cannot provide the basis for a *general* theory of underdeveloped economies, but only of those economic realities beset by it. This is concrete illustration of the point which one of the preceding sections sought to bring home.

To regard the notion of overpopulation as a myth is undoubtedly a Marxist residual. And precisely because that notion still meets with opposition in some circles, a few further remarks seem in order. If in the "so-called" overpopulated economies the marginal productivity is zero — a critic may ask — how can we explain the fact that in such economies there is a greater need for skilled labor than in the other countries? Certainly — he may continue — you are not going to say that the marginal productivity of an engineer in India or Egypt is zero. But this way of looking at the

---

in spite of England's "colossal" means for saving labor. Clearly, this implies that no skilled labor has *free* leisure, a characteristic assumption of Marxist economics.

39. E.g., Warriner, *Economics of Peasant Farming*, p. 65.

40. Theodore W. Schultz, "The Role of Government in Promoting Economic Growth," in *The State of the Social Sciences*, ed. L. D. White (Chicago, 1956), p. 375. (My italics.)

41. *Ibid.*, pp. 375–376.

42. Theodore W. Schultz, *The Economic Test in Latin America*, New York State School of Industrial and Labor Relations, Bulletin 35, August 1956, p. 15.

43. Although my knowledge of the factual situation in those countries is very superficial, I would venture to suggest that some are "underpopulated" relative to the available land resources. Professor Schultz's evidence may even corroborate such a view.

problem is to intermingle evolutionary factors with static concepts and, above all, to confuse labor with capital. An evolutionary change is bound to bring about shortages of some types of skilled labor (and surpluses of some others) in *any* economy. Thus, Italy certainly feels today a shortage of technicians for her newly discovered oilfields. This, however, represents a quasi bottleneck, to borrow an expression coined in the Marshallian spirit by Professor Lewis.[44] If no further evolutionary changes occur, the quasi bottleneck will disappear just as any quasi rent will do. But, once the new equilibrium is reached, will the marginal productivity of a petroleum technican become zero? Not at all. For the equilibrium marginal productivity of such a technician represents not only the marginal productivity of his labor but also that of the capital invested in his training.[45] Obviously, this line of reasoning regards labor as a uniform, plastic quality of all human beings, and is — I believe — in the tradition of Classical, and somewhat in that of Marxist, economics. But I fail to see a better way to analyze the problems raised by *population* in its *purely quantitative* aspect. Actually this view of labor is even more necessary in the analysis of economic growth than of a stationary state, where population may very well be regarded as a frozen distribution of *qualities*.

Therefore, the statement that the marginal productivity of labor is zero means that the marginal productivity of skilled labor consists only of the marginal productivity of the capital invested in the production of skill. It is a most rational expectation that an overpopulated economy should feel a greater shortage of skilled labor than a nonoverpopulated one. Everything points to the fact that a shortage of skilled labor reflects a shortage of capital, not necessarily of labor. It is a peculiar feature of overpopulated economies that the skilled laborer is overburdened with work while the unskilled is loafing most of the time. Further still, the real economic aspect of spreading knowledge in an underdeveloped country now appears in full light: the need for additional education competes with the need for additional physical means of production, a fact which we are apt to overlook at times and underestimate often. Where resources are very scarce, free education for *all* types of skills is as *uneconomic* as haphazard production of capital equipment. Some countries, like Soviet Russia, seem to have grasped this truth; others, such as Italy, apparently have not.

It is a simple matter of definition to observe now that in any economy, whether overpopulated or not, there is only one way to measure the marginal productivity of labor: at the margin, i.e., where labor appears un-

44. W. Arthur Lewis, "Economic Development with Unlimited Supply of Labor," *The Manchester School*, XXII (1954), 145.
45. Another part may reflect the "rent" of his personal talents, but that is a side aspect of the problem.

adulterated by capital. The marginal productivity of labor in any economy then is the marginal productivity of its unskilled labor. It is a mere factual coincidence that agricultural labor generally is unskilled labor. But this fact throws a new light upon the constant correlation of overpopulation with the agricultural conditions in the economic literature. For clearly, if the marginal productivity of labor in a country is zero, so must be the marginal productivity of the peasant.

2. *A theoretical schema.*[46] To make the argument as simple as possible, let us assume that the national *product*, represented aggregatively by x, is produced by an atomistic "industry" (an assumption fully justified in overpopulated agrarian economies). This means that the production function of the entire economy is homogeneous of the first degree,

$$(1) \qquad\qquad x = F(L, T) = TG(L/T)$$

where L stands for labor, and T for a composite variable of land and capital. To obtain this function, for each proportion of the factors of production Op (Fig. 11–1), we determine the optimum size, U, of the

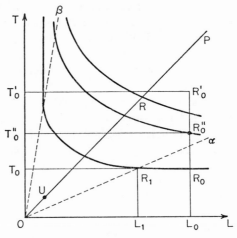

FIGURE 11–1

production unit, from the system

$$(2) \qquad\qquad T = pL, \qquad L\frac{\partial f}{\partial L} + T\frac{\partial f}{\partial T} = f(T, L),$$

where $f(T, L)$ is the production function of the "firm." The maximum product obtainable from a given combination of factors, R, is equal to

46. I prefer the term "schema" to the commonly used "model," for I wish thereby to emphasize the essential difference between the *blueprint-model* of natural sciences and the *simile-schema* of social sciences.

OR/OU times the output of *U*. That determines (1) for every *R*. It is important, however, to remember that to obtain the output computed from (1), the resources $R(L, T)$ must be equally divided among *OR/OU* identical production units.[47] It is also seen that given the amounts of the factors of production, the optimum scale of the "firm" is *uniquely* determined for every technological horizon. Hence, in the case where the geohistorical conditions of an economy are such that *all available resources* must be used in production as long as they increase output, the argument regarding the superiority of the large-scale production is poor economics.[48] In the causal relationship between the efficiency of large-scale production units and a high level of economic development (i.e., a high T/L) the latter is the cause of the former, not vice versa. There is no need to go over the reasons why the isoquants of any production function sooner or later become parallel to the axes. Thus, in the region $LO\alpha$, for instance, the output can be increased if and only if there is an increase in the factor land-capital. We shall refer to such a factor as limitative.[49] Clearly, in the region where a factor is limitative its marginal productivity is constant, while that of all other factors is zero.

A few definitions. Let the population, *P*, of a given economy be divided into P_w, the working class, and P_g, the "government" class. In the latter class we include all members of the economy who are not dependent on wages or salaries received from the "industry" producing x.[50] Let *s* and *s'* be the individual *minimum* standard of living of the working and government class respectively. The position taken in this paper is that both these variables are historically determined, and consequently susceptible of being changed by economic policy. They condition the *minimum* public need for *x* (public roads, armaments, capital accumulation, etc.). If this minimum is denoted by *E*, the minimum net product of the community is

(3) $$\chi = P_w s + P_g s' + E.$$

Let us also put

(4) $$F = \varphi P_w, \qquad L_0 = \delta F$$

where *F* is the size of the *potential* labor force, and δ represents the labor

47. Strictly speaking, *OR/OU* is not necessarily an integer, but for an atomistic industry this does not matter. We should also point out that *U* is placed out of scale in Fig. 11–1; otherwise it could hardly be distinguished from *O* on the drawing.

48. That is the fault of the argument advanced by Kautsky *et al.* against peasant holdings. *Supra*, note 18.

49. Not to be confused with *limitational*. A factor is limitational when its increase is a necessary but not a sufficient condition for an increase in output. See Georgescu-Roegen, "Limitationality, Limitativeness, and Economic Equilibrium," in *Proceedings of the Second Symposium in Linear Programming* (Washington, D.C., 1955), 301 [in the present volume, p. 341].

50. This class corresponds to what Veblen called "the kept class." It naturally includes all kinds of servants, public and personal.

time a worker can supply *above the biologically necessary minimum of sleep and rest.* For symmetry, the time necessary for sleep and rest will not be included in *leisure.*

The primordial economic problem of any community is to find a mode by which a national product equal at least to χ can be obtained with the available resources. One coordinate of the problem is the labor supply. Since *ex hypothesi* the working class cannot maintain itself on a smaller real income than sP_w, the supply curve of labor must start discontinuously from an end point M of coordinates L_0, sP_w/L_0 (Fig. 11–2). Since men

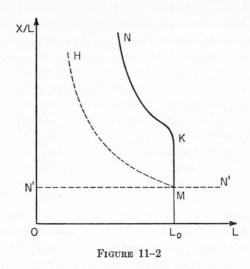

FIGURE 11–2

generally live in families, the labor supply of one individual depends upon the income of his family. To get around the difficulty of circular dependence, we may assume that the labor supply of the community is obtained by summing up the supplies of all families. With this remark, let us consider first the situation in which people can freely sell *leisure* at the market wage rate, and let MKN represent the amount of labor supplied at various wage rates. Clearly, MKN is the short-run supply of labor in a wage economy.[51] Its relation with the preference-field is shown more clearly if we refer to a familiar map of indifference curves. In Fig. 11–3, $OE = \delta$ and $OF = s$. The economic problem of the worker is in fact a discontinuous jump: from E, where his *natural* income places him, at any point in

51. Classical economics argued that the long-run equilibrium wage rate is constant and equal to ON', so that $N'N'$ represents the long-run supply curve of labor. For Marxist economics, however, $N'N'$ represents both the short- *and* the long-run supply of labor, a direct consequence of its assumption of a permanent reserve army. This is the analytical expression of the distinctive feature of Marxist economics in refusing to accept any relation whatever between economic and demographic factors. On this point, see a letter of Engels in Marx and Engels, *Correspondence, 1846–1895*, p. 199.

the area $XFF'E'$ in Fig. 11–3(a). Where he will finally land depends upon the type of economic system in which he lives. If this is a free wage market, his labor supply is given by a Hicksian "price-consumption" curve FF'', and all is well. A very frequent pattern of behavior is that for which the supply of labor is inelastic for wage rates just above the minimum possible (i.e. the slope of FE). Be this as it may, we can safely assume that the supply curve of labor has everywhere an elasticity smaller than unity. The literature on underdeveloped countries, however, often mentions a very curious pattern of behavior, that of the individual who after earning the minimum of subsistence becomes interested exclusively in leisure. Such behavior, understandably, would make any policy maker despair: the individual seems to refuse to be developed. The behavior leads to the indifference map shown by Fig. 11–3(b), and to a supply curve represented by a branch of an equilateral hyperbola, MH (Fig. 11–2).[52] Whatever the pattern of behavior, the curve MH constitutes a relevant element for the analysis of distribution: it represents the minimum average share per unit of labor time for each amount of employment.

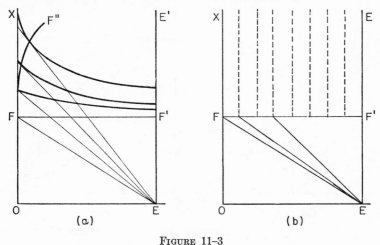

FIGURE 11–3

The second coordinate of the economic problem is the productivity of labor under the assumption that all land-capital resources, T_0, are used in the production of x. The curve, μA, representing the average productivity of labor obviously varies with T_0, but its shape presents some constant features (Fig. 11–4). All these curves begin by a horizontal segment

52. I feel that for this type of behavior we must assume a zero marginal rate of substitution between real income and leisure. But if I am wrong, policy makers confronted with this hopeless reaction to a wage scheme should be able to get around the difficulty by imposing a *corvée* simultaneously with a very high wage rate for freely contracted work. A simple diagram will show that in this way they can induce the individual to move inside the area $XFF'E'$.

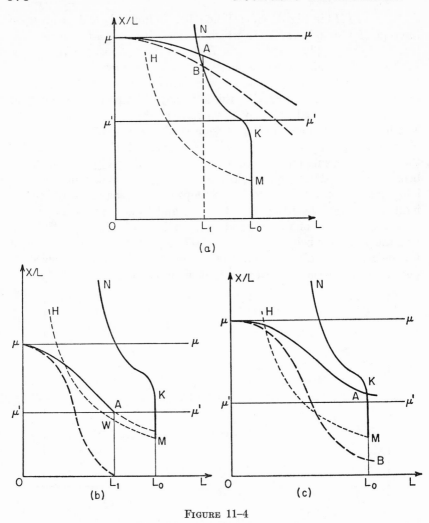

FIGURE 11–4

at the same level $O\mu$, because x/L remains constant in the domain $TO\beta$. They also coincide with equilateral hyperbolas below the level $O\mu'$ equal to the value of x/L on $O\alpha$.[53]

If T_0 is sufficiently large, the curve of the marginal productivity of labor μB intersects MN at B, as in Fig. 11–4(a). In such a case the stage is set for the solution of the community's economic problem according to the marginal productivity principle (provided $s'P_g + E$ is not too large, as has been the case in many countries during the World Wars). The economy may even allocate part of T to the direct use by consumers, so that although

53. $TO\beta$ and $O\alpha$ refer to Fig. 11–1.

the available resources are represented by R_0', only the amounts represented by R are used in production (Fig. 11–1). At the other extreme, for low values of T_0, we find the case where the average productivity curve of labor μA lies below $\mu'\mu'$ for $L = L_0$ as in Fig. 11–4(b). That is the case of *strict overpopulation*.[54] (This corresponds to R_0 in Fig. 11–1.) Obviously, in such a situation there can be no economic advantage in using labor beyond L_1, where its marginal productivity becomes zero. To continue existing without being beset by Malthusian forces, the economy cannot have an $s'P_g + E$ larger than $AW \times L_1$. In practice, however, this maximum is always attained, so that ordinarily in overpopulated countries $\chi = X_0$, X_0 being the maximum national product obtainable with the available resources. It is obvious, however, that the economy cannot possibly function according to the principles of marginal productivity theory. This is true also for an economy where the marginal productivity of labor is positive for $L = L_0$ without being greater than $L_0 M$; see Fig. 11–4(c). For this and other good reasons, such an economy should be regarded as overpopulated, without the qualification "strictly." [55]

The important conclusion is that overpopulation is correlated with a low T_0, more precisely with a low T_0/L_0. Overpopulation, therefore, is tantamount to poverty of nonlabor resources; the opposite situation, to "land-of-plenty." In the real world, however, most underdeveloped *agrarian* economies are poor not only because of the insufficiency of land, but also because of a chronic dearth of capital. For these countries the difference between T_0 and the size of usable land, A_0, is negligible. That justifies the use of L_0/A_0 as an ordinal index of agricultural overpopulation instead of L_0/T_0. Finally, by replacing L_0/A_0 by P/A_0, we obtain the most commonly used but crude form of that index.

3. *Further remarks.* In approaching issues concerning the economy as a whole the economist has but two choices: to use either a general equilibrium apparatus or an aggregative schema. In the first case, he must resign himself to being rather sterile on practical matters; in the second, he must accept the theoretical calamities of aggregation. For more than one reason, I have chosen the latter procedure. It is, however, possible to illustrate the conditions of overpopulation by a schema in which the national product is not completely aggregated. Let us assume that the economy produces an agricultural product, X_1, and an industrial product, X_2. Bearing in mind that in an overpopulated economy the standard of living barely covers the most elementary needs, and that these needs are highly rigid,

54. We consider only the alternative where μA intersects MN. The opposite case involves Malthusian aspects which though interesting lie beyond the scope of this paper.
55. In Fig. 11–1 this case corresponds to R''.

we may proceed on the assumption that the two products are not substitutable. If the minimum necessary product is represented by X_1^0, X_2^0, the case of overpopulation is illustrated by Fig. 11–5(a).[56] The only solution is M, where the marginal productivity of labor is zero in both productive sectors. Hence, in this case too, no advantage can be derived from using labor beyond L_1.

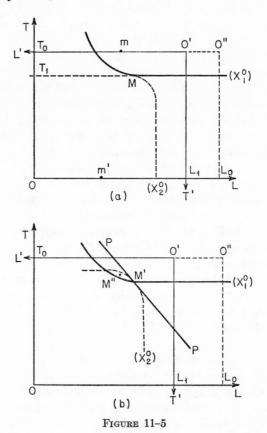

FIGURE 11–5

The schema of Fig. 11–5 forces upon us a series of highly interesting problems. We shall mention only one. A nonisolated economy has the choice between producing one kind of product at home and obtaining it through foreign trade. The question then is whether the resulting national income (X_1', X_2') could allow an overpopulated economy to move its industrial structure from M to m (or m') where the marginal productivity of labor is positive; see Fig. 11–5(a). We cannot pursue here this intricate

56. In Fig. 11–5 an "Edgeworth-box" is used with O and O' representing the origin of each system of coordinates.

problem, but we may at least remark that no agrarian country seems to have been able to escape the conditions of overpopulation by mere trading. Most probably, overpopulation will remain a local problem calling for local remedies as long as people in general neither wish nor are allowed to leave their own lands.

III. AN ANALYSIS OF THEORETICAL ISSUES

1. *Profit versus tithe.* The question of whether the Walrasian system has a mathematical solution has always been considered a crucial one for Standard theory. But no Standard economist seems to have realized that the Walrasian system raises a still more vital question: Is its mathematical solution also an *economically* valid one? That it is valid has been taken for granted by all who tackled the unusually difficult problem of the existence of the mathematical solution. To recall, Abraham Wald was content with proving that in a (simplified) Walrasian system the "equilibrium" prices are non-negative.[57] Wald was hardly an economist, but after the publication of his original paper in German (1934), no economist observed that unless we also know that the "equilibrium" price of labor is at least equal to the minimum of biological subsistence, the theorem has only meager economic relevance. The truly economic aspect of the problem is most clearly set aside in the more recent work of Arrow and Debreu. These authors start out with the assumption that *every* member of the community is endowed *ab initio* with a sufficient real income for his entire life span.[58] What we do know, however, is that man is endowed with labor of limited efficiency and can use resources of limited quantities. These limits in some cases may be such that although the economy can produce a sufficient real income for all, this *economic* solution cannot be reached by the mechanism of marginal productivity which is part and parcel of the Walrasian system.[59] We have seen that overpopulated economies are in this particular situation. The problem now is to see how production and distribution may be regulated in such an economy.

A lead for the solution of this problem is offered by the observation that agricultural overpopulation has usually been manifest in countries where

57. Abraham Wald, "On the System of Equations of Mathematical Economics," *Econometrica*, XIX (1951), 368–403.

58. Kenneth J. Arrow and Gerard Debreu, "Existence of an Equilibrium for a Competitive Economy," *Econometrica*, XXII (1954), pp. 266, 270.

59. On the surface, this statement may appear to contradict the theorem proved by Arrow and Debreu. That is not the case. Indeed, their proof assumes the economic problem already solved: the individual has already jumped from E into $XFF'E'$ (see Fig. 11–3a). In their approach, the Walrasian solution may consist of everyone conserving his initial position without any alteration.

feudalism was late in being supplanted by capitalism. To see the difference between the distribution under feudalism and capitalism, let us draw the familiar curve of the marginal productivity of labor on the available T_0, $ABCL_1$ in Fig. 11-6. Let also $abcL_1$ be drawn so that the ratio between the

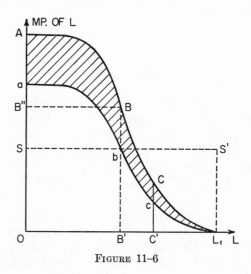

FIGURE 11-6

ordinate of the shaded area and that of $ABCL_1$, say between cC and $C'C$, be equal to the tithe ratio, ρ. It is elementary that if OB' labor is used in production, the share of the entrepreneur-landlord is $AabB$ in the feudal system, whereas the same share under capitalism is $AB''B$. The difference between the two systems is thus clear. But there is also an analogy: to the interest of the government class to maximize profit-rent in the capitalist system, in feudalism there corresponds the interest of the same class to maximize tithe.[60] Obviously, for any given value of ρ, tithe is maximum at L_1. In a strictly overpopulated country this maximum is bound to be reached as the result of workers' own necessity of securing a share at least equal to sP_w. We should keep in mind also that in the feudal system of distribution *there are workers who earn more than their specific contribution to output.* This is seen immediately if $OSS'L_1$ is drawn so that its area is equivalent to that of $OabcL_1$; all labor between B' and L_1 receives more than its *net* contribution to output. Striking historical evidence of this

60. Economic relations under feudalism presented almost an infinite gamut. However, the most frequent features of feudal economy were the *corvée* and the crop-sharing. The *corvée* usually consisted of work performed at the manor, on the land under special cultures appurtenant to the demesne (vineyards, orchards, gardens), and for public works. Both these institutions survived the legal abolition of the feudal system, as we have had occasion to observe during our own time everywhere in Eastern Europe. The use of the term "tithe" for the share of the landlord is, admittedly, improper, but convenient.

aspect of feudalism is provided by the gleaners, who received a share greater than the quantity of corn gleaned. In contrast with this, capitalism has no place for any "gleaners."

It goes without saying that a feudal government class would seek a *maximum maximorum* of tithe by maximizing also ρ. Obviously, the maximum of ρ is given by the ratio AW/AL_1 of Fig. 11–4(b), or alternatively by the relation

$$(5) \qquad\qquad X_0(1 - \rho) = sP_w.$$

This means that workers receive only their minimum of subsistence. Yet, this is not a *sine qua non* condition of the system, for at least in a strictly overpopulated economy the workers may still work up to L_1 even if ρ were equal to zero — unless their behavior conforms to Fig. 11–3(b).

The formula *"the laborer must be poor to be industrious"* came to be a key for the feudal system only as the conditions of strict overpopulation ceased to exist owing to land reclamation, to the increase in accumulated capital, and finally to the increase of labor productivity as the result of technical progress. These factors caused the productivity curves of labor to shift upwards and to the right. Ultimately the marginal productivity of labor for $L = L_0$ became positive, but still less than ML_0; see Fig. 11–4(c). Clearly, the economy in question was not yet ready for the capitalist formula. Hence, feudalism, which could still provide a solution to the problem of distribution, continued its existence although plagued by a new conflict. If the marginal productivity of labor is positive for $L = L_0$, maximization of tithe requires that workers should have no leisure. Whether they would work to that limit of their own accord in a (not strictly) overpopulated country depends upon the value of $X_0(1 - \rho)$, i.e. on ρ. To render the laborers willing to work up to L_0, the average share of labor must not exceed L_0K. This is how the above-mentioned formula came to be an undisputed economic dogma for late feudalism. Even Quesnay argued that the laborer who can buy his bread cheaply becomes "lazy and arrogant." [61] How deep-rooted this formula must have been (at least in late feudalism) is shown by its echoing in economic literature as late as the nineteenth century.[62]

61. *Oeuvres économiques et philosophiques de F. Quesnay*, ed. Auguste Oncken (Paris, 1888), p. 248. The only objection raised by Quesnay is that if abused to the extreme the formula may bring the laborer so close to the status of an animal that he will ultimately behave like one, i.e., respond only to the most elementary needs of the moment and thus lose any interest in his own economic progress (*ibid.*, p. 354). Such a behavior corresponds to that of Fig. 11–3(b). The fact that it has frequently been reported in the poorest countries of long-standing exploitation confirms Quesnay's remarks on the point.

62. Anonymous, *An Inquiry Into Those Principles Respecting the Nature of Demand as Advocated by Mr. Malthus* (London, 1821), p. 67, cited by Marx, *Capital*, vol. I, chap. xxiv, sec. 3, p. 653.

With a ratio T_0/P still too low for the curve of marginal productivity of labor to intersect the supply curve of labor, it was only natural that the feudal formula should have been used also by the nonagricultural sector during the first phase of the industrial revolution, and that it should have survived by inertia well beyond the beginning of capitalism. Only because of this circumstance was it possible for Marx to mistake the basic formula of feudalism for an essential feature of capitalism and thus to formulate a theory of surplus value which is only an elaboration of condition (5).

2. *A discussion of conduct.* Very often, some features of a system become noticeable only under the light cast by a contrasting structure. That — as we have seen — is the case of some aspects of marginal productivity theory when confronted with a feudal mechanism in an overpopulated country. The same is true of the Edgeworth-Pareto pattern of individual behavior when contrasted with the behavior of an agrarian (peasant) community.

The founders of modern utility theory all agreed that *"each individual acts as he desires,"* [63] a statement that has ever since been repeated in one form or another. But its truth is so tautological that it has no value for a student of human conduct; this student wants to know precisely what pleases the individual. In answering this question, Standard theory assumes that what pleases an individual can be expressed as a function $\psi(Y)$ involving only the quantities of commodities *in his own possession* (represented by the vector Y). That is what I meant earlier by *strictly hedonistic* behavior.

Undoubtedly, such a description is rigorously true for Robinson Crusoe, but it can hardly fit most individuals living in societies. For even *homo capitalisticus* — which Standard theory is deemed to describe — often varies his tips according to his impression of the attendant's neediness, or patronizes a shop only because its owner is hard pressed. Whether or not this conduct reflects the faculty of interpersonal comparison of needs, the fact remains that many an individual responds to changes in the income of the others. A more realistic view of the matter, therefore, leads us to regard the ophelimity

(6) $$\Omega = \psi(Y; Y_s)$$

as a function not only of Y but also of Y_s, which stands for the particular criteria by which the individual views the welfare of his community. The

63. Irving Fisher, *Mathematical Investigations in the Theory of Value and Prices* (New Haven, 1925), p. 11. Also Vilfredo Pareto, *Manuel d'économie politique* (Paris, 1927), p. 62.

individual described by (6) still reacts *hedonistically* — that is, as he desires — but not *strictly hedonistically*.[64]

The problem of individual distribution in a small group — as we know — has no determinate equilibrium in a purely mechanistic schema, whether this refers to an oligopolistic industry or to a small exchange market. In practice a solution is, however, reached only because the group follows some institutional patterns grown out of its particular historical conditions. And we should not be mistaken about it: price leadership, cartel quotas, product competition, and what not, are just as much cultural patterns as "ploughing the land of the widow" or marital dowry, for instance. Without institutional patterns concerning distribution, even the first human societies, small by necessity, could not have arrived at that modicum of stability which is the *sine qua non* condition of organic existence. Whether or not the faculty of "sympathy" for his neighbor is part of man's original nature, that faculty must have evolved before the first viable communities could be formed. Even today, only a type of conduct conforming to (6) can account for the stability of small communities. That, of course, applies first of all to the peasant communities.

A complete description of how individual distribution in a peasant community is regulated must include the institutional patterns prevailing in that particular community. That is why a mechanistic schema of peasant behavior, like the schema of Standard theory, proved to be an impossible project for all who thought of it.[65] And that is not all. If one finally decides to study the peasant institutions so as to construct a *homo oeconomicus* to represent the peasant, one soon discovers that these institutions are almost infinitely variable, a fact that precludes any relevant classification. It is natural that such a baffling and elusive problem as the conduct of the peasant should have attracted few students and resisted being caught within a simple formula.

The small community not only needs a type of conduct oriented also by Y_s, but it also provides the necessary conditions for the operationality of formula (6). In such a community the individual is bound to realize that his own actions influence his own ophelimity also indirectly, via the coordinate Y_s. In addition, everyone naturally arrives at a fairly accurate idea of the situation of everyone else. If these two conditions are not fulfilled, the coordinate Y_s cannot be an *active* agent for the conduct of the individual, even if Y_s is an element of the individual's ophelimity. The most relevant example is provided by the urban agglomerations of

64. Clearly, hedonistic behavior does not necessarily imply "altruistic" behavior; all depends upon the sign of $\partial \psi / \partial Y_s$. The pattern (6) accounts very well for the individual who wants to "keep up with the Joneses," and who consequently feels a "pain" if his neighbor's income increases.
65. Tschajanov, *Die Lehre* (note 28, above), p. 131.

an industrial society. In such large communities, the individual can no longer know the situation of all his fellows. On the other hand, he is bound to realize that by his isolated actions he can exert only an infinitesimal influence upon the variable Y_s. He is thus naturally compelled to conduct himself as if Y_s did not enter into his ophelimity function. We cannot avoid noticing the great similarity of this situation with that of the individual producer in an atomistic industry, who also is compelled by his particular situation to act as if his own offer had no influence upon the market price. There are thus two reasons that account for the success Standard theory has had with the assumption of a strictly hedonistic *homo oeconomicus*. The *theoretical* success stems from the simplicity of the atomistic structure, a reason that finds confirmation also in the completeness of the theory of competitive industry.[66] The *practical* success, on the other hand, is due to the fact that the applications of the theory have always concerned a capitalist economy.

Yet, as J. M. Clark remarked long ago, the demand (i.e., the demand derived from purely hedonistic conduct) cannot reflect any of the social purposes of the community.[67] For the fact that *homo capitalisticus* in general behaves as if his ophelimity were independent of Y_s does not mean that he is basically an egoist in comparison with *homo agricola* — as has often been argued. Yet a prolonged eclipse of the social variable, Y_s, as a coordinate of conduct is apt to cause the disappearance of that variable also from the ophelimity function (i.e. from the individual's awareness). This actually happened at the height of bourgeois liberalism in the West as economic goods in the strict sense became the only coordinate of "rational" conduct. As a Populist wrote in the 1870's, in this conduct there was no room for the principles of justice and solidarity of the village life, but only for success by "shrewdness and tricks." [68] That a society could not last long on this basis is proved by the gradual emergence of the welfare state. And, to continue our parallel, let us observe that the welfare state is a genuine cartel, the cartel of an atomistic society with the purpose of dealing with a problem for which isolated action is impotent. That is the *raison d'être* of all cartels. However, the cartel of the welfare state is as much an institution as the social patterns of peasant communities.

66. The completeness of monopoly theory also has its counterpoint in the theory of the consumer. As already remarked, in a mono-society such as Crusoe's, ψ cannot possibly contain Y_s.

67. J. M. Clark, "Economic Theory in an Era of Social Readjustment," *American Economic Review*, IX (1919), Suppl., 288–289.

68. Quoted in Mitrany, *Marx against the Peasant* (note 17, above), p. 40.

IV. CONCLUDING REMARKS

Because some conclusions of the preceding arguments have a direct bearing on practical issues, it seems appropriate to present them together in this last section.

1. The undifferentiating schema of general equilibrium can only divert our attention from the unique role that leisure plays in economics. For instance, man has always endeavored to discover labor-saving devices because in the long run leisure is an economic *summum bonum* (and for no other reason). In the short run, on the other hand, leisure may be economically *unwanted*. An advanced economy, such as that represented by Fig. 11-4(a), may very well have less leisure than a strictly overpopulated economy, Fig. 11-4(b). And indeed, visitors from the lands-of-plenty often point out reprovingly that the people of poor countries indulge in greater leisure than themselves. They seem to ignore the fact that in strictly overpopulated countries people have no choice: in those countries leisure is imposed upon them by geo-historical conditions, and is not the result of an opportunity choice between greater leisure and greater real income, as is the case in advanced economies. In a strictly overpopulated economy, leisure is not properly speaking an economic good, for it has no use but as leisure. Its value then can be but zero.[69] The peculiar characteristic of the strictly overpopulated economy, namely that leisure has no value although labor has a positive "price," bears on the definition of national income.

Walras seems to have been the first economist to include leisure in national income.[70] If x is taken as *numéraire*, and if w and r represent the prices of L and T respectively, Walras's definition of the national income amounts to

$$(7) \qquad \varphi = x + wl + rt = wL_0 + rT_0,$$

where l, t are the amounts of L, T *directly used* by the consumers. The economic relevance of (7) was further revealed by Barone's famous proposition of welfare economics. To recall, this proposition states that for given prices of L and T, optimum welfare requires that the Walrasian

69. Further proof that Marx attributed feudal features to capitalism is the fact that in his economic theory he assumes that labor power has no use-value *to its owner.* Cf. Karl Kautsky, *The Economic Doctrines of Karl Marx*, p. 60.

70. Léon Walras, *Elements of Pure Economics* (Homewood, Ill., 1954), pp. 215, 379.

national income be maximum.[71] It follows that *if national income is to be used as an index of economic progress*, its only rational definition must be that of (7). But Barone's argument is valid only for an advanced economy, where leisure is time allocated by an opportunity choice and where its price obviously is identical with that of labor. In a strictly overpopulated economy, however, leisure has a zero value. It is natural to think of eliminating it from the national income and hence to define the latter by

$$(7 \; bis) \qquad \varphi_1 = x + w(l - l_1) + rt = w(L_0 - l_1) + rT_0,$$

where l_1 represents the amount of leisure, and $l - l_1$ the amount of personal services. Yet, even the labor corresponding to personal services has no alternative use. Moreover, in overpopulated economies t is usually negligible, for such economies cannot afford hunting grounds, national parks, and the like. The conclusion is that in a strictly overpopulated economy, the most rational index of progress is the national product *stricto sensu*. We should then put

$$(7 \; ter) \qquad \qquad \varphi_2 = x.$$

In the fact that in an overpopulated economy the feudal formula leads to the maximum national product, we have an equivalent of Barone's proposition: in such an economy the feudal formula warrants maximum welfare.

It is curious that in spite of Barone's proposition the current definition of national income in advanced economies is that of (7 *bis*). Only recently, Simon Kuznets proposed to return to the Walrasian formula on grounds that recall the theoretical implications of Barone's theorem.[72] He rightly points out that by excluding leisure from national income we may obscure an important effect of technological progress. The preceding analysis yields an even stronger conclusion: in comparing the rate of economic development of one advanced and one overpopulated country, we should use in each case the appropriate definition of national income.[73] Indeed, only for the latter economies is it appropriate to define economic progress as the increase per capita net product.

71. Enrico Barone, "Il ministro della produzione nello stato collettivista," *Giornale degli Economisti*, XX (1908), 267–293, 391–414. References will be made here to the English translation in *Collectivist Economic Planning*, ed. F. A. von Hayek (London, 1935), pp. 245–290. The above-mentioned proposition is found on pp. 253–257.

72. Simon Kuznets, "Long-term Changes in the National Income of the United States of America since 1870," in *Income and Wealth, Series II*, ed. Simon Kuznets (Cambridge, 1952), pp. 63 ff.

73. This is especially important for the topical comparisons between East European economies and the advanced economies of the West. For I greatly doubt that in any of the Eastern European countries the increase in T_0 has as yet been sufficient in face of population growth to eliminate the conditions of overpopulation.

2. Ordinary statistical data may be highly misleading for a measure of leisure in the case of overpopulated countries. As surprising as this may seem, the overpopulated countries furnish the highest occupation ratios (F/P).[74] On the whole, it appears that in those countries it is hard to find someone unemployed, yet almost everyone is loafing. The paradox is easily cleared up. With an excess of labor, everyone fights to establish a solid claim to a share of the national product. This leads to a social pattern which may be labeled "splitting the job." Several persons are on a job that technically requires only one person, but each one insists on being considered a full-time employee for fear of seeing his claim challenged.

This practice has been frequently denounced as the hallmark of inefficiency if not of remissness. The scientific critic has justified this verdict on the principle that an efficient economy should pay no factor more than its marginal productivity. It is clear, however, that this argument is an unwarranted extrapolation of a law valid only in advanced economies. Indeed, as we have shown, an overpopulated economy does not operate efficiently unless some laborers earn more than their own contribution to output.

The question of the oversized bureaucracy — an unfailing characteristic of overpopulation — also has been approached from a wrong angle. Few students have realized that in overpopulated countries (and only in these) an oversized bureaucracy is a normal economic phenomenon. With labor used up to its technical limit, nothing can be gained from a reduction in the number of public or personal servants; such a reduction may create only social turmoil. Many an overpopulated country deserves to be censured not because it has a large bureaucracy but because its entire government class has a high standard of living amidst poverty. Undoubtedly, too high a standard of living for the government class is a deterrent to economic development, for it greatly reduces the already meager power of capital accumulation of the economy. If in an advanced economy equality of standard of living answers to an ethical principle, in an overpopulated country it represents an economic imperative.

3. Glossing over academic refinements, we may regard economic development as an upward shifting of the labor productivity curves (Fig. 11–4). A fabulous amount of foreign aid apart, no economy can *jump* from the situation of Fig. 11–4(b) to that of Fig. 11–4(a). In other words, it is quasi certain that in its development a strictly overpopulated economy has to pass through a phase like that depicted by Fig. 11–4(c), i.e. through a phase where the working class has no leisure at all. This situation with

74. For example, the occupation ratio in Rumania before World War II was one of the highest in the world, *Enciclopedia României*, I, 154.

its sixteen-hour day and seven-day week is well known owing to its detailed description by the socialist literature of the last century.[75] As already pointed out, Marx erroneously took it for a basic feature of
capitalism. The period in the economic history of the West that served
him as a model for depicting "the calvary of capitalism" corresponds
rather to the growing-pains of capitalism. For capitalism, understood as an
economic system regulated by profit maximization, could really exist only
after the marginal productivity of labor had reached a sufficiently high
level so that it could be equated with the wage rate. Capitalist development proper began only after this phase had been consummated. Increasing leisure (not unwanted leisure) for the working class constitutes its
most distinctive feature. To wit, the forty-hour week is a relatively new
institution, and the idea of a four-day week is already being aired.[76]

Strictly speaking, the East European countries have never come to
know the so-called calvary of capitalism. From the middle of the nineteenth century, if not before, these countries began instead to receive the
impact of Western capitalism. Although usually regarded as a phenomenon equivalent to the "calvary of capitalism," the impact was an essentially different process. The plain fact is that the East European economies
were not yet sufficiently developed to begin the calvary. The true story
can be told in a few words. Increasing trade with the West revealed the
existence of other economic patterns and at the same time opened up new
desires for the landlords and new ambitions for the bureaucracy. Under
this influence the feudal *contrat social* began to weaken. An ever-increasing
number of landlords switched to the capitalist formula of maximizing
profit-rent, a change which, even if it did not always increase their share,
had the advantage of freeing them from their traditional obligations towards the villagers. This process was later to culminate in producing the
pure absentee. From this viewpoint, the main beneficiary of the freedom
of the serfs was the landlord, not the peasant. That is true also of the
earliest agrarian reforms (1861 in Russia, 1864 in Rumania), which in
reality sanctioned the separation of the economic interests of the landlord
from those of the peasant.

75. Even in the United States the average working week was seventy hours as late
as 1850. Undoubtedly, earlier it was even longer. Interesting also is the fact that the
first attempt to limit the work of children under 12 to a ten-hour day was that of the
Commonwealth of Massachusetts in 1842. W. S. Woytinsky and Associates, *Employment and Wages in the United States* (New York, 1956), p. 98. The ten-hour day did
not become a widespread standard for the other workers until 1860. Philip S. Foner,
History of the Labor Movement in the United States (New York, 1947), p. 218; G. Gunton,
Wealth and Progress (New York, 1887), pp. 250–251.

76. Marx failed to see that one possible synthesis of his antithesis could be precisely
this. Instead he wrote that "the relative overpopulation becomes so much more apparent in a certain country, the more the capitalist mode of production is developed in it."
Capital, vol. III, chap. xiv, sec. 4, p. 277.

Now, to regulate production by profit maximization is probably the worst thing that can happen to an overpopulated economy, for that would increase unwanted leisure while diminishing the national product.[77] To be sure, newly imported techniques alleviated the crisis, but hardly the lot of the peasant. This is the explanation of the fact often commented upon that in Eastern Europe capitalism worsened the lot of the peasant, while in puzzling contrast increasing the prosperity of other sectors. It is this situation peculiar to the countries caught lagging behind Western capitalism that gave rise to the Agrarian ideology. And this ideology remained a regional philosophy at which the West looked as a curio, precisely because in its economic development the West had not had a similar experience.[78]

4. In a nutshell, the main tenets of the Agrarian doctrine are:

(1) Because of their geographical situation some communities will always rely on agriculture as a main economic activity. And since agriculture is an intrinsically different activity from industry, such communities cannot develop along identical lines with the industrial economies.

(2) For the countries with an agricultural overpopulation, individual peasant holdings and cottage industry constitute the best economic policy.

Evolution is subject to pure uncertainty, and the most we can do in tackling an evolutionary problem is to trust the existing evidence as a basis for meeting the future. As to this evidence, we have dealt at length with the historical refutation of the law of concentration in agriculture and with the specific differences between industrial and agricultural activities. We may add, however, that nothing as yet has happened to cast any doubt upon the validity of that analysis, and hence upon the first point of the Agrarian doctrine. Besides, the economic development of Denmark and Switzerland as well as parts of Germany and Austria prove that agriculture can provide the basis of its own economic development. Moreover, both anthropology and economic history confirm that only a substantial food production (independent of its source) has led to capital accumulation.[79] Quesnay's celebrated maxim works both ways: *riches paysans, riche royaume.* The logic is surprisingly simple: Robinson Crusoe could not be

---

77. Kautsky, *Economic Doctrines*, p. 235, recognizes the difficulties created by the adoption of profit maximization, but fails to see the real explanation of the process.

78. Only very recently have Western economists come to accept the view of the Agrarians that the East European countries had suffered the impact of foreign patterns not befitting their own cultures and conditions. Cf. *Méthodes et problèmes de l'industrialisation des pays sous-développés,* United Nations (New York, 1955), p. 141.

79. V. Gordon Childe, *Social Evolution* (New York, 1951), p. 22; Bruce F. Johnston, "Agricultural Productivity and Economic Development in Japan," *Journal of Political Economy,* LIX (1951), 498.

available for forging a sickle before Friday could gather enough fruits for both. "Industrialize at all costs" is not the word of economic wisdom, at least in overpopulated agricultural countries.

The second point of the Agrarian doctrine clearly aims at using as much labor in production as is forthcoming. It also reveals that Agrarians were the first to feel intuitively that the economic forms compatible with optimum welfare are not identical for all geo-historical conditions *even if the technological horizon is the same*. We should recall that the real novelty of Barone's work mentioned above was the proof that the controlled economy of a socialist state must imitate the capitalist mechanism, i.e., it must adopt the principles of marginal productivity theory, if it is to obtain optimum welfare. However, neither Barone nor others after him seem to have been aware of one important restriction, namely, that marginal productivity principles presuppose the existence of a well-advanced economy in order to achieve optimum welfare. And thus numerous writers have felt secure in using in their arguments the converse proposition: capitalism and controlled socialism provide the best systems for developing an under-developed economy. Yet this proposition is patently false, at least for an overpopulated economy.

In this light, the intuition that led the Agrarians to their double nega-tion — not Capitalism, not Socialism — proves to have been surprisingly correct. But then, what is the theoretical schema of the Agrarian doc-trine? Because Agrarians have hardly bothered with theoretical schemata, one can only attempt an *ex post* rationalization and thereby accept the risks of misinterpreting their own rationale.

5. The arguments presented in this paper unmistakably lead to the conclusion that the Agrarian schema is the feudal formula under a new form. Capitalism — as we have explained — came to Eastern Europe not as a natural phase in economic development, but as the result of cultural contamination. In the light of economic dynamics and positive welfare theory, there can be no doubt that this was a move against the grain. For feudalism was thus displaced before the respective economies had a chance to reach the calvary phase of capitalism, i.e. the *normal* gateway to the advanced stage of economic development. Only a tithe system can efficiently carry an overpopulated economy through that phase.

But, from the Marxist viewpoint, the premature disappearance of feu-dalism was a step in the right direction, for it represented an earlier fulfilment of what must inexorably come (a view probably shared also by most Standard economists). Into the Agrarian ideology we can read, however, a different position, even more Hegelian in spirit: no phase of economic development can be by-passed. In particular: feudalism cannot disappear before it has completely finished its job. If artificially displaced, it will return under one form or another (barring the occurrence of Malthusian holocausts). In such an alternative, the only logical attitude is then to plan rationally for the continuance of feudalism in such a way as to make it work even better. The policy of radical agrarian reforms in overpopulated countries, by which the head of each peasant family is turned into a feudal entrepreneur, responds precisely to this logic.[80]

A very interesting question now comes up: Which of the two Hegelianisms, the Agrarian or the Marxist, is supported by history? The answer must be sought in what happened in the overpopulated countries after a Communist regime took them over. Unfortunately, our knowledge of what happened is very incomplete. Because the marginal productivity formula cannot possibly work efficiently in an overpopulated economy, it is fair to assume that no Communist regime would use it in this situation. If, however, they use a formula equivalent to a tithe system, the Agrarians are fully vindicated. To be sure, in a Communist regime the distribution of income between the government and the working "groups" may follow an entirely new formula. "From each according to his abilities, and to each according to his needs" has no operational value. Only when we have learned the theoretical schema of this new formula in concrete concepts (labor productivity and labor supply) shall we be able to answer the question in a more complete way.

6. By chance, history supplies us with proof and counter-proof examples regarding the impact of capitalism. The consequences of the premature decay of the feudal formula are best illustrated by the economic situation of the Russian and Rumanian peasant which in relative terms continued to deteriorate all through the hundred years or so preceding World War I. The few timid agrarian reforms were not able to ameliorate the situation, a fact reflected by the frequent peasant jacqueries, some of exceptional intensity. The counter-proof example is provided by Hungary, the well-known bastion of feudalism. In comparison with all her neighbors also plagued by overpopulation, Hungary stood out by virtue of a better fate of the peasant (in most regions) and a conspicuous economic development in all fields. To a great extent this difference can be attributed to the fact that the Hungarian magnate did not succumb to the capitalist formula as his Polish and Rumanian colleagues had done on a large scale.[81] The praises uttered by the apologists of the paternalistic character

80. Engels implicitly recognized the merit of feudalism when in a letter of 1892 (Marx and Engels, *Correspondence*, p. 501) he stated that "an agrarian revolution in Russia will ruin both the landlord and the small peasant," but he clearly ignored the fact that most landlords had long since ceased to follow the feudal formula.

81. The earlier liberation from the economic yoke of a foreign power most probably constitutes another important factor of the difference. But the exploitation of national minorities for which Hungary set a paradigm cannot have been a very important

of Hungarian feudalism were more often than not a *pro domo sua* argument, but they were not entirely without basis. Its undisturbed existence, brought to an end only in 1945 by extra-economic forces, is another proof of the success of that feudalism. But precisely because of this progress, Hungarian feudalism had undoubtedly ceased to represent a necessary economic formula long before 1945.

7. The rather surprising intuition of the Agrarians, however, failed them in one important respect. They were unable to realize that in order to obtain the maximum output from given amounts of resources, the production unit must be of optimum size. Consequently, they could not foresee the danger of determining the size of the peasant holding according to extra-economic criteria. The principle "a holding for every peasant family" naturally led to a suboptimum size of the production unit. And this prevented the crystallization of the existing capital in the most efficient form compatible with the prevailing factor ratio and the available techniques. The unmistakable symptom of this situation was a relative excess of farm equipment. In Rumania, for instance, before the radical reform of 1918, there was a plow for every 26 acres; after the reform, there was one plow for every 15 acres.[82] The Agrarians discovered their error only *post partum*, and when they did, it was much too late. For in East Europe at least, the historical changes of the early 1930's prevented the Agrarian parties from forming a government again.

The facts just mentioned do not justify the prejudice of Stalinist governments in favor of large and highly mechanized farms of the North American type. This prejudice errs in the opposite direction: it leads to a size far greater than the optimum compatible with overpopulation, and hence it uses labor inefficiently.[83]

8. Poor theorists though they have been, Agrarians have never lost sight of the most elementary principle of economic development, which is that no factor should remain *unnecessarily* idle. In overpopulated economies, this may mean using labor even to the point where its marginal

---

element, for even the lot of these minorities improved in some proportion. We should also remark that, until her dismemberment in 1918, Hungary had almost as high an agricultural density as Poland, Rumania, Yugoslavia, or Bulgaria; this can be seen on the map in Moore, *Economic Demography* (cited in note 32, above), p. 73.

82. See Georgescu-Roegen, "Inventarul agricol," *Enciclopedia României*, III, 339. Instances of "inefficient" forms of capital were quite common. To mention one more: the cows represented less than 70 per cent of the entire cattle stock. This was a consequence of the fact that every peasant holding needed a pair of oxen for draught purposes, while only a few could keep more than two animals.

83. Calvin B. Hoover, *The Economic Life of Soviet Russia* (New York, 1932), p. 88, reports of farms having ten times as many workers and twice as much machinery as a farm of equal size in the United States. See also Warriner, *Economics of Peasant Farming* (cited in note 34, above), p. 169.

productivity becomes zero. From the viewpoint of positive welfare economics, we cannot do better than to cling to this principle. The question, however, is whether we can always comply with it in practice. A small farm, a small shop, can easily be run by a family or a true cooperative, and hence follow the feudal formula. On the other hand, many current products can be produced only by large plants. And large production units requiring numerous employees who have no other ties with each other than working together, lend themselves poorly, if at all, to the feudal formula. First, it is hard to see how a manager could use labor beyond the point where its marginal productivity is equal to the wage rate and still be able to prove the efficiency of his management. Secondly, once the principle that one can earn more than his own contribution to output has been accepted, the question of everyone's doing his duty becomes a thorny problem. In the continuity and closedness of the village, these problems are easily solved by the emergence of cultural patterns in which loafing is one of the worst sins.[84] There, efficiency does not need bookkeeping in order to be recognized.

The baffling problem of reconciling the requirements of modern technology with the basic principle of economic welfare is no reason for throwing the latter overboard. In every overpopulated country, at least, there are numerous sectors which either by their nature or by their tradition permit labor to be used according to the feudal formula. Agriculture is almost everywhere in this category.[85] It would be the worst economics to change the production structure of such a sector because of an ill-advised development fever. No one would dispute the truth that peasant institutions and modern industry do not fit together, but it would be a great error to sacrifice all these institutions on the altar of that truth. And if a scapegoat for failing economic policies is needed, one should find a less expensive one. Many of these institutions will still be needed if the largest output is wanted from the sustaining sectors of economic development. Besides, the iconoclast may live to regret his haste, for we should not be surprised at all if the fight of the Communist regimes against the "bourgeois spirit" [86] in reality aims at creating a "socialist man" with a peasant type of conduct.

84. Some primitive communities — we are told — hold work in such high esteem that they produce more than they need and then destroy the excess. See Richard Thurnwald, *Economics in Primitive Communities* (London, 1932), p. 209.

85. By destroying the peasant holdings altogether and replacing them by bookkeeping operated *kolkhozi*, Stalinism certainly made a losing deal with the basic welfare principle. A truly cooperative form of production on units of optimum size with *product ownership* and *tithe* paid in kind to the government would be by far the best welfare solution. Titoism seems to have realized the Stalinist error when it renounced collectivization.

86. See the resolution of the 1931 All-Russian Congress of Workers in *Report of the Ad Hoc Committee on Forced Labor*, United Nations, I.L.O. (Geneva), 1953, pp. 456–457.

9. To assume that a process that sustains the progress of advanced economies necessarily befits an overpopulated economy is an unwarranted extrapolation. The ever-growing literature on economic development, however, abounds in such extrapolations. Probably the most patent is the use of marginal productivity principles in formulating economic policies for underdeveloped economies.[87] But only a few of these economies do not suffer from overpopulation. And it is an obvious feature of overpopulated countries that enterprises operated by feudal formula exist side by side with others managed according to capitalist rules. In such circumstances, price lines are not tangent to the isoquants of every sector and, hence, the isoquants themselves are not tangent to each other. This is easily seen in Fig. 11–5(b) which represents an economy where X_1 is produced according to the feudal formula and X_2 according to capitalist rules.[88] The price line PP is tangent in M' only to the isoquant of X_2. (And only the workers of X_2 earn *wages*.) Of course, M' does not satisfy the elementary condition of positive welfare. The reason, however, is not that the national product would be definitely greater at M'' than at M', but that optimum welfare is represented by M where employment is maximum; see Fig. 11–5(a). The important fact is that neither M nor M'' can be obtained if industry X_2 works according to the capitalist formula, for in either situation the marginal productivity of labor falls below the minimum of subsistence.

That is not all: the price line may not be tangent to the X_2 isoquant either. For, in contrast with what happened in the West during the early phase of industrialization, the cities of overpopulated countries have grown to pathological sizes through a continuous immigration of rural population. The rural exodus brings into those cities not only an enormous excess of labor but also the germ of the feudal economic spirit to which practically no sector can remain immune. The social pressure of people seeking employment for their worthless leisure is so irresistible at all times that even the most convinced "marginalist" of the entrepreneurs has to yield to it and hire more people than he should according to his own rule. In these circumstances, prevailing factor prices may be proportionate to

87. Because of the momentous importance of the problem of underdeveloped countries in world affairs, the consequences of these extrapolations may well exceed those of mere academic licenses. All the more so when the source is as high an authority as the United Nations. In *Measures for the Economic Development of Under-Developed Countries* (New York, 1951), p. 49, the U.N. urges the use of the principle of marginal productivity which — it complains — "is frequently ignored in practice."

88. This schema is not a mere theoretical concoction. It actually corresponds to those economies with a high density of agricultural population living on family holdings and with industry operated more or less on the capitalist formula. Bulgaria, Rumania, and Yugoslavia — to mention only the cases with which I am thoroughly familiar — were precisely in this situation before the advent of Communism. It is precisely because agriculture lends itself easily to operation by the feudal formula that "too many farmers" is a rather general phenomenon, not peculiar to overpopulated economies.

anything except the corresponding marginal productivities. To compute the money equivalent of the marginal productivity of an investment on the basis of the prevailing prices is pure nonsense. The criteria of investment priority based on the results of such computation are therefore baseless.[89] Still worse: such criteria point in exactly the wrong direction. Indeed, except for the correction for external economies, these criteria are identical with those used by the private investor. And the result of private investment is a well-known paradox: although labor-intensive techniques are the only ones indicated for overpopulated countries, the industries developed there have generally been capital-intensive. The explanation is obvious: in an overpopulated country the ratio between wages and the price of other factors is greater than the ratio between the corresponding marginal productivities.

Economic development does not mean only pure growth; in the first place it means a growth-inducing process. Investment in capital-intensive industries is a wrong move in an overpopulated country not because it fails to bring about growth — for it generally does — but because they are not growth-sustaining. The power to sustain growth then is the only valid criterion of investment in undeveloped countries. The marginal productivity principles reflect this criterion very poorly, if at all. Even for a capitalist system they cannot explain more than distribution through allocation.

The path followed by the West in its economic development can help us in seeking a policy for the development of those areas that have remained behind. But it cannot show us *the* way. For, clearly, by this policy we cannot possibly aim to follow precisely the same route that the West followed. It would take us too long to reach the goal. Still more important: it would not even be feasible, for the opportunities the West has had at one time or another cannot be reproduced. The essential distinction between an historical and a dynamic process does not need to be re-examined here. But, at bottom, this is what Marxists and Agrarians quarreled about. Can we, as Standard economists, learn something from this quarrel?

89. For the investment criteria based on marginal productivity one may refer to A. E. Kahn, "Investment Criteria in Development Programs," *Quarterly Journal of Economics*, LXV (1951), 36–61; H. B. Chenery, "The Application of Investment Criteria," *ibid.*, LXVII (1953), 76–96, among others. Because these criteria are endorsed by some economists serving as consultants to various economic development agencies, one may infer that they are used as a guide for public policy (e.g., G. di Nardi, " 'Criteri' e 'Indicatori' per la scelta degli investimenti," *Rassegna Economica*, July 1957).

12

Mathematical Proofs of the
Breakdown of Capitalism

The old Marxist thesis that capitalism shall break down of its own accord is all too familiar. We know also that among the converging arguments used to support this thesis a prominent place is occupied by the theme of the inadequacy of the accumulation process in the capitalist system. Of late, some Marxists have endeavored to add to this particular argument the prestige of the mathematical demonstration. Apparently, the first attempt in this direction was made by Otto Bauer in 1936. More recently, Paul Sweezy improved Bauer's model and also offered a more polished proof.[1] This improved version, however, also starts out with serious mathematical errors which completely invalidate the proof. The presence of these errors has been pointed out by Domar.[2] Yet even Domar does not seem to have realized precisely where the errors lie. Moreover, in his reworked solution he uses a schema of accumulation entirely different from that assumed by Marxist analysis. We are thus still confronted with the problem of whether or not the Bauer-Sweezy conclusions rigorously follow from the Marxist assumptions about the functioning of the capitalist system.[3] This fact alone would suffice to justify the interest

NOTE: This paper is reprinted by courtesy of the publisher from *Econometrica*, XXVIII (1960), 225–243.

1. Paul M. Sweezy, *The Theory of Capitalist Development* (New York, 1942), Appendix to Chapter X, pp. 186–189.
2. Evsey D. Domar, "The Problem of Capital Accumulation," *American Economic Review*, XXXVIII (1948), 792–793.
3. Paul M. Sweezy, in "A Reply to Critics," reprinted in his *The Present as History* (New York, 1953), pp. 352–362, rightly points out that in Domar's amended scheme the problem of underconsumption — i.e., the very basis of the Bauer-Sweezy analysis — "simply disappears." In the same article (pp. 354–360), Sweezy, reflecting upon his mathematical appendix (cited in note 1) states that it was a failure because he attempted to deal with the consumption factor without using Marx's departmental scheme. Undoubtedly, an aggregative model fails to reflect some problems that only a general equilibrium scheme — be it reduced to two departments — can reveal. But, as I hope to prove, that is not the reason why the argument of the appendix misses its target. In blaming the aggregative model for this, Sweezy implicitly takes the position that his theory of underconsumption is nevertheless substantially correct.

in some probing of that argument, even if the problem of capital accumulation were not in the center of the current preoccupations of theoretical economists and policy advisers as well.

Such probing must ascertain, before anything else, whether the mathematical model used by the argument under scrutiny constitutes a *correct* translation of the Marxist scheme of expanded reproduction. It does not take long to see that the model used by Sweezy in the mathematical appendix referred to (note 1) does not correspond at all to the verbal description of that scheme as presented within the appendix itself as well as elsewhere in the same work. Beyond this point, it is most natural that the analysis should proceed by reworking the argument on the basis of a mathematical model of the Marxist scheme. The real difficulty of the analysis begins precisely here. For, as surprising as this may seem, the Marxist scheme of expanded reproduction cannot be cast into a *mathematically* correct model. Indeed, as we shall see in due time, this scheme sins against a most elementary principle, that of dimensional homogeneity.[4] The probing of the Bauer-Sweezy argument could therefore end on this hopeless note. There is, however, a more fruitful procedure: to construct first a scheme free from dimensional contradictions but embodying as many essential points of Marxist rationale as possible, and then to see whether the new dynamic model entails the Bauer-Sweezy conclusions. This is what I propose to do in the present paper. Although I am aware that retouching a Marxist scheme is Marxist anathema, I hope that the results presented here will be useful in two respects: in evaluating the argument of the inadequacy of capitalist accumulation, and in laying bare the purely logical difficulties in the Marxist formulation of expanded reproduction.

I. A DYNAMIC MODEL OF CAPITALISM

A preliminary word about notation. Although this may conflict with many conventional notations of income analysis, we propose to reserve capital letters for denoting *stocks*, and lower-case letters for denoting *flows*. By this method we hope to keep the difference between the two concepts present at all times and to allow a quick verification of dimensional homogeneity of all formulae.

Let us refer to the dynamic system we are going to describe as the

---

4. Because there seems to be no little disagreement among Marxists themselves on what is the correct interpretation of Marx's analysis of expanded reproduction, it is necessary to add that the above statement refers to Sweezy's presentation. Nothing, however, militates against accepting this presentation as the orthodox one.

system (S). We define first the net national income, y, by the standard relation

(I) $$y = c + a$$

where c represents households' consumption, and a, net accumulation. The net national income accrues to the working class as wages, w, and to the capitalist class as surplus value, s, i.e.,

(II) $$y = w + s.$$

Following Marx we shall assume that only the capitalist class accumulates; hence

(III) $$s = l + a$$

where l represents the consumption of capitalists' households. The accumulation is in turn divided into two parts:

(IV) $$a = v + k$$

where v represents the increment of variable capital, V, and k, the increment of constant capital, K. By variable capital we understand the stock of means of subsistence necessary for the maintenance of the working class, and by constant capital the stock of means of production in the strict sense. The terms are Marxist, but the distinction between V and K may be made independently of Marxist economics.

As in any economy, we must have certain technical relations between the factors of production, on the one hand, and the output, on the other. Here too we shall follow the Marxist rationale and assume that the only *purely* technical relation is the proportionality between K and the production of consumers' goods:

(V) $$K = \lambda(w + l + v).$$

As Sweezy reminds us,[5] this relation is equivalent to "the acceleration principle."

Always conforming to the same rationale, we shall assume that a technical relation (in the broad sense) exists between wages and variable capital:

$$V = \mu w.$$

This relation is nothing other than a dimensionally correct formulation of the wage-fund theory which is implied by the Marxist concept of variable capital.[6]

5. Sweezy, *The Theory*, p. 187n. See also p. 182.
6. This statement should be compared with that of J. Steindl, *Maturity and Stagnation in American Capitalism* (Oxford, 1952), p. 243, note 3, who argues that the "weird old monster, the wages fund doctrine, which Marx killed in a brilliant attack, [was nevertheless permitted as a] ghost to muddle up his terminology." I confess I cannot see how we can preserve the notion of variable capital — as conceived and used by Marx — and throw "that fossil" out of Marxist economics.

We should emphasize that both λ and μ are *dimensional* constants, of the same dimension as t (time). Consequently, their values depend upon the choice of the time unit. Either λ or μ can then be made equal to unity by a proper choice of the time unit. If this unit is chosen so that $μ = 1$, we have

(VI) $V = w.$[7]

It is extremely important to note also that *no explicit relation is assumed to exist between the output and the wage bill (i.e., labor), for according to Marxist rationale the latter is determined solely by the behavior of capitalists.*

To this behavior we now turn. According to Marxist economics, "it is a fundamental feature of capitalism that an increasing proportion of surplus value tends to be accumulated and an increasing proportion of accumulation tends to be invested." [8] Capitalists' behavior, therefore, can be expressed by the following relations

(VII) $a = a(s), \quad \dfrac{da}{ds} > 0, \quad \dfrac{d}{ds}\left(\dfrac{a}{s}\right) > 0,$

and

(VIII) $k = k(a), \quad \dfrac{dk}{da} > 0, \quad \dfrac{d}{da}\left(\dfrac{k}{a}\right) > 0.$

These relations imply that $a(s)$ and $k(a)$ have a first derivative everywhere. We need not assume that the same functions have also a second derivative. But to be realistic we must assume that $a(s)$ and $k(a)$ are *smooth* functions, and this requires that they should have at least a second derivative to the right and to the left everywhere.[9] We know that in this case the points where $a(s)$ and $k(a)$ do not have a second derivative are isolated.

Finally, we must add the dynamic relations

(IX) $\dfrac{dV}{dt} = v; \quad \dfrac{dK}{dt} = k.$

The system (S) then involves ten unknowns: V, K, y, c, a, w, s, l, v, k, and ten relations: (I)–(IX). In this particular case we can do a little better than comparing the number of unknowns with that of equations. From (III), (IV), (VII) and (VIII), we obtain the inverse functions

(1) $l = l(k), \quad v = v(k),$

7. It is a peculiarity of Marx's system to assume that λ and μ can be made equal to unity simultaneously, a fact which entitles him to write $K + V + s$ for total value. In this formula, however, there is no violation of the principle of dimensional homogeneity, for both K and V are multiplied by $1/λ$, $1/μ$ whose numerical values happen to be unity.

8. Sweezy, *The Theory*, p. 187; also p. 181.

9. To illustrate: $y = x^3/|x|$ is a smooth function without a second derivative at $x = 0$. However, for $x = 0$, the second derivative to the left is -1, and to the right $+1$.

and from (V), (VI), and (IX)

(2) $$k = \lambda v + \lambda \left(\frac{dv}{dk} + \frac{dl}{dk}\right)\frac{dk}{dt}.$$

This differential equation determines $k(t)$. The other unknown functions of the dynamic system are then derived from the other equations by straightforward operations.

There are two features of (S) that make it differ from the Marxist scheme of expanded reproduction. For the reasons already explained, we shall refer to Sweezy's presentation of that scheme as a basis of comparison.

The first difference concerns the composition of surplus value. According to (III) and (IV), for (S) we have

(3) $$s = l + v + k$$

whereas in the Marxist scheme

(3bis) $$\bar{s} = l + v + k + \frac{dl}{dt}.$$

Indeed, Sweezy's explanation of this point carefully aims at leaving no room for misunderstanding. We are told in very explicit terms that the surplus value consists of *four* parts, which in comparative notations are: 1) S_{ac}, corresponding to our k; 2) S_{av}, corresponding to our v; 3) S_c, corresponding to our l; and 4) $S_{\Delta c}$, the increment of l itself, i.e., dl/dt.[10] Moreover, Sweezy sharply criticizes N. Bukharin (the outstanding Marxist theorist liquidated during the Great Purge of the 1930's) for having used (3) instead of (3bis) in presenting the scheme of expanded reproduction. According to the same author, the omission of the term dl/dt from the analysis of surplus value proves that Bukharin was "incapable of imagining an increase in capitalists' consumption." [11] It is elementary, however, that the absence of dl/dt in (3) does not mean at all that l is necessarily a constant, for l like all other variables of the system is determined by all equations together, not by one relation alone.

On the contrary, it is formula (3bis) which is absurd, not *economically*, but in a sense independent of any material interpretation of it. Indeed, (3bis) violates the principle of dimensional homogeneity, which is essentially an arithmetical principle. As long as the letters in that formula stand for measurable material concepts and not for some Hegelian ideals, l and dl/dt cannot be added, any more than can *total* and *average* cost, for instance. I hope to be forgiven for stressing an elementary point that has been, as we know, the source of many economic fallacies. This seemed necessary for reaching the root of the difficulty of translating the Marxist

10. Sweezy, *The Theory*, p. 163, re-emphasized on p. 181.
11. *Ibid.*, p. 164n.

scheme into an arithmetically correct model. For the arithmetical incongruity of (3bis) is not accidental, but reflects a vital aspect of Marxist economics. And that aspect is the notion that *a material flow can be the source of its own growth.*

This position makes also for the second point of difference between (S) and the scheme of expanded reproduction. For (S), from (II)–(IV) and (VI) we derive

$$(4) \qquad\qquad y = V + s = V + l + v + k$$

whereas in Marxist dynamics

$$(4^{bis}) \qquad\qquad \bar{y} = V + v + \bar{s} = V + l + 2v + k + \frac{dl}{dt}.^{12}$$

This time, too, we are strongly warned against an "unguarded haste" that may lead to confusing the concept of national income with formula (4).[13]

The line of reasoning behind (4bis), however, is not as simple and obvious as in the case of (3bis). First, the *flow* increment v is added to the *stock* V during the same period in which the flow is produced. Then the stock $V + v$ is used as a basis for equating the wage bill to $V + v$. This is how it comes about that (4bis) contains one more v than (4). At bottom, this means that the flow v can be both *consumed* and passed over into the next period as an *increment* of a stock, or, in other words, that "the growth of variable capital constitutes an outlet for accumulation and *at the same time* signifies a growth in consumption."[14] In explicit terms, this means that

$$(5) \qquad\qquad a = v + k \qquad \text{and} \qquad \bar{c} = V + v + l + \frac{dl}{dt}.^{15}$$

Since Marxist practice sees nothing wrong in adding to a flow its own dynamic increment during the very period in which this increment is produced, it is difficult to see why the surplus value should not be given by

$$(3^{ter}) \qquad\qquad \bar{\bar{s}} = l + \frac{dl}{dt} + v + \frac{dv}{dt} + k + \frac{dk}{dt}.$$

For one may explain that in each period the surplus value is divided into six parts: the value of l plus its increment, and so forth.[16] Actually,

12. *Ibid.*, p. 63. In explicit mathematical terms the formula is given in Sweezy's Appendix A (p. 373 combined with p. 368n). That appendix is written by Shigeto Tsuru, but Sweezy refers to it for every question pertaining to the composition of national income.

13. *Ibid.*, pp. 248n, 371, and *passim.*

14. *Ibid.*, p. 222.

15. Cf. p. 372, *ibid.* Tsuru's justification of the double counting of v in (4bis) is highly instructive: the double counting is the natural result of three "metamorphoses" of money, whatever this may mean.

16. To be sure, we have

$$s(t + \Delta t) = s(t) + \Delta l + \Delta v + \Delta k$$

but, clearly, this is not what is meant by (3bis) and (4bis).

once the principle of dimensional homogeneity is rejected, there is no reason for not continuing to add the increments of increments of increments. . . .

But the difficulty, nay, the impossibility, of casting the Marxist scheme of expanded reproduction into a mathematical model can be illustrated in a more concrete fashion. Let us suppose that such a model has been constructed and that the corresponding system has been solved for its ten unknown functions. Let us also assume that the solution gives $l = A + Bt$, for instance. If now to the question "what is the value of the flow of capitalists' consumption at $t = t_0$?" one answers $A + Bt_0$, then, according to Marxist rationale, the answer is wrong: for this answer does not include the "increment" dl/dt, which is B. If the answer is that the value of the flow is $A + B + Bt_0$, then by all logic it follows that $l = A + Bt$ does not represent the consumption of capitalists. What does it represent then? And if we say that capitalists' consumption is given by $A + B + Bt$, what shall we do with the term dl/dt in (3bis)? If we drop it, we depart from strict Marxist rationale; if we retain it, we shall never know the amount of capitalists' consumption.[17]

Finally, let us observe that the position that a *material* flow can be the source of its own growth is tantamount to the belief in the existence not only of perpetual motion but of perpetual *accelerated* motion as well. But if a flow cannot be the source of its own growth, one may ask, what is the source of economic growth? The answer to the apparent puzzle is not difficult. Since human economy is not an isolated system, economic growth is the result of a continuous tapping of other stocks: the stocks of natural deposits, of various forms of free energy, and above all of that peculiar energy which is accumulated in the body of living organisms. The economic *process* consists precisely in this tapping. To be sure, this *process* grows without any counterbalancing decrease in something else, just as physical entropy grows without any decrease in the total energy of the universe. Only in this sense can we speak of the economic *process* being Hegelian, i.e., containing the source of its own development. But the *material* elements involved in the process must obey the universal laws of matter and energy.

II. THE ARGUMENT OF THE INADEQUACY OF CAPITALIST ACCUMULATION

We have already mentioned that the Bauer-Sweezy mathematical proof of the inadequacy of capitalist accumulation uses a model different from the Marxist scheme of expanded reproduction. But the model has a more

17. The same remarks apply to the solution $V = V(t)$.

general shortcoming. Indeed, the proof proceeds from the definition of national income by the formula

$$(6) \qquad\qquad y^* = w^* + l^* + k.$$

In the absence of an explicit statement by Sweezy, it is rational to assume that he adheres to the practice of adding the corresponding incremental flows to both wages and capitalists' consumption, i.e.,

$$(7) \qquad\qquad w^* = V + v, \qquad l^* = l + \frac{dl}{dt}.$$

With this, (6) becomes

$$(8) \qquad\qquad y^* = \bar{y} - v.$$

If, however, w and l have their normal meaning,

$$(8^{\text{bis}}) \qquad\qquad y^* = y - v.$$

It is thus seen that according to any interpretation, whether Marxist or not, (6) fails to include the accumulation of variable capital in the national income.[18]

The second observation about the Bauer-Sweezy argument concerns the mathematical inaccuracies to which the introduction of this paper alluded. Sweezy states[19] that from "the fundamental feature of capitalism," i.e., from (VII) and (VIII), it follows that

$$(9a) \qquad\qquad 0 < \frac{dw^*}{dk} < 1, \qquad \frac{d^2w^*}{dk^2} < 0,$$

$$(9b) \qquad\qquad 0 < \frac{dl^*}{dk} < 1, \qquad \frac{d^2l^*}{dk^2} < 0.$$

But (VII) and (VIII) together yield

$$(VII^{\text{bis}}) \qquad -1 - \frac{dv}{dk} < \frac{dl^*}{dk} < \frac{l^*}{v+k}\left(1 + \frac{dv}{dk}\right),$$

18. This seems the proper place to mention a position taken by Steindl, in *Maturity and Stagnation*, p. 243, note 3, which, if valid, would upset the entire argument developed so far. Steindl objects to Sweezy's analysis of expanding surplus value on the ground that s cannot even include such a term as v. To assume that it does "implies that some part of [the] national income flow is *wages*, and at the same time is also *surplus value* (profits) *in the same period;* that some part of the value created in *a given year* is unpaid labor and at the same time also paid labor!" On the surface, this argument sounds identical to that advanced against (4bis) above. But on closer examination it shows itself to be based upon how stocks grow from flows. Steindl's position implies that V either grows *by itself* or remains constant. At bottom, all this shows that it is impossible to get rid of the "fossil" and still have a consistent Marxist scheme of expanded reproduction (cf. *supra*, note 6). A two-department scheme offers no escape from the dilemma (if there is one). So that one can heartily agree with Sweezy that his problem is "a standing challenge to Marxian economists" (Sweezy, "A Reply," p. 360).

19. Sweezy, *The Theory*, p. 187.

(VIIIbis) $$-1 < \frac{dv}{dk} < \frac{v}{k}.$$

Rigorously speaking, according to (VIIbis) dl^*/dk may even be negative, while d^2l^*/dk^2 may have either sign. The supposition that (VIIIbis) entails (9a), however, reveals a more essential error of the argument.

According to either of the two possible interpretations of w^*, dw^*/dk depends on dV/dk. But the latter's value cannot be determined or restricted by the behavior of capitalists alone, i.e., by (VII) and (VIII). This value is determined only by the system as a whole, *including the technical relation* (V).[20]

The Bauer-Sweezy thesis, however, may be perfectly valid in spite of the false start in the mathematical proof. But this question cannot be elucidated without examining the same thesis in the light of a consistent model. And since (S) comes as close as possible to the Marxist rationale, we shall proceed on this basis.[21]

The Bauer-Sweezy argument can be summarized as follows:

(1) Capitalists' behavior being that described by (VII) and (VIII), if $dy/dt > 0$, then $dk/dt > 0$;

(2) On the other hand, the technical condition (V) together with capitalists' behavior require $dk/dt < 0$ if $d^2y/dt^2 < 0$;

(3) Therefore, in the case where the national income grows at a decreasing rate the *behavior* value of dk/dt is greater than the *equilibrium* value. Hence, "the output of consumption goods will display a continuous tendency to outrun demand." [22]

It can be proved, however, that both premises (1) and (2) on which the conclusion (3) rests are false. The following theorem shows this for (1).

THEOREM 1. *There are functions of t satisfying* (VII) *and* (VIII) *and such that* $\dot{y} > 0$ *and* $\dot{k} < 0$.[23]

PROOF: Let us consider the following functions of t for $t \geq 0$:

$$(10) \quad k = Ae^{-(\alpha+\beta)t}, \quad v = e^{-\alpha t}(1 - Ae^{-\beta t}), \quad l = B - \gamma e^{-\alpha t}$$

20. Domar's criticism failed to realize this aspect of the problem. After remarking that the fundamental feature of capitalism entails neither (9a) nor (9b), Domar ("The Problem," p. 793) asserts that by this feature it is given that $d(k/y)/dt > 0$. But this expression involves dV/dk and V/k. Domar's conclusions regarding Sweezy's argument need, therefore, to be re-examined.

21. Another alternative would be to proceed *per absurdum* by accepting (3bis) and (4bis) as relations in *pure numbers*. I want to stress the fact that even in this alternative the mathematical truth of the subsequent theorems is not in the least invalidated. The choice to proceed otherwise aims only at avoiding the incongruities described in Section 1.

22. Sweezy, *The Theory*, p. 189.

23. For the sake of compactness, we shall hereafter use the dot notation for the derivatives with respect to time: $\dot{x}$ for dx/dt, $\ddot{x}$ for d^2x/dt^2, etc.

where

(11) $0 < \alpha < 1$, $1 < \beta$, $0 < A < 1 + \alpha(\gamma - 1)$,
          $0 < \gamma < 1$, $\gamma < B$.

Since $\dot{k} < 0$, from (VII) and (VIII) we first obtain

(12) $\dot{k}\dfrac{da}{dk} = \dot{a} < 0$, $\dot{a}\dfrac{ds}{da} = \dot{s} < 0$.

These conditions are clearly fulfilled by (10), for

(13) $\dot{a} = \dot{k} + \dot{v} = -\alpha e^{-\alpha t}$, $\dot{s} = \dot{a} + \dot{l} = -\alpha(1 - \gamma)e^{-\alpha t}$.

The other conditions of capitalists' behavior are also satisfied:

(14) $\begin{aligned} a\dot{k} - \dot{a}k &= -\beta A e^{-(2\alpha+\beta)t} < 0 \\ \dot{a}l - a\dot{l} &= -\alpha B e^{-\alpha t} < 0. \end{aligned}$

Yet, we have

(15) $\dot{y} = v + \dot{s} = e^{-\alpha t}(1 - \alpha + \alpha\gamma - Ae^{-\beta t}) > 0$.

                                                                  Q.E.D.

The falsity of the second premise of the argument is shown by Theorem 2. Before enunciating this theorem, however, we need to introduce some definitions and prove some useful lemmas.

DEFINITION I. *If all functions v, k, l, y as well as their first derivatives with respect to t are positive, then we call* (S) *a growing system.*

From (V), (VI) and (IX) we obtain

(16) $\dot{k} = \lambda(v + \dot{v} + \dot{l})$, $\ddot{k} = \lambda(\dot{v} + \ddot{v} + \ddot{l})$

and

(17) $\dot{y} = \dfrac{\dot{k}}{\lambda} + \dot{k}$, $\ddot{y} = \dfrac{\ddot{k}}{\lambda} + \ddot{k}$.

For a growing system, relations (VII) and (VIII) become

(18) $\dfrac{\dot{v} + \lambda(\dot{v} + \ddot{v} + \ddot{l})}{v + \lambda(v + \dot{v} + \dot{l})} > \dfrac{\dot{l}}{l}$, $\dfrac{\ddot{v} + \ddot{l}}{\dot{v} + \dot{l}} > \dfrac{\dot{v}}{v}$.

REMARK I. To have a growing system (S), we need only to determine v and l as positive and increasing functions of t and such that both expressions (16) be positive, and inequalities (18) be satisfied. We should observe, however, that according to our strictly necessary assumptions, the functions $v(t)$ and $l(t)$ do not necessarily possess a second derivative everywhere.[24] That raises the problem of the meaning of the second formula (16) and of (18). According to (2), however, $k(t)$ has a derivative everywhere (except where the sum $dv/dk + dl/dk$ would be zero). It follows that even if $\ddot{v}$, $\ddot{l}$ may not exist everywhere, the integral of (2) is such that it makes $\ddot{v} + \ddot{l}$ a continuous function of t wherever $\dot{k}$ is continuous.

24. *Supra*, note 9.

Since only the sum $\ddot{v} + \ddot{l}$ enters in the formulae (16) and (18), we can use the notation $\ddot{v}, \ddot{l}$ even though these derivatives may not exist separately. We need, therefore, to choose v and l such that $\ddot{v} + \ddot{l}$ be continuous (in addition to the conditions mentioned previously in this Remark).

DEFINITION II. *If a growing system* (S) *satisfies the inequalities*

$$(18^{\text{bis}}) \qquad \qquad \frac{\ddot{v} + \ddot{l}}{\dot{v} + \dot{l}} > \frac{\dot{v}}{v} > \frac{\dot{l}}{l}$$

we shall call it a strongly growing system.

LEMMA 1. *The system*

$$(19) \qquad \qquad v = A + Be^{\alpha t}, \qquad l = A' + B'e^{\alpha t}$$

where

$$(20) \qquad \qquad A, B, A', B', \alpha > 0, \qquad A'B - B'A > 0$$

is a strongly growing system.

The proof is immediate.

LEMMA 2. *Given a strongly growing system* (v, l) *that is not of the form* (19), *we can derive another strongly growing system* (v^*, l^*) *different from* (v, l).

PROOF: We first choose arbitrarily the origin of t. Then we determine A, B, A', B', α by the system:

$$A + B = v(0), \qquad A' + B' = l(0),$$
$$(21) \qquad \qquad \alpha B = \dot{v}(0), \qquad \alpha B' = \dot{l}(0),$$
$$\alpha^2(B + B') = \ddot{v}(0) + \ddot{l}(0).$$

Because (l, v) is a strongly growing system, the unknowns of the system are easily seen to satisfy (20). Hence the system

$$(22) \quad v^*(t) = \begin{cases} v(t) \text{ for } t \leq 0, \\ A + Be^{\alpha t} \text{ for } t \geq 0, \end{cases} \qquad l^*(t) = \begin{cases} l(t) \text{ for } t \leq 0, \\ A' + B'e^{\alpha t} \text{ for } t \geq 0, \end{cases}$$

is a strongly growing system different from (l, v).

Q.E.D.

THEOREM 2. *There exist strongly growing systems* (S) *such that* $\ddot{y}$ *is negative for some values of* t.

PROOF: Let (v, l) be a strongly growing system such that $\ddot{y}$ is positive for all values of t where $\ddot{y}$ exists, and let us assume that $t = 0$ is such a value. Let us put

$$v_1(t) = v(0) + \dot{v}(0)t + v_2 \frac{t^2}{2!} + v_3 \frac{t^3}{3!},$$
$$(23)$$
$$l_1(t) = l(0) + \dot{l}(0)t + l_2 \frac{t^2}{2!} + l_3 \frac{t^3}{3!},$$

where

(24) $$v_2 + l_2 = \ddot{v}(0) + \ddot{l}(0)$$

and v_3, l_3 are some parameters to be determined later. Because $v_1(0) = v(0)$, $\dot{v}_1(0) = \dot{v}(0)$, $l_1(0) = l(0) = \dot{l}_1(0) = \dot{l}(0)$, $\ddot{v}_1(0) + \ddot{l}_1(0) = \ddot{v}(0) + \ddot{l}(0)$, and because (v, l) is a strongly growing system, the continuity of the functions (23) warrants the existence of a non-null interval $0 \leq t \leq t'$ in which (v_1, l_1) satisfy the conditions of a strongly growing system. If v_1, l_1 are introduced in (16) and (17), we obtain:

(25) $$\ddot{y}_1(t) = \dot{v}(0) + \ddot{v}(0) + \ddot{l}(0) + \lambda(v_2 + v_3 + l_3)$$
$$+ (v_2 + v_3 + \lambda v_3 + l_3)t + v_3 \frac{t^2}{2}.$$

We can choose v_2, v_3, l_3 such that $\ddot{y}_1(0) < 0$. There is then an interval $0 < t \leq t''$ where $\ddot{y}_1(t) < 0$. Let T be the smaller of t', t''. But at T the values of v_1 and l_1 satisfy the conditions of a strongly growing system. According to Lemma 2, at T we can splice the functions $v^*(t)$, $l^*(t)$ of the type (19).

Thus the system corresponding to

(26) $$\bar{v}(t) = \begin{cases} v(t) & \text{for } t \leq 0, \\ v_1(t) & \text{for } 0 \leq t \leq T, \\ v^*(t) & \text{for } T \leq t, \end{cases} \qquad \bar{l}(t) = \begin{cases} l(t) & \text{for } t \leq 0, \\ l_1(t) & \text{for } 0 \leq t \leq T, \\ l^*(t) & \text{for } T \leq t \end{cases}$$

is a strongly growing (S), for which $d^2\bar{y}/dt^2 < 0$ for $0 < t \leq T$.

Q.E.D.

The inadequacy of capitalist accumulation is derived by Sweezy from another argument as well: that capitalists' behavior can lead only to a decreasing ratio between the rate of consumption and the rate of capital investment, whereas the technical conditions require this ratio to be a constant.[25] Clearly this argument, if correct, would be far stronger than that summarized on page 233 [p. 406 in the present volume], for then capitalism could never be in equilibrium, whether $\ddot{y} > 0$ or $\ddot{y} \leq 0$. It can be shown, however, that this argument, too, is fallacious.

THEOREM 3. *There are functions of t satisfying the behavior conditions* (VII) *and* (VIII), *such that the ratio between the rate of growth of output and the rate of growth of constant capital be increasing.*[26]

25. Sweezy, *The Theory*, p. 182. The technical ratio is $1/\lambda$, from relation (V).
26. Since Sweezy (*ibid.*, p. 187) defines consumption as $w^* + l^*$, and since $w^* = w + v$, his "consumption" is in fact "output" in our sense (save for the term dl/dt). Strictly speaking, it is much more realistic to relate constant capital to output than to consumption.

PROOF: Using the functions of (10), we have

(27) $$\frac{v + \dot{v} + \dot{l}}{k} = \frac{1}{A}\left[(1 - \alpha + \alpha\gamma)e^{\beta t} + A(\alpha + \beta - 1)\right] > 0.$$

Q.E.D.

III. A FUNDAMENTAL PROPERTY OF THE SYSTEM (S)

Lemma 1 proves that a "capitalist" system (S) may be growing in such a way that the net national income be continuously increasing at an *increasing* rate. On the other hand, Theorem 2 proves that there are growing systems (S) for which the net national income increases at a *decreasing* rate in some intervals. The question now is whether the symmetrical situation of Lemma 1 exists, namely, whether a growing system exists for which $\ddot{y}$ may remain negative after a certain value of t. The answer to this question is negative, as shown by the following:

THEOREM 4. *No growing system* (S) *exists such that* $\ddot{y} \leq 0$ *for all* $t \geq t_0$.

PROOF: Because the details of the proof vary according to whether $\lambda \gtrless 1$ (without however requiring a different method of demonstration) we are going to give the proof only for the case $\lambda > 1$.[27] The proof will show that the assumption $\ddot{y} \leq 0$ for all $t \geq 0$ is in contradiction with the properties of (S).

Let

(28) $$\ddot{y} = -e^{t/\lambda}p(t)$$

where $p(t) \geq 0$. To avoid unnecessary complications we shall assume that $p(t)$ exists everywhere except for a number of isolated values.

From (17) we obtain

(29) $$e^{t/\lambda}\left(\frac{k}{\lambda} + \ddot{k}\right) = -p(t)$$

which yields

(30) $$e^{t/\lambda}\dot{k} = C' - \int_0^t p(t)dt.$$

But since $\dot{k} > 0$, we must have

(31) $$C' \geq \lim_{t = +\infty} \int_0^t p(t)dt, \qquad \lim_{t = +\infty} p(t) = 0.$$

Let us put

(32) $$\chi(t) = \int_t^\infty p(t)dt.$$

27. In view of the fact that the value of λ is determined by the condition $\mu = 1$, I doubt that on *a priori* grounds we can hold that $\lambda > 1$ is the only realistic case.

Because of (31),

(33) $$\lim_{t=+\infty} \chi(t) = 0.$$

From (30), we obtain

(34) $$k = [C + \chi(t)]e^{-t/\lambda}, \qquad C \geq 0$$

and further

(35) $$k = \lambda[C_0 - Ce^{-t/\lambda} - \gamma(t)]$$

where

(36) $$\gamma(t) = \frac{1}{\lambda} \int_t^\infty e^{-t/\lambda}\chi(t)dt.$$

Because of (33), this integral exists. And since k is a positive and increasing function of t, we must have

(37) $$k(0) = \lambda[C_0 - C - \gamma(0)] > 0, \qquad k(+\infty) = \lambda C_0 > 0.$$

Now, because v must grow less rapidly than k, $v(+\infty)$ must be finite. Therefore, $a = k + v$ also is bounded. But l must grow less rapidly than a; hence l, too, must be bounded. Since $l > 0$, it follows that

(38) $$\lim_{t=+\infty} \int_0^t l dt < +\infty, \qquad \lim_{t=+\infty} l = 0.$$

From (V) we get

(39) $$v + \dot{v} = C_0 - Ce^{-t/\lambda} - \gamma(t) - l(t)$$

which, by integration, yields

(40) $$v = C_0 - \frac{C\lambda}{\lambda - 1} e^{-t/\lambda} + [v_0 - \Gamma(t) - L(t)]e^{-t}$$

where

(41) $$\Gamma(t) = \int_0^t e^t\gamma(t)dt, \qquad L(t) = \int_0^t e^t l(t)dt.$$

To see whether (35) and (40) are compatible with a growing (S), we shall examine this problem for t sufficiently large. In this case, the corresponding formulae of k and v can be replaced by some asymptotic expressions, $\bar{k}$ and $\bar{v}$.

To determine these asymptotic expressions, let us observe that since $\lambda > 1$, $e^{-t/\lambda}$ tends more slowly towards zero than e^{-t}, and more slowly than both $\gamma(t)$ and $\Gamma(t)$. Indeed, from (36) and (41) we have

(42) $$\lim_{t=+\infty} e^{t/\lambda}\Gamma(t) = \lim_{t=+\infty} e^{t/\lambda}\gamma(t) = 0.$$

There are several alternatives to be considered.

(A) $C \neq 0$. In this case $e^{-t/\lambda}$ represents the first order infinitesimal, and we have

$$(43) \qquad \tilde{k} = \lambda(C_0 - Ce^{-t/\lambda}),$$

$$\tilde{v} = C_0 - \frac{\lambda C}{\lambda - 1} e^{-t/\lambda} - L(t)e^{-t}.$$

Now, if

$$(44) \qquad \lim_{t=+\infty} e^{t/\lambda}\tilde{l} = M < +\infty,$$

we have

$$(45) \qquad e^{-t}L(t) \cong Me^{-t/\lambda}$$

which, introduced in (43), yields

$$(46) \qquad \tilde{v} = C_0 - \left(\frac{\lambda C}{\lambda - 1} + M\right) e^{-t/\lambda}.$$

It is easy to see that $\tilde{v}$ grows faster than $\tilde{k}$, and hence the assumption (44) is incompatible with the structure of (S).

If in (44) $M = +\infty$, then $e^{-t/\lambda}$ tends faster toward zero than Le^{-t} and we have

$$(47) \qquad \tilde{k} = \lambda C_0, \qquad \tilde{v} = C_0 - L(t)e^{-t}.$$

This case must be rejected for the same reason as above.

(B) $C = 0$. In this case $\gamma(t) \neq 0$. There are several alternatives.

(B1) *Both integrals (41) converge for $t = +\infty$.* It is not possible to have

$$(48) \qquad v_0 - \Gamma(+\infty) - L(+\infty) \geq 0,$$

for then

$$(49) \qquad v_0 - \Gamma(t) - L(t) > 0$$

and $\tilde{v}$ would be negative. And if

$$(50) \qquad v_0 - \Gamma(+\infty) - L(+\infty) < 0$$

we have

$$(51) \qquad \begin{aligned} \tilde{k} &= \lambda C_0, \\ \tilde{v} &= C_0 + [v_0 - \Gamma(+\infty) - L(+\infty)]e^{-t}. \end{aligned}$$

This must be rejected because v would grow faster than k for t sufficiently large.

(B2) *Only Γ converges.* In this case,

$$(52) \qquad \tilde{k} = \lambda C_0, \qquad \tilde{v} = C_0 - e^{-t}L(t),$$

which gives a decreasing ratio $\tilde{k}/\tilde{v}$.

(B3) *Only L converges.* If Γ diverges while

$$(53) \qquad \lim_{t=+\infty} e^t\Gamma(t) < +\infty,$$

we have

(54) $$\tilde{k} = \lambda C_0, \qquad \tilde{v} = C_0 - e^{-t}\Gamma(t),$$

which must be rejected for the same reason as in (B2).

We should then examine the alternative

(55) $$\lim_{t=+\infty} e^t \Gamma(t) = +\infty.$$

If this alternative is true, $\dot{\gamma}(t)e^t$ cannot tend toward zero as t tends toward $+\infty$. On the other hand, we have

(56) $$\Gamma(t) = \gamma(t)e^t - \gamma(0) - \int_0^t \dot{\gamma}(t)e^t dt.$$

And, since by (36) $\dot{\gamma}(t)$ is negative, it follows that

(57) $$\tilde{\Gamma}(t) = \epsilon(t)\gamma(t)e^t, \qquad \epsilon(t) > 1.$$

Therefore,

(58) $$\tilde{k} = \lambda(C_0 - \gamma), \qquad \tilde{v} = C_0 - \epsilon\gamma,$$

which again gives a decreasing ratio $\tilde{k}/\tilde{v}$.

(B4) *Both Γ and L diverge.* The only case which cannot be immediately reduced to one of the types examined under (B2) and (B3), is that where

(59) $$\lim_{t=+\infty} L(t)/\Gamma(t) = \delta, \qquad 0 < \delta < +\infty.$$

The case where (53) would be true is disposed of by the same argument as in (B3). If, however, (55) prevails, then

(60) $$\tilde{k} = \lambda(C_0 - \gamma), \qquad \tilde{v} = C_0 - (\epsilon + \delta)\gamma,$$

which also gives a decreasing $\tilde{k}/\tilde{v}$.

With this, all alternatives have been examined and proved to contradict some property of (S).

Q.E.D.

REMARK 2. It is important to note that $\ddot{y}$ may end by remaining negative in a system (S) where v and k grow but l decreases. This is shown by the following example:

(61)
$$k = \lambda(C_0 - Ce^{-t/\mu}), \qquad v = C_0 - Be^{-t/\mu},$$
$$l = l_0 + (\mu C - \mu B + B)e^{-t/\mu}$$

where

(62) $$\mu < \lambda, \qquad B > C > C_0, \qquad 0 \le t.$$

IV. CONCLUDING REMARKS

We shall now summarize the salient points of the preceding analysis and tie them together with a few additional remarks.

1. As we have seen (Theorem 3), from the mere knowledge of the behavior of capitalists as described by Marxist economics, it is not possible to deduce that the ratio between the means of production, K, and the output of consumers' goods, $w + l + v$, is decreasing.

2. We have also seen that an economic system corresponding to the Marxist description of capitalism (corrected for dimensional absurdities), can be *growing* — even *strongly growing* — although its net national income increases only at a *decreasing* rate (Theorem 2).

Assuming therefore that the Marxist description of capitalism is epistemologically valid, a period in which the net national income of a capitalist economy increases at a decreasing rate does not justify any prediction of the breaking down of capitalism. Moreover, the number of periods in which the net national income of a growing capitalist economy increases at a decreasing rate may be unlimited (Theorem 2). Therefore, even the repetition of such phases does not justify that prediction.

3. It is true, however, that a phase in which the national income of a growing capitalist economy increases at a decreasing rate cannot last indefinitely (Theorem 4). Thus, it seems that we could proclaim the end of the capitalist system provided we knew that we have actually entered into such an everlasting phase. But how can we know in practice that the phase in point is everlasting and not a temporary one?

4. On the other hand, it is not very clear why in a growing capitalist economy, capitalists' consumption should be increasing. Actually, according to the very Marxist theory of concentration the number of capitalists should continually decrease. The speedier the concentration, the more likely it is that the consumption of the capitalist class should decrease. If this happens, then neither is a lasting phase in which the national income increases at a decreasing rate incompatible with a growing economy in all other respects (Remark II).

5. At the appropriate time we emphasized the fact that, according to Marxist rationale, no technical relation exists between output and variable capital, i.e., between output and employment.[28] The aggregate production function of the capitalist system is thus reduced to a relation between output and constant capital alone. This is entirely in agreement with the position that the magnitude of employment depends solely upon the interest of capitalists in having a greater or smaller recourse to the reserve army. In other words, the amount of employment is determined by the amount of variable capital capitalists are willing to set up, and not by a technical condition.[29] Therefore, given that the behavior of capitalists is that of (VII) and (VIII), we could not possibly assume also an *independent* relation between output and employment. For if we added such

28. *Supra*, p. 401.
29. K. Marx, *Capital*, vol. I (Chicago, 1932), chap. xxv, sec. iii.

a relation, the number of equations of (S) would increase by one unit and would thus become greater than that of the variables of the system. The capitalist system would then be an impossible system *ab initio*, as impossible as a square with five sides.

It is hard to see then how one can reconcile Marxist economics with the assertion that capitalism produces more consumers' goods than the demand for them. For if there is no technical relation between employment and output, there also is no demand equation in the system. The *employed* workers have no demand: they always receive and consume exactly what results from capitalists' behavior.[30] This peculiarity of the Marxist position has puzzled many followers of Marx who — like Rosa Luxemburg, especially — kept on asking: "Where does the demand come from [in a capitalist system]?"[31]

Of course, capitalists may not hit at once upon their *true* preferences when they are confronted with entirely new personal situations. All economic decisions in an evolutionary system are subject to this type of error. Clearly, such errors of appreciation introduce some *shocks* into the functioning of the system, but it is highly improbable for these shocks to accumulate in the same direction so as to produce a lasting deviation from the trend determined by the system of equations itself.

6. In the preface of an admirable little book, Erwin Schrödinger expressed the thought that the difficulty of analyzing the process of life does not reside in the complication of mathematics, but in the fact that the process is too complicated for mathematics.[32] This remark applies admirably also to the problem of the future of capitalism. For capitalism, like all other economic systems that preceded it and that will be produced by the continuous evolution of human society, is a form of life. Some aspects of its functioning lend themselves perfectly to mathematical analysis. Yet, when we come to the problem of its *evolution*, of its mutation into another form, mathematics proves to be too rigid and hence too simple a tool for handling it. Mathematical proofs of future evolutionary changes in any domain should, therefore, be viewed with skepticism, even if, unlike those analyzed in this paper, they are logically irreproachable.*

30. And to recall, one of the highlights of Marx's teachings is that capitalism produces always less than the demand, if "demand" is to be interpreted as "need." See the categorical statement on this point in K. Marx and F. Engels, *Correspondence, 1846–1895* (New York, 1935), p. 199.

31. Rosa Luxemburg, *The Accumulation of Capital* (New Haven, 1951), p. 19.

32. Erwin Schrödinger, *What Is Life?* (Cambridge, Eng., 1955), p. 1.

* A postscript to this article is provided by my replies to the criticism advanced in Georges Bernard's paper, "Quelques réflexions sur les modèles de croissance," *Econometrica*, XXXI (1963), 219–229. See Georgescu-Roegen, "Some Thoughts on Growth Models: A Reply," *ibid.*, 230–236; "A Rejoinder," *ibid.*, 239.

Analytical Index

(Index of Names begins on page 431)

Index of Names

Alchian, Armen A., 185, 186, 205
Allais, Maurice, 262n
Allen, R. G. D., 134, 139, 140, 141n, 142n, 150, 186, 218–221
Anaxagoras, 33n
Anaximander, 9
Aristotle, 5n, 6, 9, 10, 16, 30, 32, 33, 69n, 108, 117, 184, 191, 192, 196, 360
Armstrong, W. E., 190n, 201n, 203n, 204n, 208, 262n
Arrow, Kenneth J., 316n, 338–339, 381
Ayer, A. J., 43n, 104n
Azaïs, Pierre Hyacinthe, 250, 253, 264n

Bacon, Francis, 16
Bailey, Cyril, 16n
Banfield, T. C., 194, 195n, 201
Barone, Enrico, 387–388, 392
Bauer, Otto, 398–399, 404n, 405–406
Bauer, S., 17n
Bergmann, G., 265n
Bergson, Henri, 33, 34, 66n, 75n, 82n
Bernard, Georges, 415n
Bernoulli, Daniel, 184
Bertalanffy, L. von., 19n, 55n, 62n, 72n, 81n, 89n
Bertrand, Joseph, 250, 264n
Birkhoff, Garrett, 194n, 199n, 201n, 225n, 244n, 245n
Birkhoff, G. D., 32n, 78n
Black, John D., 366n
Black, Max, 20, 21n, 23n, 32n
Blake, William, 184n
Blum, Harold F., 89n
Bohm, David, 64
Bohr, Niels, 13, 86n, 104n, 361
Boltzmann, Ludwig, 76n, 82n, 86
Bolzano, Bernhard, 4
Bondi, H., 30n
Boole, G., 252n
Borel, Emile, 248, 251
Born, Max, 78n
Bowley, Sir Arthur, 187n, 203n, 262n

Breshkovskaia, Katerina, 368n
Bridgman, P. W., 4n, 13, 15n, 18n, 22n, 28, 29, 30n, 33n, 34n, 37, 42, 44, 48n, 54n, 62, 69n, 70n, 72n, 75n, 78n, 81n, 86, 90, 94, 110, 116n, 119
Broglie, Louis de, 64
Bromwich, T. J. I'A., 166n
Buffon, Georges L. L., 247n
Bukharin, N., 402
Burnet, John, 9n, 12n, 15n, 29n, 33n, 211n

Caird, E., 27n
Cantor, George, 32
Carnap, Rudolph, 210n, 243n, 254–257, 265, 269
Carnot, Sadi, 68, 92
Cassel, Gustav, 286n
Cattell, J. M., 151n
Chamberlin, E. H., 172n
Chenery, H. B., 397n
Childe, V. Gordon, 127n, 391n
Church, Alonzo, 247n
Clapham, J. H., 110n
Clark, J. M., 386
Clausius, R., 68, 75
Cobb, J. A., 102
Condorcet, Marie Jean, 275
Corlett, W. J., 188n
Courant, R., 204n
Cournot, A., 118
Cramer, Gabriel, 184
Croce, Benedetto, 15
Cuvier, Georges, 6, 16

Dalton, H., 30
Darmois, G., 161n
Darwin, Charles, 16, 25, 31n, 42, 44n
Debreu, G., 337–339, 381
Dedekind, Richard, 32
Delage, Yves, 39
Delewski, J., 368n
Demaria, G., 118n